Process Design and Simulation

Computer Applications to Chemical Engineering

Computer Applications to Chemical Engineering

Process Design and Simulation

Robert G. Squires, EDITOR
Purdue University

G. V. Reklaitis, EDITOR
Purdue University

Based on a symposium sponsored by the Division of Industrial and Engineering Chemistry at the 178th Meeting of the American Chemical Society, Washington, D. C., September 11–13, 1979.

ACS SYMPOSIUM SERIES **124**

AMERICAN CHEMICAL SOCIETY
WASHINGTON, D. C. 1980

Library of Congress CIP Data

Computer applications to chemical engineering process design and simulation.
(ACS symposium series; 124 ISSN 0097-6156)

Includes bibliographies and index.

1. Chemical process control—Congresses.
I. Squires, Robert G., 1935- . II. Reklaitis, G. V., 1942- . III. American Chemical Society. Division of Industrial and Engineering Chemistry. IV. Series. American Chemical Society. ACS symposium series; 124.

TP155.75.C65 660.2'81 79-27719
ISBN 0-8412-0549-3 ACSMC8 124 1–511 1980

PRINTED IN THE UNITED STATES OF AMERICA

Process Design and Simulation

ACS Symposium Series

M. Joan Comstock, *Series Editor*

FOREWORD

The ACS SYMPOSIUM SERIES was founded in 1974 to provide a medium for publishing symposia quickly in book form. The format of the Series parallels that of the continuing ADVANCES IN CHEMISTRY SERIES except that in order to save time the papers are not typeset but are reproduced as they are submitted by the authors in camera-ready form. Papers are reviewed under the supervision of the Editors with the assistance of the Series Advisory Board and are selected to maintain the integrity of the symposia; however, verbatim reproductions of previously published papers are not accepted. Both reviews and reports of research are acceptable since symposia may embrace both types of presentation.

CONTENTS

APPLICATIONS

PREFACE

This volume contains selected papers from a five-session symposium on "Computer Applications to Chemical Engineering Process Design and Simulation" sponsored by the I&EC Division of ACS held in Washington, D.C. in September of 1979. Although shorter symposia on special topics in chemical engineering computation have been held under the auspices of the AIChE, this was the first symposium devoted to the entire field to be held in the United States. The European Federation of Chemical Engineers has held regular symposia on Computer Applications in Chemical Engineering but the proceedings of these meetings have enjoyed only limited circulation in the United States. This volume thus represents the only collection of works on computer applications to appear in the United States since the C.E.P. Symposium Series volumes on computational topics which appeared in the middle sixties.

The papers comprising this volume are subdivided into four categories: reviews of four major areas of computation research, reviews of several key computational topics within these areas, papers discussing specific new advances in methodology, and papers demonstrating the effective use of computation in modeling, design, and control covering a broad range of applications.

The first of the broad computational area reviews discusses the general direction of research in steady-state process simulation and summarizes the new ideas in computational architecture to have emerged since 1975. This is followed with a review of the main thrusts in control theory and on evaluation of the relevance to chemical engineering applications. Next the significant developments in numerical methods for minimizing nonlinear constrained and unconstrained functions are traced. The new developments in recursive quadratic programming methods for general nonlinear programs should be of particular interest to chemical engineers since they appear to offer a significant advance over the generalized reduced gradient techniques that have dominated the field for some ten years. Finally, research in computer-aided synthesis is appraised, and a summary is given of the significant results in six problem areas: heat exchanger networks, separation systems with and without heat integration, reaction paths, total flowsheets, and control systems.

These wide-ranging reviews are followed by analyses of progress in several specialized problem categories: chemical and physical equilibrium computations; vapor–liquid equilibrium computations including single and multistage VLE separations, multiliquid phase systems, and VLE systems with reaction and electrolytes; treatment of measurement errors in process networks and computations of choking flows in gas pipe networks.

Next follows a series of reports on important developments in computational methods or program packages incorporating novel computational features.

Finally, the volume is capped with papers discussing computer applications involving modeling, design, and control spanning a wide range from microbial conversion to industrial reactor modeling to drug therapy control.

The scope and quality of these contributions have made the symposium a milestone in chemical engineering computation and ensure that this volume will be of permanent significance to those involved or interested in this area. Both the papers and the symposium as a whole have benefitted substantially from the anonymous contributions of a large number of conscientious referees for whose efforts we are indebted. Finally, the organization and smooth functioning of the symposium as well as the successful assembly of this volume are in large part due to the commendable work of the following session chairmen: D. A. Mellichamp and R. G. Rinker of the University of California–Santa Barbara, G. Blau of Dow Chemical USA, and J. Zemaitis of OLI Systems, Inc.

School of Chemical Engineering
Purdue University
West Lafayette, IN 47907
October, 1979

G. V. Reklaitis
R. G. Squires

STATE-OF-THE-ART REVIEWS

1

Steady State Chemical Process Simulation: A State-of-the-Art Review

E. M. ROSEN

Monsanto Company, 800 North Lindbergh, St. Louis, MO 63166

Perspective. The use of a mathematical model on a computer to simulate a chemical process is now approximately two decades old. The field, which has been referred to as steady state chemical process simulation, flowsheeting or computer aided chemical process design to emphasize various shadings and meanings has had a major impact on moving chemical process design from essentially an art form of the 1950's to an accepted engineering science today.

The field, which of necessity has always attempted to merge the areas of chemical engineering, physical chemistry, thermodynamics and the various disciplines of computer science, has been especially dynamic the last several years. This is no doubt due in part to the increasing pressure to make better use of energy, minimize operating costs and increase the productivity of the chemical processes studied as well as the chemical engineer himself.

A determination of the state-of-the-art in a particular field can probably best be viewed by understanding the motivation of the contributors. Academic work is motivated by a desire to explain nature, a desire to solve unsolved problems and, for pragmatic reasons, a desire to attract funding. Academic work is usually found in the literature.

Industrial work is motivated by profit, which in turn leads to a desire to increase productivity and a desire to increase robustness of solutions. Industrial organizations judiciously choose among competing ideas and programs. The implementations carried out to solve their problems are not generally found in the literature.

0-8412-0549-3/80/47-124-003$08.50/0

Reviews, Books and Projects. The general field was reviewed in 1975 by Motard, Shacham, and Rosen (1) and in a comprehensive fashion in 1977 by Hlavacek (2). A first book on the subject is scheduled to be released in the latter half of 1979 (3).

An indepth evaluation of the field was afforded by the ASPEN project at MIT sponsored by the U. S. Department of Energy. The project was started June 1, 1976 and is entitled, "Computer-Aided Industrial Process Modeling". Its quarterly and annual reports are available from the National Technical Information Service (4).

The User Interface. A wide variety of stand alone steady state simulation programs and flowsheet systems are available to the process engineer. These have been reported in a series of articles by Peterson, Chen and Evans in 1978 (5) and by Chen and Evans in 1979 (6). Some practical advice on the use of the computer in design is reported by Weismantel (7). A course in the use of several commercially available systems is given in the AIChE Today Series (8). A report on the use of networks to share chemical engineering programs among educators was recently issued (9).

The use of on-line systems to edit the input data for simulation systems is widely used. However, interacting with the program during its execution is now being carried out industrially. Its advantages (or disadvantages) have not yet been discussed in the literature.

The continuing decline in costs of graphical devices and the broadening availability of easy-to-use graphical software has made computer graphics a feasible tool in flowsheeting presentations and analysis (10).

General Direction of the Field. The characteristics of early flowsheeting systems and their limitations were defined by Evans and Seider in 1976 (11). They also attempted to define the criteria for an advanced computing system.

Several trends have been noted, however, in this field over the last few years:

1. Use of flowsheeting systems has become widespread. Many have been developed to meet the particular needs of their environments (12, 13, 14) and often serve as a repository of the company's or developer's expertise.

2. There has been a trend toward integrating flowsheeting systems into much larger systems for project engineering (15, 16, 17, 18). The same physical property data used in flowsheet simulations is being increasingly applied to other project engineering programs.

3. There has been a broadening acceptance of the UNIFAC program for the determination of activity coefficients from molecular structure when no data is available (19, 20). Systems increasingly are storing both pure component and mixture data.

4. The capability to handle different physical property correlations for different pieces of equipment are being added (4).

5. An effort to develop new algorithms for difficult or complex calculations, often not attempted before, were undertaken.

6. A major academic effort has been mounted to reevaluate system architectures. This has been motivated by the limitations of the sequential modular method for design and optimization (21). This in turn has led to a strong research effort in equation solving methods tailored to meet the needs of process simulation.

Trends 5 and 6 will be explored further after noting progress in some of the scientific and technological foundations of this subject.

Scientific and Technological Foundations

Sparse Matrix Methods. In order to get around the limitations of the sequential modular architecture for use in design and optimization, alternate approaches to solving flowsheeting problems have been investigated. Attempts to solve all or many of the nonlinear equations simultaneously has led to considerable interest in sparse matrix methods generally as a result of using the Newton-Raphson method or Broyden's method (22, 23, 24).

The field was comprehensively reviewed by Duff (25) in 1977. The design features of sparse matrix codes are discussed by Duff and Reid (26) and the need for suitable user interfaces is the subject of a paper by George and Liu (27). Generally the goal of the work is to set up the equations directly in such a manner so that the effort to solve the resulting systtems, which are generally sparse, is minimized. Alternately given a sparse system the goal is to minimize the work to get a solution realizing the basic system will need to be solved many times.

Stadherr (28) and Westerberg and Berna (29) discuss the reduction of a flowsheet to a form that can utilize sparse methods and Hildago, Correa, Gomez and Seader (30) show how the linearized equations resulting from interlinked separators can be arranged.

Lin and Mah (31) discuss data structures and processing techniques for solving nonlinear equations and Hernandez and Sargent (32) consider the use of automatic algebraic manipulation in the general case of using a mixed system of equations and subroutines.

Physical Properties. A symposium held at Asilomar and subsequently published as an ACS Symposium Series volume Phase Equilibria and Fluid Properties in the Chemical Industry (33) addressed the state-of-the-art in this field. Motard and Winter (34) discussed the history and implementation of physical property subsystems in flowsheeting systems and indicated the future needs of such systems.

In order to gain the flexibility to change physical property methods between modules or add new methods, Evans, Joseph and Seider (35) and Seider, Evans, Joseph, Wong, and Jirapongphan (36) suggested using a pointer system (plex data structure) in the design of physical property subsystems in flowsheeting systems. This method, in fact, has been adopted in the ASPEN system (4). Kaijaluoto (37) confirms the advantages of the plex data structure but points out its negative effects. There are complexities in understanding the code, difficulties in debugging and testing as well as difficulty of integrating plex data based programs with non-plex programs. Space savings were reported to be wiped out by the added administrative overhead required.

The time that simulation programs spend in the physical property routines has long been a point of concern. Both Leesley and Heyen (38) and Barrett and Walsh (39) suggest the use of approximations to the thermodynamic functions during the simulation to save time. Mah (40) reports on the effects that poor thermophysical data can have on a design.

The UNIFAC computer program (UNIversals quasichemical Functional group Activity Coefficients) has received broad acceptance throughout the profession

(19, 20). The program estimates the liquid phase activity coefficients at specified temperatures and liquid compositions from molecular structure of the molecules. Binary or multicomponent solutions can be considered.

A major limitation in many flowsheeting systems is the inability to handle electrolytes. New work in this area is now being reported (41, 42) and two commercial systems are available (43, 44). The ASPEN system (4) has incorporated a solids handling capability to overcome another common deficiency.

A new cooperative effort in the physical property area was initiated in November 1978, named the Design Institute for Physical Property Data (DIPPR) under the American Institute of Chemical Engineers. The effort will focus on the development of a single comprehensive source of evaluated data for chemicals of industrial importance (45).

For routine industrial work the Technical Data Book of the American Petroleum Institute is currently generally used for physical properties of petroleum compounds and the book by Reid, Prausnitz and Sherwood (The Properties of Gases and Liquids, McGraw Hill, 1977) is used for other organic compounds and their mixtures.

Algorithmic Foundations

Flash Calculations. The ability to carry out vapor-liquid equilibrium calculations under various specifications (constant temperature, pressure; constant enthalpy, pressure; etc.) has long been recognized as one of the most important capabilities of a simulation system. Boston and Britt (46) reformulated the independent variables in the basic flash equations to make them weakly coupled. The authors claim their method works well for both wide and narrow boiling mixtures, and this has a distinct advantage over traditional algorithms (47).

Attention has also focused on the reliable solution to the three phase (liquid-liquid-vapor) flash problem. That the solution is difficult is attested by the fact that few flowsheeting systems have this capability. Three papers (48, 49, 50) recently appeared which proposed solution algorithms. Generally infinite dilution activity coefficients are used to generate starting compositions.

Independent testing of these algorithms has not yet been reported.

Distillation Calculations. Work done with flash calculations and sparse matrix methods was extended to distillation calculations. Holland and Gallun (51) explored the use of Broyden's method coupled with sparse updating procedures to distillation calculations with highly non-ideal solutions. Shah and Boston (52), and Ross and Seider (53) discuss the case of multiple liquids phases on a tray.

Brannock, Verneuil and Wong (54) and Boston (55) considered equality and inequality constrained distillation calculations.

Distillation calculations with reactions taking place in the column are cases for which algorithms were developed by Bentzen, Irarraraz, Anthony and Holland (56) as well as by Tierney (57).

As with flash calculations, independent testing and/or comparison of these algorithms has not yet been reported.

Flowsheet Architectures

Steady state process simulations require the solutions to large sets of generally nonlinear equations subject to a variety of constraints on input flows, output flows and/or internal unit variables. Generally, the equations are sparse (most equations have only a small number of the unknowns) and many equations may be linear.

Historically, many of the naturally grouped equations (flash calculations, distillations, etc.) were solved together as a module utilizing specially designed algorithms which were robust (rarely failing) and efficient. With the advent of flowsheeting systems the most natural method of solving these equations was module by module. Systems utilizing this approach are called sequential modular.

With the rise of more complex designs, the needs of optimization and designs with greater integration and interaction between modules, the advantages (great flexibility) of the sequential modular structure has been brought into question. Solving many or all of the equations (or modules) simultaneously has been suggested as a viable alternative to increase the speed of solution and allow the easy integration of design constraints.

Many of the recently suggested architectures for flowsheeting systems have been reviewed by Motard (58). Table 1 is a list of selected system architectures and some representative systems.

TABLE I

FLOWSHEET ARCHITECTURES AND SOME REPRESENTATIVE SYSTEMS

ARCHITECTURE	SYSTEM	REFERENCE
SEQUENTIAL	CAPES	14
	CONCEPT	62
	DESIGN	60
	FLOWTRAN™	59
	FLOWPACK II	93
	GPS	63
	PROCESS™	61
LINEAR	MPB-II	78
	SYMBOL	77
SIMULTANEOUS	JUSE-GIFS	84, 85
	MULTICOL	92
	QUASILIN	91
SIMULTANEOUS MODULAR	DIS	13
	TISFLO	12

Sequential Modular. By far the most experience with flowsheeting systems has been with the sequential modular architecture (59-63). It is this architecture that is most easily understood by the process engineer. Each module calculates all output streams from input streams subject to module parameters. Generally, the stream variables consist of component flows, temperature (or enthalpy) and pressure as the independent variables. Other dependent variables such as total flow, fraction vapor and total enthalpy (or temperature) are often carried in the stream.

There are two major problems which arise in sequential modular architectures which seriously affect their solution efficiency:

1. Recycle loops
2. Design specifications

Partitioning and the Cycle Matrix. Sequential modular systems require an order of calculation (precedence order) be given to the modules. There are generally four steps taken to determine this ordering.

1. Partitioning of the flowsheet into recycle loop nets.
2. Precedence ordering of the nets.
3. Tearing of the recycle loops in the nets.
4. Precedence ordering of the modules (nodes) in the recycle loops.

The basic idea in partitioning procedures is a successive identification and condensation of recycles into pseudonodes with the same inputs and outputs as the included recycles. Methods for identification of recycles for partitioning purposes can be divided into two main classes: Powers of adjacency matrix (e.g. Kehat and Shacham (64)) and path tracing (e.g. Henley and Williams (65)). These are reviewed by Gros, Kaijaluto and Mattsson (66) and Venkatesh (67).

Figure 1 gives an example (66) of a network partitioned into two recycle loop nets, consisting of the loops S001, S002, S003 and S005, S006, S007. A net is a group of modules that cannot be solved without tearing (guessing) a stream(s) to make the net solvable. The precedence order is solve net A then net B.

Figure 2 is a system that cannot be partitioned and consists of a single net. While net A in Figure 1 can be solved first to determine S004 and then net

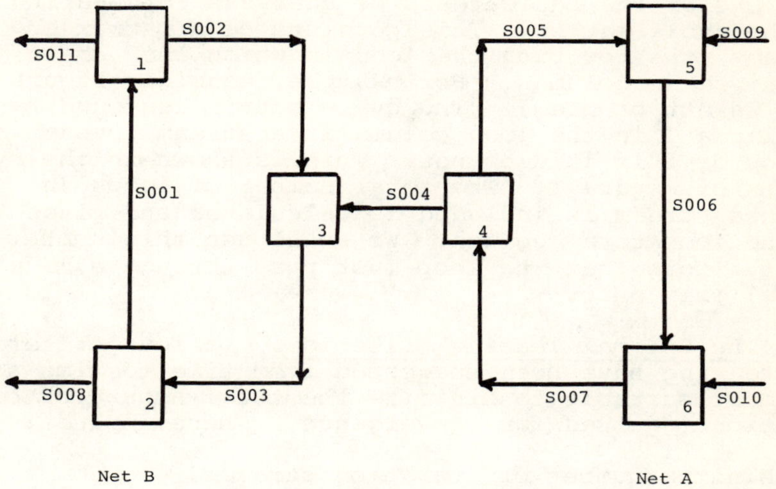

Figure 1. A network with two nets

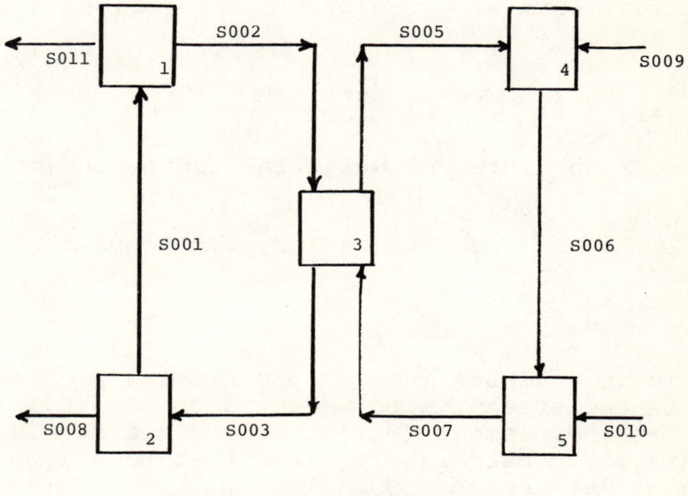

Figure 2. A system with a single net

B can be solved, the entire system in Figure 2 cannot be partitioned and all streams must be solved simultaneously.

One of the basic tools of analysis for tearing is the cycle matrix. This consists of a matrix of streams (the row) and the loop in which they are contained in columns. For example, consider Figure 3 (the Cavett problem). The cycle matrix is found by placing a 1 in the loop column if a stream appears in a loop or 0 if it does not. This is shown in the cycle matrix of Figure 4. The total number of loops in which a stream is included is calculated and placed in the loop total column. We shall use this number later. Note that the loop must pass through a module (node) only once.

<u>The Optimal Tearing Criterion</u>. Various criteria for tearing have been suggested from time to time as the most suitable to minimize the total amount of calculation required for convergence. These include:

1. Minimum number of iteration streams.
2. Minimum number of stream variables (iteration stream parameters).
3. Weighting according to preferred streams.
4. Minimum total number of loops torn.

Each of these criterion can be put into a single mathematical form as given by Pho and Lapidus (<u>68</u>):

$$\text{Min} \sum_{j=1}^{n} \rho_j x_j \qquad j = 1,2,\ldots,n \text{ streams} \qquad (1)$$

where ρ_j is an (integer) weighting factor subject to

$$\sum_{j=1}^{n} a_{ji} x_j \quad 1 \qquad i = 1,2,\ldots m \text{ loops}$$

$$x_j = 0 \text{ or } 1$$

This is a linear integer programming problem where the x_j is the stream to be selected or not (0,1) and the a_{ji} is the element (0,1) in the cycle matrix. Note that the constraint requires that each loop must be torn at least once.

Criterion (1) assigns each $\rho_j = 1$. Criterion (2) sets ρ_j equal to the number of unknowns in each stream. Criterion (3) would allow the engineer to select the

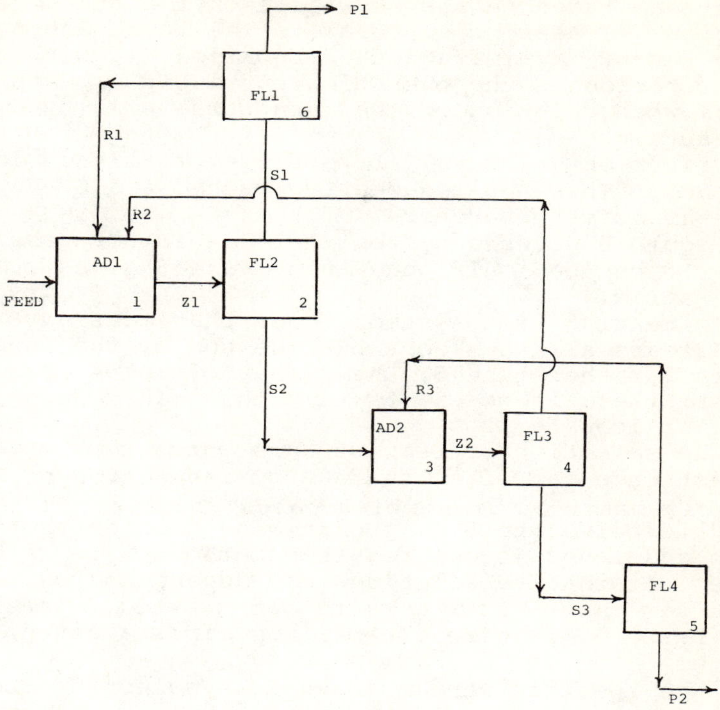

Figure 3. Cavett problem

Stream	Loop 1	Loop 2	Loop 3	Loop Sum
Z1	1	0	1	2
S2	0	0	1	1
S1	1	0	0	1
Z2	0	1	1	2
S3	0	1	0	1
R1	1	0	0	1
R3	0	1	0	1
R2	0	0	1	1

Figure 4. Cycle matrix for Cavett problem

tear stream based on his knowledge of the streams. Criterion (4) assigns ρ_j equal to the total loops each stream tears. Minimization of Equation (1) with this last criterion is the one which currently appears optimal when solution of the torn streams is by direct substitution.

Criterion (4) says, if possible, tear each loop only once. This was deduced by Upadhye and Grens (69) and separately by Barchers (70).

Figure 5 gives a system where it is not possible to tear each loop only once and Figure 6 gives its cycle matrix.

In tearing this system we see that tearing S2, S4, S6 tears all the loops but the sum of the loops torn is 6 rather than 5, the number of loops present. However, there is no way to tear that can reduce this number of loop tears.

The selection of tear streams which tear each loop at least once can often be done by inspection of the cycle matrix. In complex cases, however, a method to solve the 0, 1 programming problem is needed. Branch and bound is one feasible method. The selection between alternative solutions may depend on how well the stream can be estimated to get started. Generally it is best to minimize the maximum times a given loop is torn.

Once the tear streams have been selected, the problem is reduced to solving the simultaneous non-linear equations:

F(X)	=	Φ(X)	-	X	(2)
Residual		Calculated by sequential pass through the flowsheet		Estimated value	

where X is the unknown tear stream variable vector and Φ(X) is the calculated values of the variables resulting from calculating through the loops. If the largest absolute element (i.e., the cubic norm) of the residual using direct iteration is plotted versus iteration on semilog paper then the slope of the line in the linear region gives the maximum eigenvalue of the Jacobian of Φ(X) at the solution (71). This assumes the maximum eigenvalue is positive, the most common case.

Evidence that tearing each loop once, if possible, is the "best" tearing criterion has been presented by Gros, Kaijaluoto and Mattsson (66) independent of the diagonal method (see below) used for convergence. Figure 7 taken from Rosen and Pauls (59) gives a plot

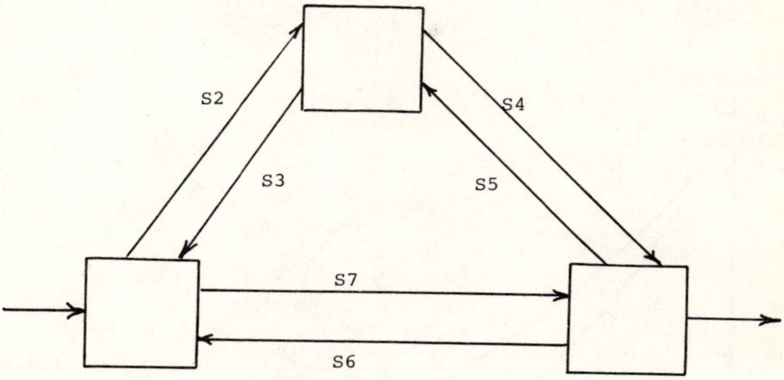

Figure 5. Tearing loops more than once

Stream	Loop 1	Loop 2	Loop 3	Loop 4	Loop 5	Loop Sum
S2	1	0	0	1	0	2
S3	1	0	0	0	1	2
S4	0	1	0	1	0	2
S5	0	1	0	0	1	2
S6	0	0	1	1	0	2
S7	0	0	1	0	1	2

Figure 6. Cycle matrix for system

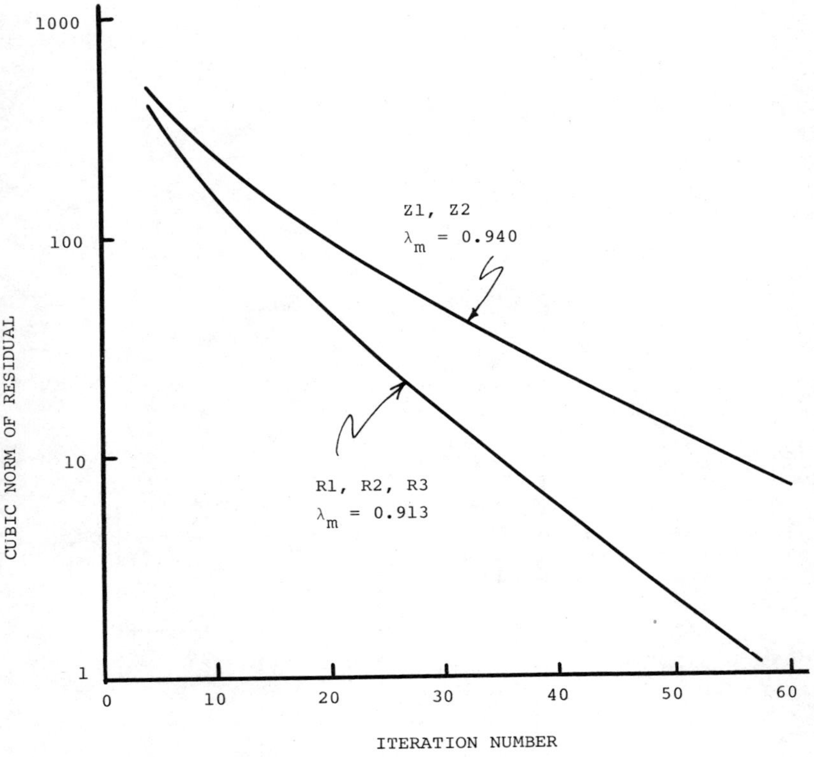

Figure 7. Convergence of the Cavett problem by direct substitution

of the cubic norm of the residual for different tears to the Cavett problem using direct substitution. Note the R1, R2, R3 three stream tear (which is the minimum loop tear) leads to a smaller maximum eigenvalue than the Z1, Z2 two stream tear which tears one loop more than once. All equivalent solutions to Equation (1) would have direct substitution slopes which are the same.

It is unfortunate that the cubic norm of the residual has been called error. Hence we prefer the word residual to distinguish it from error, the latter which implies the solution is known.

Equation (2) is of the size equal to the number of tear streams times number of variables. Since this can easily be equal to 50 or 60, methods to solve Equation (2) have been sought which avoid as much as possible evaluations of the flowsheet, i.e. $\Phi(X)$. The most popular methods have been direct substitution (which is slow but stable) or acceleration methods such as bounded Wegstein method (72) the Dominant Eigenvalue method (71) or Broyden's method (22). Each of these methods can be thought of as a means of approximating the Jacobian (J(X)) of F(X) in the solution of Equation (2) by the Newton-Raphson method. (In this sense these methods are all quasi-Newton methods though the term is generally used only when the full matrix or its inverse is approximated.) In the case of the Dominant Eigenvalue method, the diagonal elements of J(X) are a constant equal to the estimate of the maximum eigenvalue minus one. In the case of Wegstein's method the diagonal is the secant slope approximation to the derivative minus one. In the case of Broyden's method, the full matrix is approximated.

Kluzik (73) compared diagonal versus full matrix approximations to J(X) and concluded that:

1. Broyden's method converges rapidly in the linear region (near the solution) to a tight tolerance. It is slow in the nonlinear region.

2. In the nonlinear region, far from the solution, diagonal methods make more rapid progress.

This would suggest that initial approximations to J(X) using Wegstein's method should be followed by using Broyden's method.

Another combination method shown to be effective is the delayed Wegstein method (59). Here Wegstein's method is applied at fixed intervals between direct iterations. Orbach and Crowe (71) used the Dominant

Eigenvalue acceleration between direct substitution only when the estimate of the dominant eigenvlaue changed by some small amount between direct substitutions.

The Problem of Design. Since the modules in sequential modular systems calculate outputs from inputs, one cannot easily specify an output (often design) specifications. This problem is generally handled by placing a root finding procedure (i.e., a "control block") around the modules and adjusting a parameter in a module such that the desired specification is met (see 47). From an equation solving point of view, this leads to a loop within loop structure. (It should be emphasized, however, this has the great advantage of stability and subsequently, solution reliability). Since many design specifications may be required, one often ends up with many loops within loops.

Consider for example the problem of converging a recycle loop within a control loop as shown in Figure 8. If units 1, 2 and 3 were, say, flash units and it was desired to adjust the temperature in unit 2 to achieve a specified component flow in stream S5, then the recycle loop (S4) could be solved within the control loop:

1. Estimate T in unit 2
2. Estimate S4
3. Calculate units 1, 2, and 3 to get new estimate of S4
4. Compare calculated S4 with estimated S4
5. Evaluate component flow in S5
6. Compare design specification with observed value

Other loop within loop orderings are possible.

Metcalfe and Perkins (74) and Perkins (75) combined the recycle calculations with the design specifications to solve simultaneously equations of the form

$$\begin{aligned} F(X,P) &= \Phi(X,P) - X \\ G(X,P) &= H(X,P) - D \end{aligned} \tag{3}$$

where P are the system parameters, D are design specifications and X are the recycle loop variables. Broyden's method was used on the equations with the modification that if a newly predicted point led to a much worse (order of magnitude) functional evaluation (sum of squares residuals) then a step length factor would be reduced by 10 until a step length would be found

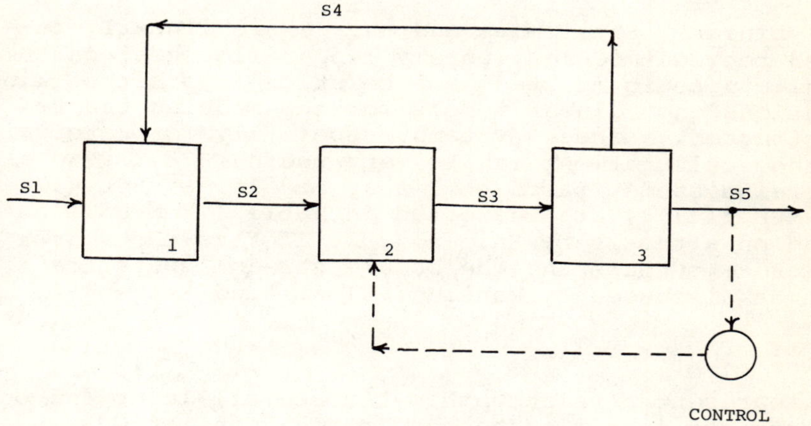

Figure 8. Control with recycle loop

that led to a less than order of magnitude functional increase. This new point only then would be used to update the Jacobian inverse. Metcalfe and Perkins showed if a (greater than order of magnitude) poor point is used to update the matrix, then it would lead to a nearly singular matrix. In addition, a near singular matrix, the authors indicate, is an indication of a badly posed problem.

It may be commented that the loop identification methodology and tearing criterion does not include control loops. With control loops present, one ordering deduced from a minimum loop tear may be vastly more efficient than an equivalent solution ordering. Just how to incorporate control loops in the tearing criterions does not appear to be addressed in the literature.

Linear. Since mass and energy are linearly related between modules, purely linear flowsheet calculations can be formulated as a solution to a set of linear equations once linear models for the modules can be constructed. Linear systems, especially for material balance calculations can be very useful (76). Two general systems, based on linear models, SYMBOL (77) and MPB II (78) are indicated in Table 1. MPB II is based on a thesis by Kniele (79). If Y is the vector of stream outputs and the module stream inputs are X, then as discussed by Mahalec, Kluzik and Evans (80)

$$Y = A\,X + B \tag{4}$$

can represent a relationship between all input and output streams in a flowsheet. In addition, if C is a connection matrix which indicates how output streams are connected to input streams then

$$X = CY + F \tag{5}$$

where F is a vector of external feed streams. Knowing the C matrix from the flowsheet, the A matrix, the B and F vectors Equations (4) and (5) may be solved simultaneously to find the X and Y vectors. Alternnately, Equations (4) and (5) can be combined to give either

$$X = [I - CA]^{-1}\,(CB + F) \tag{6}$$

or

$$Y = [I - AC]^{-1}\,(AF + B) \tag{7}$$

Table 2 gives a simple flowsheet and calculations indicating how these equations are used. Generally, simple modules such as split modules, add modules and fixed extent of reaction modules may be utilized within this approach. Note that for fixed extent of reaction modules

$$Y = X + B$$

where B is a vector of component reaction productions. This is of the form

$$\sum_{i=1}^{n} \alpha_{ij} e_j \qquad j = 1,2\ldots\text{number of reactions},m$$

and the α_{ij} is the stoichiometric coefficient for component i in reaction j. The extent of reaction j is e_j. Certain types of design specifications can often be included directly into linear systems. If any input or output stream is fixed then a system parameter would have to be adjusted (i.e., become a variable). For example in the Table 2 example if Y were fixed then the B vector (reactor production) could become the independent variable. Hutchison (81), Sood, Reklaitis, and Woods (82) and Sood and Reklaitis (83) discuss linear systems.

Simultaneous. In order to circumvent the inefficiencies associated with loop within loop structures forced by the module design and sequential modular approach, there has been considerable academic effort to investigate how to perform all computations simultaneously. The potential advantages of this global (or "equation oriented") approach are generally recognized but acceptance of the approach has been slow due to a number of reasons:

1. The complexity of the executive in setting up the equations to be solved.

2. The potential space required for such a solution is large, though this problem is disappearing.

3. The numerical problems associated with the methods.

4. If the solution fails the user may be left with little useful information.

TABLE II

LINEAR SYSTEM WITH FIXED EXTENT REACTOR

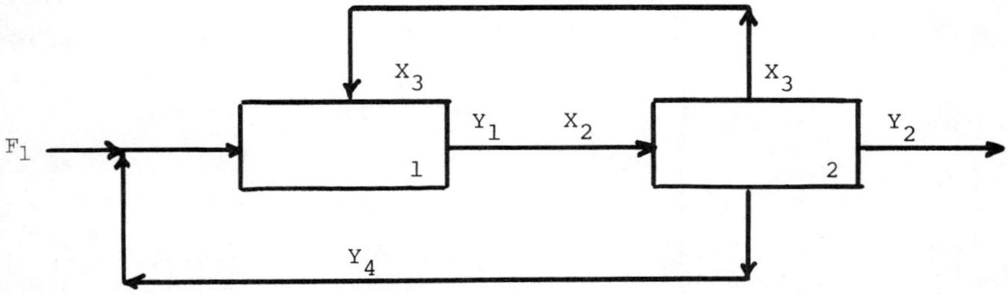

			STOICHIOMETRIC COEFFICIENTS		$\sum_{j=1}^{2} \alpha_{ij} e_j =$	B_i
No.	COMPONENT	F_1	1	2		
1	CH_4	7	-1		$-e_1$	-1
2	H_2O	15	-1	-1	$-e_1 - e_2$	-3
3	CO	4	1	-1	$e_1 - e_2$	-1
4	H_2	4	3	1	$3e_1 + e_2$	5
5	CO_2	1		1	e_2	2
	SUM	31				2

UNIT 1 Reactor

$$CH_4 + H_2O \rightleftharpoons CO + 3H_2 \qquad e_1 = 1$$

$$CO + H_2O \rightleftharpoons CO_2 + H_2 \qquad e_2 = 2$$

UNIT 2 Splitter

$$Y_2 = 0.3\ X_2 \qquad Y_3 = 0.5\ X_2 \qquad Y_4 = 0.2\ X_2$$

TABLE III

SOLUTION TO LINEAR SYSTEM OF TABLE 2

VECTOR/MATRIX SIZES

$$\frac{X_i,\ Y_i,\ F_i,\ B_i}{S \times 1} \qquad \frac{A}{20 \times 15} \qquad \frac{C}{15 \times 20} \qquad \frac{F}{15 \times 1} \qquad \frac{B}{20 \times 1}$$

$$\begin{bmatrix} Y_1 \\ Y_2 \\ Y_3 \\ Y_4 \end{bmatrix} = \begin{bmatrix} I & 0 & I \\ 0 & .3 & 0 \\ 0 & .5 & 0 \\ 0 & .2 & 0 \end{bmatrix} \begin{bmatrix} X_1 \\ X_2 \\ X_3 \end{bmatrix} + \begin{bmatrix} B_1 \\ 0 \\ 0 \\ 0 \end{bmatrix}$$

$$Y = A \quad X + B$$

$$\begin{bmatrix} X_1 \\ X_2 \\ X_3 \end{bmatrix} \quad \begin{bmatrix} 0 & 0 & 0 & I \\ I & 0 & 0 & 0 \\ 0 & 0 & I & 0 \end{bmatrix} \begin{bmatrix} Y_1 \\ Y_2 \\ Y_3 \\ Y_4 \end{bmatrix} \quad \begin{bmatrix} F_1 \\ 0 \\ 0 \end{bmatrix}$$

$$X = C \quad Y + F$$

SOLUTION

$$X_1 = \begin{bmatrix} 11 \\ 23 \\ 6 \\ 10 \\ 3 \end{bmatrix} \qquad X_2 = \begin{bmatrix} 20 \\ 40 \\ 10 \\ 30 \\ 10 \end{bmatrix} \qquad X_3 = \begin{bmatrix} 10 \\ 20 \\ 5 \\ 15 \\ 5 \end{bmatrix}$$

SUM = 53 SUM = 110 SUM = 55

$$X_1 = \frac{5}{3} F_1 + \frac{2}{3} B_1 \qquad X_2 = \frac{10}{3}\left[F_1 + B_1\right] \qquad X_3 = \frac{5}{3}\left[F_1 + B_1\right]$$

Despite these potential difficulties, efforts to attack this problem have been undertaken and some progress has been made. The nonlinear equations are generally attacked by methods (e.g. Newton-Raphson) which require periodic solution of linear equations.

Equation Solvers. This approach may be implemented in a number of ways. One approach is to pass the residuals of the equations and the independent variables to the executive for solution. In this way the nature of the modules can be preserved. JUSE-L-GIFS (84, 85) appears to use this type of architecture.

Kubicek, Hlavacek and Prochaska (82) applied the Newton-Raphson method to the equations resulting from interconnected distillation columns.

The authors reported nonconvergence when nonideal vapor liquid equilibria was used, slow convergence at other times and non-feasibility for more than two "controlled simulation" loops.

Berna and Westerberg (87) indicate how some of the multiple root problems encountered in equation solving approaches in process simulations can be overcome.

Quasi Linearization. This approach attempts to put the nonlinear equations in the form

$$A(X)\ X = B(X) \qquad (8)$$

The A matrix and B vector is generally a function of X. Once X is solved from Equation (8) it is used to regenerate a new value of A. This is repeated until convergence.

Equation (8) is of the form of the Newton-Raphson method. The A(X) matrix, however, is not necessarily the Jacobian, J(X). Just how the A(X) is set up depends on the application. Bending and Hutchison (88) developed the method for pipe flow networks. Hutchison and Shewchuk (89) applied the method to multiple distillation towers. Gorczynski and Hutchison (90) detail the method for flowsheeting systems. Quasilin (91) is a flowsheeting system based on this approach. MULTICOL (92) appears to solve interconnected columns by means of this approach as well.

Simultaneous Modular. There has been an almost continuum of architectures suggested to take advantage of the better features of sequential modular, linear and simultaneous architectures. Most of these suggestions seek to retain the calculation modules (since millions of dollars have been invested in sequential modular software) and thus the name simultaneous modular has been applied. FLOWPACK II (93) apparently has some simultaneous modular features.

Simultaneous modular architecture can probably be further broken down into two categories.

1. Those architectures which attempt to solve recycle and control loops simultaneously.

2. Those architectures which use a "two tiered" approach (Figure 9) using a fully linearized system alternately with a rigorous modular calculation.

Rosen (94) suggested this latter approach alternating between a split fraction model of the system and rigorous flowsheet modules to regenerate new split fractions. The split fractions were initially estimated to begin the iterations and the system converged when the split fractions changed by a small amount. Weisenfelder and Olsen (95) reported success with this method for interlinked distillation columns but Mahalec, Kluzik and Evans (80) indicated split fraction models tend to be unstable.

A number of variations are possible with such two tiered sytems. Tearing can take place in the conventional way and the torn streams can be estimated. Each module in turn can be calculated as in the sequential modular systems. A linearized model of each module can then be generated which in turn can be used in the linearized flowsheet model. From Equation (1)

$$\underset{\text{Residual}}{F(X)} = \underset{\substack{\text{Calculated}\\\text{from}\\\text{linearized}\\\text{models}}}{\Phi(X)} - \underset{\text{Estimated}}{X} \qquad (9)$$

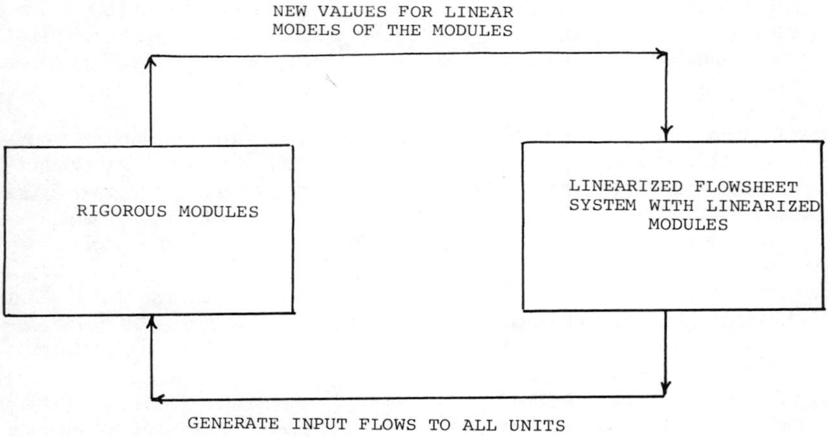

Figure 9. Two-tier approach

Here X is the initial estimate of the tear stream and $\Phi(X)$ is calculated from the linearized model. Alternately all streams can be torn and then be reestimated from the linearized model.

Kehat and Shacham (96) used split fraction models to estimate the Jacobian when the Newton-Raphson method is used to solve Equation (1). The authors concluded that their method is very efficient for systems with more than one tear stream and when there is only a weak interaction between variables in the tear stream.

Sood, Khanna and Rekalitis (97) and McLane, Sood and Reklaitis (98) discuss multiple tier systems and strategies to use for their solution.

Umeda and Nishio (99) using fully linearized models compared the sequential modular and simultaneous modular approaches and concluded each architecture had its area of applicability.

Lin (100) suggested breaking the process flowsheet into one or more blocks of modules. Each block of modules contains one or more modules and all of the modules in the same block are solved simultaneously. The whole process flowsheet is then solved by conventional sequential modular approach by treating each block as a module.

The Future

Flowsheeting systems have become and will remain a routine tool used in the design and analysis of chemical processes. The spurt in new algorithms and architectures over the last three years will probably result in a period of digestion and evaluation over the next several years. Current systems will probably remain in place as long as they are providing useful results for their users. Nevertheless, there will continue to be pressures to generate more robust algorithms, improvements to speed up the calculation and integrate flowsheeting systems in more comprehensive systems for project engineering.

Literature Cited

1. Motard, R. L.; Shacham, M.; Rosen, E. M., "Steady State Chemical Process Simulation", AIChE J(1975) 21 No. 3; 417-436.

2. Hlavacek, V., "Analysis of Complex Plant - Steady State and Transient Behavior", Comp. & Chem. Eng. (1977) 1, No. 1; 75-100.

3. Westerberg, A. W.; Hutchison, H. P.; Motard, R. L.; R. L.; Winter, P., "Process Flowsheeting", Cambbridge University Press, Cambridge, England (In Press).

4. ASPEN Project - 1st Annual Report-MIT-2295T9-4 June 15, 1977; 2nd Annual Report-MIT-2295T9-9 June 15, 1978; Available from: Contract E(49-18)-2295 Task No. 9, National Technical Information Service, U. S. Dept. of Commerce, 5225 Port Royal Road, Springfield, VA 22161.

5. Peterson, J. N.; Chen, C. C.; Evans, L. B., "Computer Programs for Chemical Engineers: 1978 - Part 1" Chem. Eng. June 5, 1978; Part 2 - July 3, 1978; Part 3 - July 31, 1978.

6. Chen, C. C.; Evans, L. B., "More Computer Programs for Chemical Engineers", Chem. Eng., May 21, 1979.

7. Weismantel, G. E., "Smoothing Out Wrinkles in Computer-Aided Design", Chem. Eng., July 17, 1978.

8. AIChE Today Series, "Computer Aided Process Design and Simulation", 345 E. 47th St., New York.

9. EDUCOM/CACHE - "Utilization of Networks for the Sharing of Computer Based Resources Within Chemical Engineering", Report to National Science Foundation Workshop, September 28-29, 1978, Washington, D.C., Grant No. MCS78-18288.

10. Carnahan, B.; Mah, R. S. H.; Fogler, H. S., "Computer Graphics in Chemical Engineering Education", CACHE Corporation Report, Cambridge, Mass. (1978).

11. Evans, L. B.; Seider, W. D., "The Requirements of an Advanced Computing System", (1976) Chem. Eng. Prog. 72, No. 6; 80-83.

12. de Leeuw den Bouter, J. A.; Swenker, A. G., "TISFLO, a Flowsheet Simulation Program Based on New Principles", Paper to EFCE Conference, "Computer Application in Process Development", April 1974, Erlangen, Germany.

13. Klemes, J.; Lutcha, J.; Vasek, V., "Recent Extension and Development of Design Integrated System-DIS" CACE '79 EFCE Montreux, April 1979.

14. Maejima, T.; Shindo, A.; Umeda, T., "Computer Aided Process Engineering System - CAPES", Chem. Economy and Eng. Review (1973) 5, No. 2 (No. 58); 34-41.

15. Niida, K.; Yagi, H.; Umeda, T., "An Application of Data Base Management System (DBMS) to Process Design", Comp. & Chem. Eng. (1977) 1, No. 1; 33-40.

16. Rodriguez-Miaja, F. E.; Leesley, M. E., "Computer Aided Project Evaluation for Chemical Process Plants", Computer Aided Design (1979) 11, No. 1; 5-11.

17. Tsubaki, M.; Motard, R. L., "Data Based Process Simulation", CACE '79, EFCE Montreux, April 1979.

18. Waligura, C. L.; Motard, R. L., "Requirements for Data Management in Engineering and Construction", Paper presented at AIChE Meeting, Houston, March 1977.

19. Fredenslund, A.; Gmehling, J.; Rasmussen, P., Vapor-Liquid Equilibria Using UNIFAC, Elsevier Scientific Publishing Co., New York (1977).

20. Fredenslund, A.; Gmehling, J.; Michelsen, M. L.; Rasmussen, P.; Prausnitz, J. M., "Computerized Design of Multicomponent Distillation Column Using the UNIFAC Group Contribution Method for Calculation of Activity Coefficients", I&EC Process Des. Dev. (1977) 16, No. 4; 450.

21. Parker, A. L., "Chemical Process Optimization by Flowsheet Simulation and Quadratic Approximation Programming", Ph.D. Thesis in Chemical Engineering, U. of Wisconsin, 1979.

22. Broyden, C. G., "A Class of Methods for Solving Nonlinear Simultaneous Equations", Math. Comp. (1965) 19, No. 92; 577-593.

23. Broyden, C. G., "The Convergence of an Algorithm for Solving Sparse Nonlinear Systems", Math. Comp. (1971) 25, No. 114; 285-294.

24. Schubert, L. K., "Modification of a Quasi-Newton Method for Nonlinear Equations with Sparse Jacobian", Math. Comp. (1970) 24; 27-30.

25. Duff, I. S., "A Survey of Sparse Matrix Research", Proc. of the IEEE (1977) 65, No. 4; 500-535.

26. Duff, I. S.; Reid, J. K., "Some Design Features of a Sparse Matrix Code", ACM Trans. on Math Software (1979) 5, No. 1; 18-35.

27. George, A.; Lui, J. W. H., "The Design of a User Interface for a Sparse Matrix Package", ACM Trans. on Math Software, (1979) 5, No. 2; 139-162.

28. Stadherr, M. A., "A New Sparse Matrix Method for Process Design", Paper presented at Miami AIChE Meeting, November 1978.

29. Westerberg, A. W.; Berna, T. J., "Decomposition of Very Large Scale Newton-Raphson Based Flowsheeting Problems", Comp. & Chem. Eng., (1978) 2, No. 1; 61-63.

30. Hildalgo, R. S.; Correa, A. V.; Gomez, A. M.; Seader, J. D., "An Optimal Arrangement of Simultaneous, Linearized Equations for General Systems of Interlinked, Multistaged Separators", Paper presented at Houston AIChE Meeting, April 1979.

31. Lin, T. D.; Mah, R. S. H., "A Sparse Computation System for Process Design and Simulation: Part I. Data Structures and Processing Techniques; Part II. A Performance Evaluation Based on the Simulation of a Natural Gas Liquification Process", AIChE (1978) 24, No. 5; 830-848.

32. Hernandez, R.; Sargent, R. W. H., "A New Algorithm for Process Flowsheeting", CACE '79, EFCE, Montreux, April 1979.

33. Storvick, T. S.; Sandler, S. I., Ed., "Phase Equilibria and Fluid Properties in the Chemical Industry", ACS Symposium Series (1977) 60, American Chemical Society, Wash. D. C.

34. Motard, R. L.; Winter, P., "Physical Property Needs in Computer-Aided Process Design", Proceedings of the Fifth Biennial International CODATA Conference, (1977) Pergamon Press, Oxford.

35. Evans, L. B.; Joseph, B.; Seider, W. D., "System Structures for Process Simulation", AIChE J (1977) 23, No. 5; 658-666.

36. Seider, W. D.; Evans, L. B.; Joseph, B.; Wong, E.; Jirapongphan, S., "Routing of Calculations in Process Simulation", Ind. Eng. Chem. Process Design Dev. (1979) 18, No. 2; 292-297.

37. Kaijaluoto, S., "Experiences of the Use of Plex Data Structure in Flowsheeting Simulation", CACE '79, EFCE, Montreux, April 8-11, 1979.

38. Leesley, M. E.; Heyen, G., "The Dynamic Approximation Method of Handling Vapor-Liquid Equilibrium Data in Computer Calculations for Chemical Processes", Comp. & Chem. Eng. (1977) 1, No. 2; 109-112.

39. Barrett, A.; Walsh, J. J., "Improved Chemical Process Simulation Using Local Thermodynamic Approximations", CACE '79, EFCE, Montreux, April 8-11, 1979.

40. Mah, R. S. H., "Effects of Thermophysical Property Estimation on Process Design", Comp. & Chem. Eng. (1977) 1, No. 3; 183-189.

41. Cruz, J. L.; Renon, H., "A New Thermodynamic Representation of Binary Electrolyte Solutions Nonideality in the Whole Range of Concentrations", AIChE J (1978) 24, No. 5; 817-829.

42. Gautam, R.; Seider, W. D., "Computation on Equilibrium in Electrolyte Solutions" CACE '79 EFCE, Montreux, April 8-11, 1979.

43. Ottmers, D. M. Jr., "A Description of the Radian Chemical Equilibrium Program", Technical Note 200-403-69, Radian Corporation.

44. Zemaitis, J. F., Jr.; Rafal, M., "Automatic Program Generation Applied to Chemical Equilibria and Reactions in a Fractionation Tower Design - FRACHEM", Paper presented to AIChE Meeting in Chicago, November 1976.

45. Chem. & Eng. News, "AIChE Forms Group for Physical Property Data", November 27, 1978; 23-24.

46. Boston, J. F.; Britt, H. I., "A Radically Different Formulation and Solution of the Single-Stage Flash Problem", Comp. & Chem. Eng. (1978) 2, No. 2/3; 109-122.

47. Henley, E. J.; Rosen, E. M., "Material and Energy Balance Computations", John Wiley, New York, (1969).

48. Tinoco-Garcia, L.; Cano-Dominquez, J. L., "A New Technique for Solving Multi-Phase Equilibria", Paper presented at Houston AIChE Meeting, April 1979.

49. Gautam, R.; Seider, W. D., "Multiphase Chemical Equilibrium in Process Design", Paper presented at Houston AIChE Meeting, April 1979.

50. Boston, J. F.; Fournier, R. L., "A Quasi-Newton Algorithm for Solving Multiphase Equilibrium Flash Problems", Paper presented at Miami AIChE Meeting, November 1978.

51. Holland, C. D.; Gallun, S. E., "Modifications of Broyden's Method for the Solution of Distillation Problems Involving Highly Non-Ideal Solutions", Paper presented at Houston AIChE meeting, April 1979.

52. Shah, V. B.; Boston, J. F., "An Algorithm for Rigorous Distillation Calculations with Two Liquid Phases", Paper presented at Houston AIChE meeting, April 1979.

53. Ross, B. A.; Seider, W. D., "Simulation of Three Phase Distillation Towers", Paper presented at Houston AIChE Meeting, April 1979.

54. Brannock, V. F.; Vernevil, V. S.; Wong, Y. L., "Rigorous Distillation Simulation with Equality and Inequality Process Specifications", Paper presented at AIChE Meeting in Chicago, November 1976.

55. Boston, J. F., "Algorithms for Distillation Calculations with Bounded-Variable Design Constraints and Equality-or Inequality-Constrained Optimization", Paper presented at Houston AIChE Meeting, April 1979.

56. Bentzen, G. W.; Izarraraz, A.; Anthony, R. G.; Holland, C. D., "Algorithm for Simultaneous Distillation and Reaction", paper presented at Houston AIChE Meeting, April 1979.

57. Tierney, J. W., "Calculation Method for Distillation Systems with Reaction", Paper presented at Houston AIChE Meeting, April 1979.

58. Motard, R. L., "Computational Architectures in Process Simulation", AIChE CAST Division Newsletter, 1978.

59. Rosen, E. M.; Pauls, A. C., "Computer Aided Chemical Process Design: The FLOWTRAN System", Comp. & Chem. Eng. (1977) 1, No. 1; 11-21.

60. ChemShare Corporation, "Distill, Design, Refine", 2500 Transco Tower, Houston, Texas 77027.

61. Brannock, N. F.; Vernevil, V. S.; Wang, Y. L., "Process™ Simulation Program - An Advanced Flowsheeting Tool for Chemical Engineers", CACE '79, EFCE, Montreux, April 8-11, 1979.

62. CONCEPT Mark III, Computer Aided Design Centre, Madingley Road, Cambridge, CB10HB, England.

63. McDonnell Douglas Automation Co., "General Process Simulation Program"; Box 516, St. Louis, MO 63166.

64. Kehat, E.; Shacham, M., "Chemical Progress Simulation Programs 2", Process Technology International (1973) 18, No. 3; 115-118.

65. Henley, E. J.; Williams, R. A., "Graph Theory in Modern Engineering", Academic Press, New York (1973).

66. Gros, H.; Kaijaluoto, S., Mattsson, I., "Some New Aspects on Partitioning and Tearing in Steady-State Process Simulation" in Computer Applications in the Analysis of Chemical Data and Plants, Science Press, Princeton (1977).

67. Venkatesh, C. K., "Computational Precedence Ordering in Modular Cascade Systems", M. S. Thesis in Chemical Engineering, University of Houston, August 1978.

68. Pho, T. K.; Lapidus, L., "Topics in Computer Aided Design. Part I - An Optimum Tearing Algorithm for Recycle Streams", AIChE J (1973) 19, No. 6; 1170.

69. Upadhye, R. S.; Grens, A. E., II, "Solution of Decompositions for Chemical Process Simulation", AIChE J (1975) 21 No. 1; 136.

70. Barchers, D., "Optimal Convergence of Complex Recycle Process Systems", Ph.D. Thesis in Chemical Engineering, Oregon State University, 1975.

71. Orbach, O.; Crowe, C. M., "Convergence Promotion in the Simulation of Chemical Processes with Recycle - The Dominant Eigenvalue Method", Can. J. of Chem. Eng. (1971) 49; 503-513.

72. Kliesch, H. C., "An Analysis of Steady State Process Simulation: Formulation and Convergence", PhD Thesis in Chemical Engineering, Tulane University, 1967.

73. Kluzik, H. A., "A Study of the Simultaneous Modular Convergence of Chemical Process Flowsheets", M. S. Thesis in Chemical Engineering, MIT, Cambridge (January 1979).

74. Metcalfe, S. R.; Perkins, J. D., "Information Flow in Modular Flowsheeting Systems" Trans I. Chem. E. (1978) 56; 210-213.

75. Perkins, J. D., "Efficient Solution of Design Problems Using a Sequential Modular Flowsheeting Programme", CACE '79, EFCE Montreux, April 1979.

76. Agarwal, J. C.; Klumpar, I. V.; Zybert, F. D., "A Simple Material Balance Model", Chem. Eng. Prog. (1978) 74; 68.

77. SYMBOL-Computer Aided Design Centre, Madingly Road, Cambridge CB10HB, England.

78. Sood, M. K.; Reklaitis, G. V., "Material Balance Program - II", School of Chemical Eng., West Lafayette, Indiana (December 1977).

79. Kneile, R. G., "Solution of Material Balance Problems for Process Design", Ph.D. Thesis in Chemical Engineering (December 1975).

80. Mahalec, V.; Kluzik, H.; Evans, L. B., "Simultaneous Modular Algorithm for Steady State Flowsheet Simulation and Design", CACE '79, EFCE Montreux, April 8-11, 1979.

81. Hutchison, H. P., "Plant Simulation by Linear Methods", Trans. Instr. Chem. Engrs. (1974) 52; 287-290.

82. Sood, M. K.; Reklaitis, G. V.; Woods, J. M., "Solution of Material Balances for Flowsheets Modelled With Elementary Modules: The Unconstrained Case" AIChE J (1979) 25, No. 2; 209.

83. Sood, M. K.; Reklaitis, G. V., "Solution of Material Balances for Flowsheets Modelled with Elementary Modules: The Constrained Case" AIChE J (1979) 25, No. 2; 220.

84. JUSE-L-GIFS - Generalized Interrelated Flow Simulation Program. Technical Brief - Paper presented at JAPAN/U.S. Joint Seminar, June 23-27, 1975, Kyoto, Japan.

85. IRI, M.; Tsunekawa, J.; Yajima, K., "The Graphical Techniques Used for A Chemical Process Simulator JUSE GIFS", Information Processing 71 - North Holland Publishing Company (1972).

86. Kubicek, M.; Hlavacek, V.; Prochaska, F., "Global Modular Newton-Raphson Technique for Simulation of an Interconnected Plant Applied to Complex Rectification Columns" Chem. Eng. Science (1976) 31; 277-284.

87. Berna, T. J.; Westerberg, A. W., "Polynomial, Chao-Seader and Newton Raphson - The Use of Partially Ordered Pivot Sequences" DEC-06-1-79 Dept. of Chem. Eng., Carnegie-Mellon University, Pittsburgh, Penn. 15213 (January 1979).

88. Bending, M. J.; Hutchison, H. P., "The Calculation of Steady State Incompressible Flow in Large Networks of Pipes", Chem. Eng. Sci. (1973) 28; 1957.

89. Hutchison, H. P.; Shewchuk, C. F., "A Computational Method for Multiple Distillation Towers", Trans. I Chem E (1974) 52; 325.

90. Gorczynski, E. W.; Hutchison, H. P., "Towards a Quasi-Linear Process Simulator - I. Fundamental Ideas", Comp. & Chem. Eng. (1978) 2, No. 4; 189-196.

91. Gorczynski, E. W.; Hutchison, H. P.; Wajih, A. R. M., "Development of a Modularly Organized Equation - Oriented Process Simulator". CACE '79, EFCE, Montreux, April 1979.

92. MULTICØL, Computer Aided Design Centre, Madingly Road, Cambridge CB10HB, England.

93. Bluck, D.; Hughes, P.; Mallin-Jones, A. K.; Perris, F. A.; Sheppard, A. J., "FLOWPACK II - A Third Generation Flowsheeting System", Paper B-6 to EFCE Conference "Design Congress '76", Aston, England, September 1976.

94. Rosen, E. M., "A Machine Computation Method for Performing Material Balances", Chem. Eng. Prog. (1962) 58, No. 10; 69-73.

95. Weisenfelder, A. J.; Olson, R. E., "Solution of Recycle Streams in Multicolumn Distillation", Paper presented at AIChe Meeting in Houston, April 1979.

96. Shacham, M.; Kehat, E., "The Fraction Separation Method I-For Calculation of Process Flowsheets", Dept. of Chem. Eng., Technion, Haifa, Israel, Report CE-73/74.

97. Sood, M.; Khanna, R.; Reklaitis, G. V., "A Two Level Approach Exploiting Sparsity in Flowsheeting Material Balancing", Paper presented at AIChE Meeting in Houston, April 1979.

98. McLane, M.; Sood, M. K.; Reklaitis, G. V., "A Hierarchial Strategy for Large Scale Process Calculations", CACE '79, EFCE, Montreux, April 1979.

99. Umeda, T.; Nishio, M., "Comparison Between Sequential and Simultaneous Approaches in Process Simulation", Ind. Eng. Chem. Proc. Design & Dev. (1972) 11; 153.

100. Lin, T. D., " A Simultaneous Modular Simulator and A Sequential Block - Modular Simulator for Process Design or Simulations", Paper presented at AIChe Meeting, Houston, April 1979.

RECEIVED November 5, 1979.

2

A Review of Optimization Methods for Nonlinear Problems

R. W. H. SARGENT

Department of Chemical Engineering and Chemical Technology, Imperial College of Science and Technology, London SW7, England

The field of optimization is vast and all-embracing. Relevant papers are published at a rate of more than 200 per month, spread over more than 30 journals, without counting the numerous volumes of conference proceedings and special collections of papers. The Tenth International Symposium on Mathematical Programming held in August this year has alone added 450 papers to the list. Applications are equally varied and widespread.

This review cannot therefore hope to be comprehensive and its scope is firmly restricted to general methods for dealing with nonlinear problems, both with and without constraints, since these are the most common in chemical engineering applications. Integer programming methods are not reviewed, since most of the mathematical developments are concerned with mixed integer-linear problems which are of limited interest to chemical engineers. Branch-and-bound techniques are still the basic tools for nonlinear integer problems, and since heuristics play such an important role the techniques can only be considered in relation to specific applications.

Many specialized techniques exploiting particular problem structures are ignored, and fields which involve considerations outside the question of the optimization techniques themselves are also excluded. Thus for example the whole field of function approximation and model parameter fitting has been left out.

Although there have been significant theoretical advances in recent years, particularly in connection with stability, sensitivity and convergence analysis, these also are largely ignored. The emphasis is on algorithmic developments because to the user the theoretical advances are of no account until they are embodied in implementable algorithms.

0-8412-0549-3/80/47-124-037$05.00/0

Unconstrained Minimization.

The quasi-Newton or variable-metric methods introduced by Davidon {1} have now become the standard methods for finding an unconstrained minimum of a differentiable function f(x), and an excellent review of the basic theory has been given by Dennis and Moré {2}.

These are iterative methods of the form

$$\left.\begin{aligned} x_{k+1} &= x_k - \alpha_k S_k g_k , \\ S_{k+1} &= S(S_k, p_{k+1}, q_{k+1}) , \end{aligned}\right) \quad (2.1)$$

where x is an n-vector, $\{x_k\}$ k=0,1,2, ... is a sequence of iterates with an arbitrary starting point x_o, g_k is the gradient of the function f(x) at x_k, $p_{k+1} = x_{k+1} - x_k$, $q_{k+1} = g_{k+1} - g_k$, and S_k is a local approximation to the inverse of the Hessian matrix of f(x). Classically, the scalar α_k is chosen to minimize the function $f(x_k - \alpha S_k g_k)$ with respect to α.

The methods differ in the formula used to generate the sequence S_k, k=0,1,2, ..., and after Fletcher and Powell's {3} analysis of Davidon's method a whole spate of formulae were invented in the sixties. Broyden {4} introduced some rationalization by identifying a one-parameter family, and recommended a particular member, now commonly referred to as the BFGS (Broyden-Fletcher-Goldfarb-Shanno) formula. Huang {5} widened the family, but by the end of the sixties numerical experience was producing a consensus that the BFGS formula was the most robust of the formulae available. The formula is

$$S_{k+1} = S_k + \frac{1}{\sigma_{k+1}} \{ p_{k+1} - S_k q_{k+1}) p_{k+1}^T + p_{k+1} (\tau_{k+1} p_{k+1} - S_k q_{k+1})^T \}, \quad (2.2)$$

where $S_o = S_o^T$, $\sigma_{k+1} = p_{k+1}^T q_{k+1}$, $\tau_{k+1} = p_{k+1}^T q_{k+1} / q_{k+1}^T S_k q_{k+1}$.

A turning point came with a theorem of Dixon {6}, who showed that all quasi-Newton formulae (those for which $S_{k+1} q_{k+1} = p_{k+1}$) in Huang's family generate identical steps even for general functions, and this directed attention to a choice based on numerical stability rather than on theoretical properties, such as maintenance of positive-definiteness of the S_k {7}. In fact Broyden {4}, Fletcher {8} and Shanno {9} all arrived at the choice of the BFGS formula from consideration of conditioning of the resulting matrices. Shanno and Kettler {10} specifically considered a quantitative criterion for optimal conditioning, while Fletcher {8} was the first to suggest varying the update formula from step to step in the light of such a criterion. The idea was further developed by Davidon {11} and by Oren and

Spedicato {12}, but later Spedicato {13} noted that the criteria used by these authors were identical.

Clearly S_k is related to the function f(x), and in particular it must be scaled in inverse proportion to any scaling of f(x). This led Oren and Luenberger {14} to investigate the symmetric members of Huang's family for which $S_{k+1}q_{k+1}=\rho_{k+1}p_{k+1}$, with the scalar ρ_{k+1} chosen to adjust the scaling of S_{k+1}. This "self-scaling" idea was further developed by Spedicato {15} who considered formulae which were invariant to a scalar nonlinear transformation of f(x), and this also generalizes other attempts to approximate f(x) using more general classes than quadratic functions {16,17,18,19}.

Numerical comparisons of the optimal conditioning and self-scaling ideas with the classical formulae have been published by Spedicato {15,20}, Brodlie {21}, Shanno and Phua {22}, Zang {23} and Schnabel {24}. The evidence is not conclusive, but it seems that the classical BFGS formula is hard to beat. Optimal conditioning involves more arithmetic at each iteration, which pays off only on seriously ill-conditioned problems. There seem to be special types of functions for which self-scaling gives a marked improvement but in general its performance is inferior, and the same seems to be true of the methods based on nonlinear transformations.

The early analysis of Fletcher and Powell {3} interpreted Davidon's method as one which generates conjugate directions, which naturally gives rise to the idea of minimization along these directions. However it was soon realized that minimization to high precision is an unnecessary expense, and indeed is not implied if the formulae are interpreted as secant approximations to the inverse of the Hessian matrix. In fact true minimization must be abandoned in favour of a "descent test" to guarantee convergence in a practical algorithm {25}, and various step-length rules are given by Sargent and Sebastian {7} who showed how algorithms can be designed to ensure global convergence to a stationary point. Numerical experience also shows that the simple Armijo rule {26,25} coupled with a descent test is more efficient than minimization, provided that step-length expansion is also used if the test is satisfied immediately.

For years everyone has been content with algorithms which produce a descent path to a stationary point, which can of course be a saddle-point rather than the desired local minimum. However McCormick {27} has put forward an idea, later developed by Moré and Sorensen {28}, for the use of directions of negative curvature coupled with descent directions to ensure convergence to a local minimum.

The goal of achieving the global minimum rather than just a local minimum still has its attractions. Various approaches are given in the collections of papers edited by Dixon and Szëgo{29}, while the recent "tunnelling algorithm" of Levy and Montalvo {30} seems to be an effective version of the function-modification approach to the problem. An excellent discussion of the issues and the different approaches is given by Griewank {31}.

As computers become more powerful the problems tackled become ever larger, and inevitably storage problems arise. This has revived interest in the conjugate gradient methods, which require storage of only a few n-vectors rather than an nxn matrix. Powell {32} gives an interesting analysis yielding new insight into the working of these methods. He extends the work of Beale {33} and Calvert {34}, giving evidence for favouring a particular conjugate-gradient formula and providing an automatic test for restarting. Even so, conjugate-gradient methods remain less efficient and less robust than quasi-Newton methods, providing an incentive to apply sparse-matrix techniques to the latter.

Now if the Hessian matrix is sparse its inverse is likely to be dense, so instead of (2.1), we use

$$\left.\begin{aligned} x_{k+1} &= x_k - \alpha_k s_k \quad , \\ H_k s_k &= - g_k \quad , \\ H_{k+1} &= H(H_k, p_{k+1}, q_{k+1}) \quad , \end{aligned}\right\} \qquad (2.3)$$

where H_k is an approximation to the Hessian matrix itself, and in order to solve for s_k we store and update the triangular factors of H_k. The techniques for updating sparse triangular factors are given by Toint {35}.

There has been little recent work on methods for differentiable functions which avoid explicit evaluation of derivatives. Powell's conjugate direction method {36} is still used, but the generally accepted approach is now to use standard quasi-Newton methods with finite-difference approximations to the derivatives. On the other hand there has been considerable interest in methods for nondifferentiable functions, as shown by the collection of papers edited by Balinski and Wolfe {37}, in which the technique described by Lemarechal is of particular interest. Other contributions in this difficult field are due to Shor {38}, Goldstein {39}, Clarke {40}, Mifflin {41,42}, Auslender {43} and Watson{44}. In general these problems are much more difficult to solve than those involving differentiable functions, but they are becoming increasingly relevant to optimum design problems involving tolerances {45,46}.

Nonlinear Programming.

The general nonlinear programming problem is

$$\left.\begin{aligned} &\text{Minimize} && f(x) \quad , \\ &\text{subject to} && \phi(x) \geq 0 \quad , \\ & && \psi(x) = 0 \quad , \end{aligned}\right\} \qquad (3.1)$$

where f(x) is a scalar function of the n-vector x, $\phi(x)$ is an m-vector and $\psi(x)$ is a q-vector.

The state of the art in 1974 in dealing with such problems is admirably summarized in the collection of papers edited by Gill and Murray {47}. At that time the middle ground was held by feasible-point projection or reduced-gradient methods, with a strong challenge from augmented Lagrangian methods. Fletcher himself was disenchanted with his "exact penalty-function" method and tended to favour the augmented Lagrangian approach, and there were still strong protagonists for the original penalty-function approach.

The classical penalty-function methods have now finally become part of history, the early promise of the augmented Lagrangian approach has faded, and there has been a coalescence of the approach used in the projection methods with the exact penalty-function approach.

The classical penalty-function idea was to convert the original constrained problem into an unconstrained one by increasing the objective function artificially if the constraints were violated, adding a penalty term reflecting the magnitude of the constraint violations. The method originated with Frisch {48} and Carroll {49} but was mainly developed by Fiacco and McCormick {50}. Good reviews are given by Lootsma {51} and Ryan {47,pp175-190}. The difficulty with the approach is that it is by definition approximate, and to obtain good approximations the constraint violations must be heavily weighted in relation to the objective function, yielding an ill-conditioned unconstrained problem. The practical solution was to solve a sequence of unconstrained problems with steadily increasing weight of the constraint violations, and methods were devized for extrapolating the sequence to infinite weight. In 1968, Powell {52} likened the process to shooting at a target in a strong wind and suggested it was better to "aim off" rather than wheel up heavier and heavier guns; he therefore introduced a shifting parameter for each constraint, adjusted so that the minimum of the penalty function actually satisfied the constraint. A sequence of minimizations is still necessary to adjust the shifting parameters, but these subproblems are much easier to solve.

The "exact penalty-function" idea was to devize a penalty function which has an unconstrained local minimum exactly coinciding with the constrained minimum of the original problem (3.1). This goal seems to have been consciously sought independently by Fletcher {53} and Pietrzykowski {54}, but the idea was already implicit in the work of Arrow and Solow {55} and Zangwill {56}.

The Zangwill-Pietrzykowski penalty function for problem (3.1) is

$$P(c,x)=f(x)+c\{ \sum_{j=1}^{q} | \psi^j(x) | + \sum_{j=1}^{m} \max (0, - \phi^j(x)) \}. \quad (3.2)$$

This function is indeed an exact penalty function for all values of the scalar c above a certain finite threshold value. However it is nondifferentiable, and hence its minimization presents even more severe difficulties than that of the classical penalty functions. The general methods for nondifferentiable functions referred to in Section 2 could be used, but specific methods for (3.2) have been proposed by Conn and his coworkers {57,58,59}, Bertsekas {60} and Chung {61}. More recently Charambalous {62,63} has proposed the use of the more general l_p-norm for the penalty term instead of the l_1-norm used in (3.2), and points out some advantages for a choice $1<p<\infty$, but the penalty function is still nondifferentiable.

It is well known that when (3.1) contains no inequality constraints the Lagrangian function

$$L(x,\mu) = f(x) - \sum_{j=1}^{q} \mu^j \psi^j(x) \qquad (3.3)$$

has an unconstrained stationary point with respect to x and μ at the constrained minimum. Unfortunately however, if the functions $\psi^j(x)$ are nonlinear there is no guarantee that this stationary point is a local minimum - it could be a saddle-point or even a maximum. Hence Arrow and Solow {55} suggested "convexifying" $L(x,\mu)$ in the neighbourhood of the stationary point to make this a local minimum by adding a quadratic penalty term:

$$L(c,x,\mu) = f(x) - \mu^T\psi(x) + \tfrac{1}{2}c\, \psi^T(x).Q.\psi(x), \qquad (3.4)$$

where c is a scalar and Q a positive definite matrix. For a given Q this function has a local minimum for all values of c above a certain threshold, and hence is a differentiable exact penalty function. Moreover since c is finite the unconstrained problem is not usually ill-conditioned. In fact Arrow and Solow considered only Q=I, and they proposed a continuous descent method for the minimization; they also showed that inequality constraints could be dealt with by the use of slack variables.

Independently of this work, Fletcher {53} started with (3.4) and sought to make μ and Q continuous functions of x which would converge to the required values at the stationary point. Later {64} he generalized the approach to deal with inequality constraints, and showed that the Lagrangian function for (3.1):

$$L(x,\lambda,\mu) = f(x) - \lambda^T\phi(x) - \mu^T\psi(x) \qquad (3.5)$$

is itself an exact penalty function if the multipliers λ,μ are obtained at each iteration by solving the quadratic programme:

$$\left.\begin{aligned} &\text{Minimize} \quad f_x(x).\delta + \tfrac{1}{2}c\delta^T\delta \ , \\ &\text{subject to} \quad \phi(x) + \phi_x(x).\delta \geq 0 \ , \\ &\qquad\qquad\quad \psi(x) + \psi_x(x).\delta = 0 \ , \end{aligned}\right\} \qquad (3.6)$$

for given x, and c sufficiently large.

Unfortunately, to compute derivatives of $L(x,\lambda(x),\mu(x))$ with respect to x involves second derivatives of the functions $f(x)$, $\phi(x)$, $\psi(x)$, although Fletcher {47,pp219-239} suggests the use of approximations involving only their first derivatives. When inequality constraints are present, the exact penalty function again has discontinuities in its derivatives, but in fact these do not occur in the neighbourhood of the required stationary point and hence are not as serious as those involved in (3.2).

The idea of adding a penalty term to the Langrangian function was exploited in a different way by Hestenes {65}, who again used Q=I. He proposed a two-stage algorithm, solving a sequence of problems in which (3.4) is minimized with respect to x for fixed values of c and μ, with intermediate adjustment of these latter variables. This turns out to be mathematically identical to Powell's "aiming off" method, and is often referred to as the "Powell-Hestenes multiplier method". Later Rockafellar {66} extended the method to deal with inequality constraints, deriving a modified penalty term through the use of slack variables. In fact there are many alternatives for the methods of adjusting c and μ, and the form of the penalty term can also be varied. As a class these methods are known as "augmented Lagrangian methods", anduseful reviews have been given by Bertsekas {67} and Fletcher {47,pp219-239}. An excellent discussion of the problems of estimating Lagrange multipliers has been published by Gill and Murray {68}. Recently Rockafellar {69,70} has further extended the method by adding a second penalty term, $(\mu/2c)\,|x-x_k|^2$, where x_k is the last iterate, which can be used to force global convergence. He terms this the "proximal method of multipliers" because it is derived from the proximal point algorithm for maximal monotone operators, as described in the paper.

In fact the first effective nonlinear programming algorithm was Rosen's {71} "gradient projection" method. Starting at a feasible point, the method involved at each iteration a step in the steepest descent direction i.e. (along the projected gradient) within a current set of "active" constraints (i.e. constraints treated as equalities). If the minimum along this direction violated any of the inactive constraints, the step was made to the nearest such constraint, which was added to the active set; otherwise the step was made to the minimum. If the projected gradient was zero, indicating a constrained minimum in the current active set, the Lagrange multipliers were used to test whether further progress could be made by dropping a constraint from the active set. For nonlinear constraints, linear approximations were used in the above strategy, and at the end of each iteration a simplified Newton correction procedure was used to recover a point satisfying the active nonlinear constraints.

In 1967, Goldfarb {72} improved the method by using the projected gradients in Davidon's variable-metric formula to generate the steps. In 1968, Murtagh and Sargent {73} proposed a "variable-metric projection" version of Rosen's method, using the rank-one quasi-Newton formula to give the metric, and two years later {74} they revised the algorithm by altering the active set selection strategy to retain the smallest possible active set at each step, and by approximating the inverse of the Hessian matrix of the Lagrangian function (with respect to the x-variables) rather than the objective function, in order to allow for curvature of nonlinear constraints. They also suggested the possibility of using a penalty function to limit constraint violations and assist the correction procedure, and introduced a step-reduction procedure to ensure descent steps and hence force global convergence. An initial feasible point was obtained by applying the algorithm itself, with the penalty term alone as objective function.

Meanwhile in 1967 Wolfe {75} had proposed his "reduced gradient" method for quadratic programming, which in 1968 was generalized to deal with nonlinear programming by Abadie and his coworkers {76}, who later {77} used the reduced gradients in a conjugate-gradient method to generate the steps.

Many other variants on these basic themes are possible, but Sargent {47,pp149-174} has shown that all these methods merely involve different kinds of projection, and essentially amount to solving a sequence of quadratic programming problems, with linear approximations to the constraints but different choices of the objective function. Since in general these quadratic programmes have a non-feasible starting-point Sargent suggested their solution by Lemke's algorithm for linear complementarity problems, and later {78} described a stable implementation of this algorithm for the purpose.

In fact, in an unpublished thesis in 1963, Wilson {79} had already proposed a method based on solving a sequence of quadratic programmes, with the Lagrangian function as objective function, but he had evaluated all the required first and second derivatives of the (original) objective and constraint functions analytically, which of course involves heavy computations. In 1972, Robinson {80} proposed an algorithm which solved a sequence of linearly constrained problems, with a general nonlinear objective function, and showed that it was locally quadratically convergent; however the subproblem is itself an infinite problem so this too is not a practical algorithm. In 1976, Garcia-Palomares and Mangasarian {81} suggested using simple quadratic approximations in Wilson's or Robinson's algorithms, studied conditions for different local rates of convergence, and proposed on specific algorithm (with local R-superlinear convergence) based on a quasi-Newton update of an approximation to the complete Hessian matrix of the Lagrangian, with respect to both the x-variables and λ-multipliers. Han {82} proposed the use of a quasi-Newton update on the portion of the Hessian matrix corresponding to the x-variables (as used in the Murtagh-Sargent algorithm) and proved local Q-superlinear convergence for the resulting algorithm. All these local convergence results require very strong conditions, which were slightly weakened by Powell {83} for the Han algorithm.

These studies have shown that recursive quadratic programming methods have very satisfactory local convergence properties, with superlinear convergence in the final stages whether or not the iterates x_k are feasible (with respect to the nonlinear constraints). However they say nothing about global convergence from arbitrary starting points. In their algorithm, Murtagh and Sargent ensured global convergence by using a descent condition on the objective function, which is clearly sufficient in the first phase, and also in the second phase because the algorithm then generates feasible points. If feasibility of the iterates is not forced, then forcing a monotonic decrease of the objective function will not work, since an unfeasible iterate may have a lower value than any feasible point. Some balance is required between reduction of the objective function and of the constraint violations, and this leads us back naturally to the exact penalty -function methods. However the descent tests normally used for unconstrained problems {25} rely on differentiability of the objective function, so that even Fletcher's exact penalty function is not satisfactory for global convergence, unless the descent test can be modified.

In 1975 Han {84} resolved this impasse by showing that if an exact penalty function is merely used in association with the descent test in the step-reduction procedure, and not to replace the objective function in forming the quadratic programmes, then the required properties are obtained. He established his result for the particular penalty function given in (3.2) with a computable lower bound on c, noting however that strictly it involved exact minimization along the search direction and a more practical procedure was required. Powell {85} reported poor results with this algorithm, mainly because of the rule used for adjusting c, and suggested using the Lagrangian (3.5) instead, without however proving that it had the required properties. Later {86} he reverted to Han's algorithm using (3.2), but used a heuristic rule to adjust the value of c, improving performance on some problems but of course relinquishing the guarantee of global convergence. This algorithm gave excellent results on a few standard test-problems.

Mayne and Polak {87} point out several disadvantages of the Han algorithm. There is the problem of choosing c, and the problem that exact minimization of the penalty function is strictly required to ensure convergence, as already mentioned. Then there is the possibility that the quadratic programme may not have a solution, and if it does there is the question of ensuring that the direction generated by it is a descent direction. Finally there is the difficulty that step-reduction to ensure global convergence can destroy the fast superlinear convergence in the final stages; they give an example illustrating this last failure. They then give a modified algorithm avoiding these difficulties, together with the requisite proofs. Briefly, they give a rigorous rule for adjusting c, an Armijo-type rule to replace the exact minimization, and use a search arc rather than a search direction, which ensures ultimate superlinear convergence. If the quadratic programme fails to have a solution or fails to generate a descent direction they switch to a first-order step selection procedure which maintains global convergence, and they show that this will not be invoked in the final stages, so that it does not impair the final convergence rate.

The unanswered questions are whether the use of the penalty function, with the associated rule for adjusting c, does indeed slow down convergence, and whether the first-order "rescue procedure" is frequently invoked. They use a general secant formula to update the Hessian matrix, which could make their algorithm vulnerable to descent direction failure, but other updating formulae could be substituted. For example Powell {85,86} used the BFGS formula, modified to maintain positive-definiteness and hence ensure descent directions, but this modification would again destroy the superlinear convergence if its use persisted to the final stages.

Very recently Sargent {88} has shown that, although the Lagrangian function is not an exact penalty function, it can indeed be used in a descent test to force global convergence. Moreover the test is satisfied without step-reduction in the final stages, so the convergence is superlinear. Thus the penalty parameter and its associated problems are eliminated. The algorithm also ensures that a regular quadratic programme with a positive definite Hessian matrix is obtained at each step, provided that this is so for the initial point. Thus, although a "first-phase" procedure may sometimes be required to locate such a point, no rescue procedure is needed subsequently.

Of course proof of global convergence for any algorithm requires that the problem itself has a regular solution, and that the functions do not behave pathologically over the region covered by the iterations. For real problems the functions are usually well behaved in the feasible region, but extrapolation too far outside this may well induce pathological behaviour. For such situations there may therefore still be a case for including a correction procedure to limit constraint violations.

Thus there are still some issues to be explored, but these recent developments are all really refinements, and from the user's point of view there are now available several very efficient algorithms for solving general nonlinear programming problems. What must now be done is to address the problems of dealing with large-scale nonlinear programmes. In the sixties and early seventies there was a great deal of activity on techniques for decomposing large structured optimization problems, but little has been reported recently. Two papers however are particularly relevant to the above developments. Murtagh and Saunders {89} implemented large-scale versions of variants of the Murtagh-Sargent and Robinson algorithms, obtaining similar numerical performance on a variety of test problems. They do not use decomposition but implemented the algorithms directly, using sparse matrix techniques. Berna, Locke and Westerberg {90} implemented Powell's 1977 version of Han's algorithm, using decomposition to eliminate a large proportion of the variables, then using sparse matrix techniques to deal with the reduced problem.

The relative ease with which large-scale versions of these algorithms could be produced is an encouraging development. It means that we now have the capability of solving realistic engineering problems by standard nonlinear programming algorithms, without the need for clever exploitation of particular features of the problem to make the computations practicable.

Acknowledgement

I should like to thank Peggy Browett, our indefatigable librarian, for her efficient help in obtaining many of the references cited.

Literature Cited

1. Davidon, W.C., "Variable Metric Method for Minimization", A.E.C. Research and Development Report 1959, ANL-5990.
2. Dennis, J.E., and Moré, J.J., "Quasi-Newton Methods: Motivation and Theory", SIAM Review 1977, 10, 46-89.
3. Fletcher, R., and Powell, M.J.D., "A Rapidly Convergent Descent Method for Minimization", Computer Jnl,1963-4 6,163.
4. Broyden, C.G., "Quasi-Newton Methods and their Application to Function Minimization", Math.Comp., 1967, 21,368.
5. Huang, H.Y., "A Unified Approach to Quadratically Convergent Algorithms for Function Minimization", JOTA, 1970 6 (3), 269.
6. Dixon, L.C.W ., "Quasi-Newton Algorithms Generate Identical Points. Math. Programming, 1972, 2 (3), 383-387.
7. Sargent, R.W.H., and D.J. Sebastian, "Numerical Experience with Algorithms for Unconstrained Minimization" in F.A. Lootsma (Ed)., "Numerical Methods for Nonlinear Optimization", Academic Press 1972, pp45-68.
8. Fletcher, R., "A New Approach to Variable Metric Algorithms" Computer Jnl., 1970, 13, (3), 317-322.
9. Shanno, D.F., "Conditioning of Quasi-Newton Methods for Function Minimization", Math.Comp., 1970, 24, 647-656.
10. Shanno, D.F., and Kettler, P.C., "Optimal Conditioning of Quasi-Newton Methods", Math.Comp., 1970, 24, 657-664.
11. Davidon, W.C., "Optimally Conditioned Optimization Algorithms without Line Searches", Math.Programming, 1975, 9(1), 1-30.

12. Oren, S.S., and Spedicato, E. "Optimal Conditioning of Self-scaling Variable Metric Algorithms", Math. Programming, 1976, 10, (1), 70-90.
13. Spedicato, E., "On a Conjecture of Dixon and Other Topics in Variable Metric Methods" Math.Programming, 1978, 15, (2), 123-129.
14. Oren, S.S., and Luenberger,D.G.,"Self-scaling Variable Metric (SSVM) Algorithms", Parts I and II, Management Science, 1974, 20, 845-862, and 863-874.
15. Spedicato, E., "A Note on the Determination of the Scaling Parameters in a Class of Quasi-Newton Methods for Unconstrained Minimization", J.Inst.Maths.Applics. 1978 21,285-291.
16. Fried, I., "N-Step Conjugate Gradient Minimization Scheme for Non-quadratic Function Minimization", AIAAJ., 1971, 9,2286-87.
17. Jacobson, D.H., and Oksman, W.,"An Algorithm that Minimizes Homogeneous Functions in N Variables in N+2 Iterations and Rapidly Minimizes General Functions" J.Math.Anal. and Applics., 1972, 38, 535.
18. Kowalik, J.S., and Ramakrishnan, K.G., "A Numerically Stable Optimization Method Based on a Homogeneous Function", Math.Programming 1976, 11(1), 50-66.
19. Davison, E.J. and Wong, P., "A Robust Algorithm that Minimizes L-Functions in a Finite Number of Steps and Rapidly Minimizes General Functions, Assuming the Derivatives are not Available", University of Toronto Control Systems Report No.7313,1973.
20. Spedicato,E., "Computational Experience with Quasi-Newton Algorithms with Minimization Problems of Moderately Large Size", Report CISE-N 175, Milano,1975.
21. Brodlie, K.W., "An Assessment of Two Approaches to Variable Metric Methods", Math.Programming, 1977, 12 (3),344-355.
22. Shanno,D.F., and Phua, K.H., "Matrix Conditioning and Non-linear Optimization", Math.Programming, 1978, 14(2), 149-161.
23. Zang,I., "A New Arc Algorithm for Unconstrained Optimization" Math.Programming, 1978, 15 (1), 36-52.
24. Schnabel, R.B., "Optimal Conditioning in the Convex Class of Rank-two Updates", Math.Programming,1978,15(3),247-260.
25. Sargent, R.W.H., and Sebastian, D.J.,"On the Convergence of Sequential Minimization Algorithms". JOTA ,1973 ,12(6) 567-575.(See also JOTA, 1974, 14, 439-442).
26. Armijo, L., "Minimization of Functions having Continuous Partial Derivatives", Pacific J.Math., 1966, 16, 1-3.
27. McCormick, G.P., "A Modification of Armijo's Step-size Rule for Negative Curvature", Math.Programming, 1977, 13(1),111-115.
28. Moré,J.J. and Sorensen, D.C., "On the Use of Directions of Negative Curvature in a Modified Newton Method", Math.Programming 1979,16(1), 1-20.

29. Dixon, L.C.W. and Szëgo, G.P. (Eds.) "Towards Global Optimization", Vol.1, 1975, Vol.2, 1978, North Holland (Amsterdam).
30. Levy, A.V., and Montalvo,A., "The Tunnelling Algorithm for the Global Minimization of Functions", presented at Biennial Numerical Analysis Conference, Dundee 1977.
31. Griewank, A., "A Generalized Descent Method for Global Optimization", MSc Thesis, Australian National University, 1977.
32. Powell, M.J.D., "Restart Procedures for the Conjugate Gradient Method", Math.Programming, 1977, 12(2),241-254.
33. Beale, E.M.L., "A Derivation of Conjugate Gradients" in F.A. Lootsma (Ed) "Numerical Methods for Nonlinear Optimization", Academic Press, London 1972,pp39-44.
34. Calvert, J.McI., "A Comparison of Several Conjugate Gradient Algorithms", MSc Dissertation, Imperial College 1972.
35. Toint, P.L., "On Sparse and Symmetric Matrix Updating Subject to a Linear Equation", Math.Comp. 1977,31,954-961.
36. Powell, M.J.D., "An Efficient Method of Finding the Minimum of a Function of Several Variables without Calculating Derivatives", Computer Jnl., 1964, 7,155.
37. Balinski, M.L., and Wolfe, P., (eds.), "Nondifferentiable Optimization", Math.Programming Study 3, North Holland Amsterdam, 1975.
38. Shor, N.Z., "A Class of Almost-differentiable Functions and a Minimization Method for this Class", Cybernetics, 1974, 8(4), 599-606.
39. Goldstein, A., "Optimization of Lipschitz Continuous Functions", Math.Programming, 1977, 13(1),14-22.
40. Clarke, F., "Generalized Gradients and Applications", Trans.Am.Math.Soc. 1975, 205,247-262.
41. Mifflin, R., "Semismooth and Semiconvex Functions in Constrained Optimization", SIAM Journal on Control and Optimization, 1977, 15, 959-972.
42. Mifflin, R., "An Algorithm for Constrained Optimization with Semismooth Functions", Maths.Oper.Research 1977, 2, 191-207.
43. Auslender, A., "Minimisation de Fonctions Localement Lipschitziennes: Applications à la Programmation Mi-convexe, Mi-differentiable", in O.L. Mangasarian, R.R. Meyer and S.M. Robinson (eds)"Nonlinear Programming 3", Academic Press, New York, 1978, pp 429-460.
44. Watson, G.A., "The Minimax Solution of an Overdetermined System of Nonlinear Equations", J.Inst.Maths.Applics. 1979, 23, 167-180.

45. Mayne, D.Q., Polak, E. and Trahan,R., "An Outer Approximation Algorithm for Computer Aided Design Problems", Memorandum No.UCB/ERL M77-10, 1977.

46. Grossmann, I.E., and Sargent, R.W.H., "Optimum Design of Chemical Plants with Uncertain Parameters", AIChE Journal, 1978, 24,(6),1021-28.
47. Gill, P.E. and Murray, W. (Eds.), "Numerical Methods for Constrained Optimization", Academic Press, London, 1974.
48. Frisch,R., "The Logarithmic Potential Method for Solving Linear Programming Problems", Memorandum of the University Institute of Economics, Oslo, 1955.
49. Carroll, C.W., "The Created Response Surface Technique for Optimizing Nonlinear Restrained Systems", Oper.Res.1961 9,169-184.
50. Fiacco,A.V. and McCormick,G.P., "Nonlinear Programming: Sequential Unconstrained Minimization Techniques", Wiley, New York, 1968.
51. Lootsma, F.A., "A Survey of Methods for Solving Constrained Minimization Problems via Unconstrained Minimization", in F.A. Lootsma (Ed.) "Numerical Methods for Nonlinear Optimization", Academic Press, New York, 1972, pp313-348.
52. Powell, M.J.D., "A Method for Nonlinear Constraints in Minimization Problems", in R. Fletcher (Ed.) "Optimization" Academic Press, London, 1969, pp283-298.
53. Fletcher, R., "A Class of Methods for Nonlinear Programming with Termination and Convergence Properties", in J. Abadie (Ed.) "Integer and Nonlinear Programming", North Holland, Amsterdam, 1970, pp157-175.
54. Pietrzykowski, T., "An Exact Potential Method for Constrained Maxima", SIAM J. Numer.Anal. 1969, 6,(2),299-304.
55. Arrow, K.J., and Solow, R.M., "Gradient Methods for Constrained Maxima with Weakened Assumptions", in Arrow, K.J., Hurwicz, L. and Uzawa, H., "Studies in Linear and Nonlinear Programming", Stanford University Press, Stanford, 1958, pp.166-176.
56. Zangwill, W.I., "Nonlinear Programming via Penalty Functions", Management Science, 1967, 13(5), 344-358.
57. Conn,A.R., "Constrained Optimization Using Nondifferentiable Penalty Functions", SIAM Journal Numer.Anal., 1973 10,760-784.
58. Conn,A.R., and Pietrzykowski,T., "A Penalty Function Method Converging Directly to a Constrained Optimum", SIAM Journal Numer.Anal.1977,14,348-375.
59. Charambalous, C., and Conn, A.R., "An Efficient Method to Solve the Minimax Problem Directly, SIAM Journal Numer. Anal. 1978, 15,162-187.
60. Bertsekas, D.P., "Nondifferentiable Optimization via Approximation", Math.Programming Study 3, 1975, 3, 1-25
61. Chung, S.M., "Exact Penalty Algorithms for Nonlinear Programming", in O.L. Mangasarian, R.R. Meyer and S.M. Robinson, (Eds.) "Nonlinear Programming 3, Academic Press, New York, 1978, pp197-224.

62. Charambalous, C., "Nonlinear Least pth Optimization and Non-Linear Programming", Math.Programming, 1977, 12(2), 195-225.
63. Charambalous, C., "A Lower Bound for the Controlling Parameters of the Exact Penalty Functions", Math. Programm-ing, 1978, 15(3), 278-290.

64. Fletcher, R. "An Exact Penalty Function for Nonlinear Programming with Inequalities", Math. Programming, 1973 5(2),129-150.
65. Hestenes, M.R., "Multiplier and Gradient Methods", JOTA 1969 4,303-320.
66. Rockafellar, R.T., "A Dual Approach to Solving Nonlinear Programming Problems by Unconstrained Optimization", Math.Programming, 1973, 5 (3), 354-373.
67. Bertsekas,D.P., "Multiplier Methods: A Survey", Automatica, 1976. 12, 133-145.
68. Gill, P.E., and Murray, W., "The Computation of Lagrange-Multiplier Estimates for Constrained Minimization", Math.Programming, 1979, 17(1), 32-60.
69. Rockafellar, R.T., "Augmented Lagrangians and Applications of the Proximal Point Algorithm in Convex Programming, Maths.Oper.Res. 1976 1,97-116.
70. Rockafellar, R.T., "Monotone Operators and Augmented Lagrangian Methods in Nonlinear Programming", in O.L. Mangasarian, R.R. Meyer and S.M. Robinson (Eds.) "Nonlinear Programming 3", Academic Press, New York, 1978,pp1-26.
71. Rosen, J.B., "The Gradient Projection Method for Nonlinear Programming: Part I Linear Constraints, Part II: Nonlinear Constraints", J.Soc.Indust. Appl.Math.1960 8(1),181-217, and 1961, 9(4),514-532.
72. Goldfarb,D., "Extension of Davidon's Variable Metric Method to Maximization under Linear Inequality and Equality Constraints", SIAM J. Appl. Math. 1969,17,739-764.
73. Murtagh, B.A., and Sargent, R.W.H., "A Constrained Minimization Method with Quadratic Convergence", in R.Fletcher (Ed.), "Optimization, pp215-246, Academic Press, London, 1969.
74. Sargent, R.W.H., and Murtagh, B.A., "Projection Methods for Nonlinear Programming", Math.Programming, 1973, 4(3), 245-268.
75. Wolfe, P., "Methods of Nonlinear Programming", in R.L. Graves and P. Wolfe (Eds.), "Recent Advances in Mathematical Programming", McGraw Hill, New York, 1963, pp67-86.
76. Abadie, J., and Carpentier, J., "Generalization of the Wolfe Reduced Gradient Method to the Case of Nonlinear Constraints", in R. Fletcher (Ed.)"Optimization, Academic Press, London, 1969, pp49-64.
77. Abadie, J., "Application of the GRG Algorithm to Optimal Control Problems", in J. Abadie (Ed.) "Integer and Nonlinear Programming", 1970, North Holland, Amsterdam, pp191-214.

78. Sargent, R.W.H., "An Efficient Implementation of the Lemke Algorithm and its Extension to Deal with Upper and Lower Bounds", in Math.Programming Study 7, North Holland, Amsterdam, 1978, pp36-54.

79. Wilson, R.B., "A Simplicial Method for Convex Programming", PhD Dissertation, Harvard University, 1963.

80. Robinson, S.M., "A Quadratically-convergent Algorithm for General Nonlinear Programming Problems", Math.Programming, 1972, 3(2), 145-156.

81. Garcia-Palomares, U.M. and Mangasarian, O.L., "Superlinearly Convergent Quasi-Newton Algorithms for Nonlinearly Constrained Optimization Problems", Math. Programming, 1976, 11(1), 1-13.

82. Han, S.P., "Superlinearly Convergent Variable-metric Algorithms for General Nonlinear Programming Problems", Math.Programming, 1976, 11(3), 263-282.

83. Powell, M.J.D., "The Convergence of Variable Metric Methods for Nonlinearly Constrained Optimization Calculations" in O.L. Mangasarian, R.R. Meyer and S.M. Robinson (Eds) "Nonlinear Programming 3, Academic Press, New York, 1978, pp27-64.

84. Han, S.P. "A Globally Convergent Method for Nonlinear Programming", JOTA, 1977, 22(3),297-309.

85. Powell, M.J.D., "Algorithms for Nonlinear Constraints that Use Lagrangian Functions", Math.Programming, 1978, 14(2),224-248.

86. Powell, M.J.D., "A Fast Algorithm for Nonlinearly Constrained Optimization Calculations", in G.A. Watson (Ed) "Numerical Analysis-Proceedings Biennial Conference, Dundee 1977", Lecture Notes in Mathematics, 630, Springer Verlag, Berlin, 1978, pp144-157.

87. Mayne, D.Q., and Polak, E., "A Superlinearly Convergent Algorithm for Constrained Optimization Problems", presented at Biennial Numerical Analysis Conference, Dundee, 1979.

88. Sargent, R.W.H., "On the Parametric Variation of Nonlinear Programmes and the Convergence of Nonlinear Programming Algorithms",paper presented at "The Tenth International Symposium on Mathematical Programming", Montreal, August, 1979.

89. Murtagh, B.A., and Saunders, M.A., "An Implementation of Some Lagrangian-based Algorithms for Sparse Nonlinear Constraints", UNSW Report 1979/OR/5, University of New South Wales, 1979.

90. Berna, T.J., Locke, M.H. and Westerberg, A.W., "A New Approach to Optimization of Chemical Processes", Report DRC-06-3-79, Carnegie-Mellon University 1979.

RECEIVED December 13, 1979.

A Review of Process Synthesis

ARTHUR W. WESTERBERG

Department of Chemical Engineering, Carnegie-Mellon University,
Pittsburgh, PA 15213

The purpose of the paper is to present a review of the area of chemical process synthesis. As two earlier review articles on synthesis have already appeared (Hvalecek (1978) and Hendry, Rudd and Seader (1973)), this paper will list most but will not review all previous synthesis publications. This paper will respond to the often asked question, "Has anything really useful come from the research activity called process synthesis?"

A Definition of Synthesis

We shall start by defining the term "synthesis." The activities of design are 1) synthesis — the step where one conjectures the building blocks and their interconnection to create a structure which can meet stated design requirements, 2) analysis — the activity of modeling and then solving the resulting equations to predict how a selected structure would behave if it were constructed, 3) evaluation — the activity of placing a worth on the structure where the worth might be its cost, its safety, or its net energy consumption, and 4) optimization — the systematic searching over the allowed operating conditions to improve the evaluation as much as is possible. The design engineer moves iteratively through each of these activities, developing more and more details and/or a better understanding about the design with each iteration.

Synthesis is the inventive step in the design of a process. It is often the most enjoyable one. The primary research question is: Can the synthesis activity be automated and, if so, to what extent? An interesting paradox occurs because if "synthesis" as an activity is automated, is the activity correctly labeled "synthesis?" Using a dictionary definition, synthesis should be based on inductive (going from the particular to the general) reasoning whereas analysis is deductive (going from the general to the particular). If process design follows a series of automatic steps, leading more or less to a preordained solution, was not this solution deduced from the given problem definition? We may only be arguing semantics here, but this point

0-8412-0549-3/80/47-124-053$08.75/0

leads one to restate the important question in process synthesis: How much of the structure can be deduced from the statement of the design requirements for a problem? Stated this way the activity may no longer appear so mysterious an activity. The computer should indeed be able to invent process structure when that structure can be deduced from the problem statement. It is questionable if the computer can be used with a high degree of success to discover automatically a design rule by solving several particular "synthesis" problems and from these discovering an unanticipated rule. In contrast, the human is often very successful at this activity.

Restating, the computer should be effective in developing structure based on well-defined technology, where the heuristic rules which guide it to find successful designs quickly are provided as part of its programming. An example is for the computer to select a heat exchanger network to exchange heat among several hot and cold streams to minimize or reduce the use of utilities. The selection of such a network amounts to a clever search among a very large number of reasonably well-defined alternatives. The computer should be ineffective in automatically developing a new rule which might be obvious to an engineer after performing a large number of heat exchanger network designs. Such a rule might be that the best networks normally result when one attempts at each step to minimize the temperature driving forces within each exchanger, i.e. when one attempts to minimize irreversibilities.

Clearly the engineer and the computer can complement each other, and one should be striving for synthesis tools where they do. Such a view of synthesis suggests that the right synthesis computer programs are those which permit the designer to put new heuristics into them as he or she (and very likely not the computer) discovers them.

The computer can still be used in this manner to find unexpected solutions. For example, Johns (1977) proposed an intriguing use of synthesis which uses the strength of the computer to search quickly for alternatives using known technology but with the hope of finding new solutions. He suggested allowing the computer to use an infeasible step at no cost in developing a solution. For example, known technology may not include a method to separate component A from component B. Thus, if A and B are not to end up mixed, the computer must either avoid mixing them, or, if allowed one infeasible step, it could develop a structure which includes the infeasible step of splitting A from B. If this infeasible step reduces the cost by $7 million/yr, one could then decide if that makes it worth looking for new technology to split A from B.

Douglas (1977, 1979) has been pursuing a quite different approach to process synthesis. He has been developing techniques which permit the designer to analyze and optimize alternative configurations quickly and by hand. His "synthesis" strategy is to teach people to recognize valid approximations for the problem at hand; it has proved a rather controversial exercise since people claim it is not new in concept. Indeed, it is not since hand approximations were the tools of design before the computer. However, Douglas is attempting to push these techniques even further than before to gain problem insight as well as good first approximations to the solution of a process design problem.

We have also used the approach (Westerberg (1978)) to develop a set of hand analysis techniques, which we use in our undergraduate design course. This approach works. Valid designs almost always result. We shall not be stressing these ideas here, however, since the purpose of this paper is to examine automatic synthesis techniques. We only mention these ideas here to make people think twice about ignoring the talents of the engineer when developing a synthesis tool.

The activity of synthesis occurs throughout a design, from original process conception to construction and operation. Examples are 1) the synthesis of a control system for fixed plant configuration, 2) the synthesis of a start-up procedure for a new process, 3) the selection of materials of construction and 4) the layout of the plant to improve safety. All of these activities involve discrete decision making; this aspect is a common denominator of all synthesis activities.

A more formal definition of process synthesis is possible. It is formally a nonlinear mixed integer and continuous variable optimization problem, generally a very large one. The selection of the building blocks and how they are to be interconnected may be formulated as a set of discrete decisions represented by a set of zero/one variables. For a particular set of discrete decisions one must determine the optimal operating levels for the corresponding process structure; these levels are represented generally by continuous variables. Certain discrete decisons disallow others which may be formally stated using inequality constraints among the discrete variables. We could continue, but by now the correspondence should be evident.

The Research Problems of Synthesis

The principal research problems of synthesis were stated by Simon (1969) in an essay on design. They are 1) representation of the alternatives, 2) analysis and evaluation of each alternative and 3) the strategy of searching among the alternatives.

The representation problem can be stated briefly as: Can a representation be developed which is rich enough to allow all alternatives to be included and clever enough to exclude automatically ridiculous options? Simon argues by example that the discovery of a correct representation may often convert what appears to be a very difficult problem into one for which the solution is easily stated. To appreciate the representation problem, consider the difficulty of uniquely representing an arbitrary organic molecule. Beside giving a unique representation, it should also allow one to discover quickly functional groups on a molecule. A second example is to develop a suitable unique representation for a sequence of separation units, allowing for thermally coupled distillation columns among the alternatives. What is needed in each case is a representation which aids in solving the synthesis problem.

The analysis and evaluation problem involves modeling the alternatives in an appropriate fashion and then developing an evaluation criterion so that one can compare them to each other. Developing a suitable criterion is often a nearly impossible task. How does one compare the safety of two processes? For the problem of reaction path synthesis, how does one compare alternative chemical routes to the same molecule when one cannot predict kinetics? If it proves possible to develop a suitable evaluation function, then one must still be able to do the analysis and evaluation quickly. One is usually faced with an enormous number of alternatives in synthesis, and a trade off is necessary between analysis and evaluation speed and accuracy.

Lastly, the search strategy problem is to develop a strategy to locate quickly the better alternatives without totally enumerating all options. While synthesis problems are finite, one is faced with an enormous number of alternatives even for simple problems. To illustrate, one can enumerate 47 different structures for exchanging heat among two hot and two cold streams where one disallows the splitting of any of the streams, where no two streams may exchange heat between them more than one time, and where each match must involve heat transferring only from a hot stream to a cold stream. The structures include the "null" structure (no exchange), four structures with only a single heat exchange, eight with two exchanges which both involve the same stream, etc. If one has three cold and two hot streams, the number of alternatives is 847.

A realistic process may involve 20 streams and an incredible number of alternative structures. Industrial practice allows one to split a stream and to rematch streams so the actual synthesis problem is much larger than even these numbers indicate.

Simple separation sequences give rise to similar results. Thompson and King (1972) developed a formula for predicting the number of different structures which exist for simple separation sequences. A simple separator is one which splits a multicomponent mixture into two products, the two products having no common components. For such a problem the number of structures possible is $(2(N-1))!\ S^{N-1}/((N-1)!\ N!)$ where N is the number of components in the multicomponent mixture and S the number of different separation methods. The following table indicates the size of this problem. The two structures for N=3 components (say A, B and C) and S=1 method (distillation) are 1) separate A from BC in the first column and then separate B from C in the second and 2) separate AB from C in the first and then A from B in the second.

Thus, one must develop a strategy to locate quickly the better alternatives for these virtually infinite problems, a strategy which cannot require one to enumerate all options.

N	S	No. of Structures
3	1	2
3	2	8
5	1	14
5	3	1134
⋮	⋮	⋮
9	5	558, 593, 750

Classification of Synthesis Problems

A problem arises when one attempts to classify the various approaches to synthesis because at least three classification schemes exist. The synthesis approaches might be classified as being predominantly heuristic or predominantly algorithmic. An heuristic approach is one which relies on rules of thumb to search quickly among the alternatives. For example, one may select the separation sequence based on simple distillation columns which always makes the least costly next split at each step through the column sequence. An algorithmic approach is one which states the exact synthesis problem to be solved and then guarantees to find the best solution among the alternatives possible. The exact synthesis problem is usually a much simplified one and one whose size is often significantly reduced first using heuristics.

A second classification for synthesis schemes, and the one we shall use here, is by the nature of the synthesis problem—heat exchanger networks, separation schemes, reaction paths, control systems, fault trees, total flowsheets, etc. The third classification is based on whether the synthesis method being proposed 1) begins with a feasible flowsheet and the method seeks to improve it or 2) begins with no structure and the method seeks to find a good initial candidate. For class (1) the methods frequently encountered include evolutionary ones, where one makes small changes to the structure in the search for improvements, and embedding, where one proposes a superstructure which includes all possible alternatives. Optimization is used in embedding to find the "best" substructure which solves the synthesis problem. For class (2) one will encounter heuristic based methods and algorithmic methods based on integer programming (e.g. branch and bound) or dynamic programming, etc.

We shall now examine several areas of synthesis, discussing the key contributions for each. We shall start with the synthesis of heat exchanger networks, an area for which industrially significant results exist.

Heat Exchanger Network Synthesis. The typical problem for heat exchanger network synthesis is stated as follows. Given a set of hot streams to be cooled and another set of cold streams to be heated plus a set of available heating and cooling utilities, develop the least cost network of heat exchangers to accomplish the required heating and cooling. The costs include the annual cost of buying the utilities and the cost of purchasing the equipment. For this problem one is given the process stream flow rates and their inlet and exit temperatures, the heat transfer coefficients for all possible matches, the inlet and exit temperature for all utilities, the cost of each utility and the cost on an annual basis of a heat exchanger versus its area.

Several representations exist for a heat exchanger network problem. The temperature (ordinate) versus stream heat content (e.g. kJ/s) diagram is one of the oldest. The heat content is on a relative basis so each stream once drawn may be moved right or left relative to any other. Figure 1 illustrates.

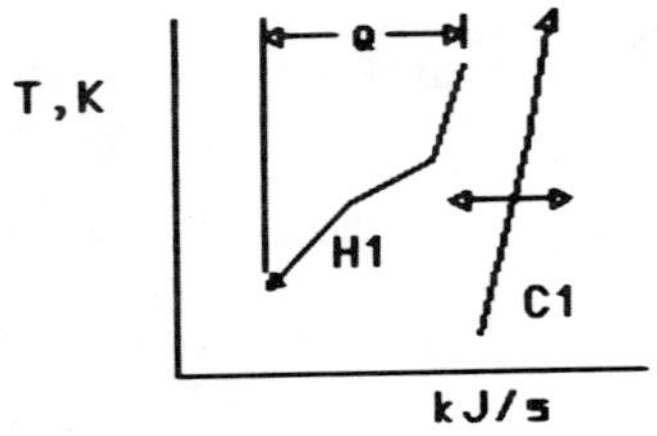

Figure 1. Stream heat content diagram

A second and very useful representation is the so-called area oriented "heat content" diagram of Nishida, Kobayashi and Ichikawa (1971). It is a plot of T (ordinate) versus heat content per degree (kJ/sK), and the heat content of each stream is represented as an area. Figure 2 illustrates. The area is equal to the amount (e.g. kJ/s) of heat available or needed by each stream. The first representation is the integral of the second. A third representation which is not related to these two is by Linnhoff and Flower (1978) and is one which shows the structure of the network as well as the relative temperatures of the streams. It does not directly show the quantity of heat in each stream except by labeling it. Figure 3 illustrates.

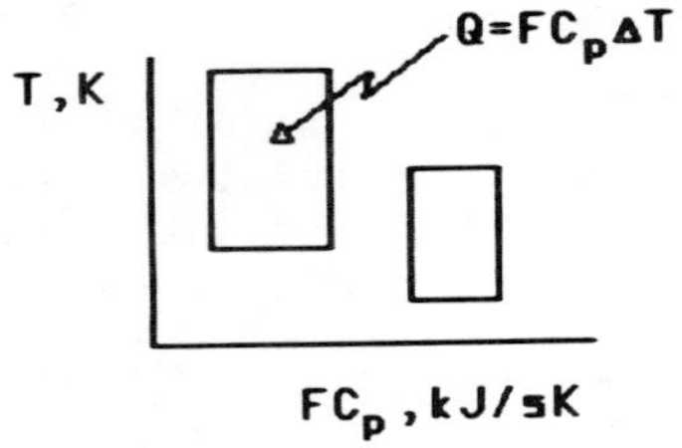

Figure 2. "Area-oriented" heat content diagram

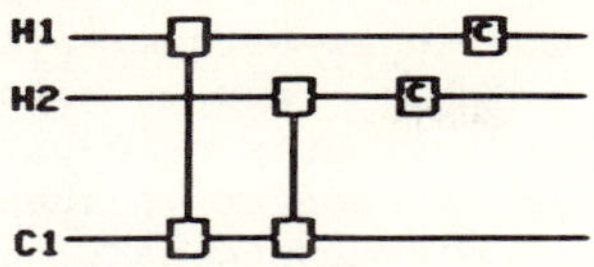

Figure 3. Linnhoff and Flower heat exchanger network representation

The first paper to discuss the heat exchanger network synthesis problem appears to be that by Hwa (1965) who used separable programming techniques.

Kesler and Parker (1969) and others since (Kobayashi et al. (1971); Cena et al. (1977)) have formulated the problem as an assignment problem in linear programming. In an assignment problem formulation, each stream is partitioned serially into a sequence of small substreams, each requiring the exchange of Q units of heat. Q is chosen so the heat requirement for each process stream is, to an adequate approximation, an integer multiple of Q. A binary (zero/one) variable is associated with each potential match of a hot substream or utility with a cold substream or with a cold utility. Undesirable or infeasible matches are readily excluded with constraints or artifically high costs. The difficulty with this approach is posing the objective function. The real objective function cannot be linear because a large heat exchanger is much less expensive per unit area than a small one. The method cannot account for this nonconvexity of cost versus area. Thus the method willingly creates networks with large numbers of heat exchangers. Nonetheless it appears to generate good results easily even for quite large problems particularly when utility costs dominate. The usual approach to reduce the number of exchangers is to make a sequence of small changes manually to the network which the program produces.

Masso and Rudd (1969) proposed the next algorithm, one in which the computer was to learn how to construct a heat exchanger network. Masso and Rudd conjectured that the better networks could be found by making a sequence of decisions about matching streams. They suggested a list of plausible heuristics which could be used to make the next stream/stream heat exchange decision. Once made, as much heat as possible was exchanged between the two selected streams. Then the list of heuristics was used again to select the second stream/stream match. Scores were generated automatically for the heuristics by solving the complete network problem repeatedly, raising the heuristic scores used for each step if the result was a better network, lowering them if a worse one. Thus a best first match heuristic was found, a best second match heuristic and so forth. The results demonstrated how difficult it is to get the computer to do pattern matching as this approach was interesting but not very successful.

Several articles followed in quick succession using a variety of algorithms—branch and bound, heuristics, etc. Some restrict the problem solution to no stream splitting and/or to no rematching of streams. Another class of algorithms, as with the linear programming approach, creates networks with an enormous number of split streams and exchangers. The heuristic based algorithms are very fast and often successful but are also often very unsuccessful.

One might classify several of these heat exchanger network synthesis algorithms into two broad classes. There are several algorithms which view the synthesis problem as one which selects the next hot process stream/cold process stream match to make. The difficult issues are the definition of a match and which match to make next.

Masso and Rudd (1969), Lee, Masso and Rudd (1970) and Pho and Lapidus (1973) start a match with the two stream inlet conditions and exchange all the heat possible, terminating when one stream reaches its target outlet temperature or when a temperature pinch occurs. Ponton and Donaldson (1974), Donaldson, Paterson and Ponton (1976) and Grossmann and Sargent (1978) start a match at the hot end of both streams (or as close to the hot end of the cold stream as possible). The idea is to introduce necessary utilities at their least cost level while exchanging as much heat as possible. Rathore and Powers (1975) do both, getting two alternative matches and obviously many more alternative networks.

To solve a problem of this first class is to perform a tree search. Ponton and Donaldson use heuristics to select each next match and find only one solution, often a good one but not always. Pho and Lapidus propose a total enumeration scheme, but, for very large problems (10 streams), suggest a fallible look-ahead strategy to eliminate branches. Lee, Masso and Rudd, Rathore and Powers, and Grossmann and Sargent propose using branch and bound strategies. The enormous number of alternatives generated by Rathore and Powers limits their algorithm. After locating the better structure using a branch and bound algorithm Grossmann and Sargent perform a continuous variable optimization, several times reducing costs by 10 to 30%. Note they have partitioned the problem into one of finding the better discrete decision variable values first, using heuristics to set the values for the continuous variables. Then, with fixed discrete variables, they optimize the continuous variables.

For problems of this first class, stream splitting is not permitted.

The second class of algorithms merge all heat sources or sinks operating at the same temperature level. Thus if parts of two or more hot streams have heat available over the same temperature range, those parts are merged and treated as one larger heat source in that temperature interval. The first or outer level decisions are to select which merged heat sources supply heat to which merged heat sinks. These first level decisions effectively partition the problem into a set of smaller ones. In each of these small problems one must select which stream parts within the merged streams are to exchange heat. With this second approach, one should expect to develop networks which are more thermodynamically reversible because they will have a more uniform average ΔT driving force throughout than results when using the first approach. Stream splitting is readily accommodated. The networks will likely have more heat exchangers too however, a disadvantage.

Algorithms falling into the second class include those by Nishida, Kobayashi and Ichikawa (1971), Hohmann and Lockhart (1976) and Linnhoff and Flower (1978a,b).

In a sense the assignment problem in linear programming based algorithms belong to this second class of algorithms.

The really significant results for the heat exchanger network synthesis problem have been given first by Hohmann (1971) and then later by Linnhoff and Flower and others (Linnhoff and Flower (1979a), Flower and Linnhoff (1978), Linnhoff, Mason and Wardle (1979)). These results are as follows.

1) One can readily determine the minimum amount of each type of utility needed to effect the needed heating and cooling <u>without</u> having to develop a heat exchanger network to accomplish it.

2) One can easily predict the likely least number of heat exchangers required in this network.

All the example problems in the literature demonstrate that the better networks, from an annualized cost point of view, use the minimum amount of utilities and the fewest exchangers. Thus it is a significant insight to see that one can calculate these bounds before developing a network.

The utility usage bounds are discovered quite easily by using the temperature versus heat content (kJ/s) representation (see Figure 1). One simply merges (see Umeda, Itoh and Shiroko (1978)) all the hot streams into one super hot stream and all the cold streams into a super cold stream. These merged super streams are each treated as a single stream and are moved to the right or left until the minimum vertical approach distance of the cold beneath the hot stream just equals the allowed minimum ΔT (say $5^{o}C$) for the network. Exchange of heat can occur between the hot and cold streams where the hot stream is directly above the cold on this diagram. The leftover portions of each must be dealt with using utilities.

The fewest exchangers usually needed is N-1, where N is the total number of both process and utility streams involved in the network.

Several schemes suggest themselves for using these bounds. Obviously one can develop any network he or she chooses subject to safety constraints, controllability constraints, etc., and then compare the results to these bounds. If the network is on these bounds, one probably has found an economic network. If far from these bounds, one should attempt to improve the network he or she has developed.

A second scheme is to use these bounds as a means to estimate quickly the potential heat exchanger network costs for a flowsheet. Thus several different process flowsheets might be compared quickly without developing the exchanger network in detail.

A third scheme developed by Flower and Linnhoff (1978) is to generate directly all networks having the two properties of minimum utility usage and fewest exchangers. Remarkably few (under 10) networks result for the example problems they use to illustrate their approach. These are considered then as prime candidates for being the optimal network.

Linnhoff and Flower (1978a) note that, if a pinch point occurs for a network, then it effectively decomposes the heat exchanger synthesis problem into two disjoint synthesis problems. No heat can be exchanged between two streams where one is above the pinch and the other is below if one wishes to use the minimum amount of utilities. Note that this partitioning may increase the fewest exchangers possible as each partition must be treated separately.

Cerda and Westerberg (1979) have generalized the solution to the minimum utility usage problem by showing how to solve it when particular hot stream/cold stream matches are forbidden in total or in part. In this form, the design engineer can discover the effect of forbidding a particular match or set of matches on the minimum utility costs. The analysis is extremely fast and is readily implemented as an interactive program.

Hohmann (1971) and later Umeda, Niida and Shiroko (1979) and Umeda, Harada and Shiroko (1979) indicate how to use the heat content diagram (see Figure 1) to find the modifications which lead to improvements in the heat integration for a given flowsheet configuration. A total cooling curve for all heat sources is drawn against a total heating curve for all heat sinks, including all utilities. These curves are then moved together until they reach an allowed minimum approach temperature. This moving together translates into a heat integration among the sources and sinks since it will show where certain sources can be used to heat certain sinks. More importantly it also shows where the curves "pinch," that is, the points on the total heating and cooling curves which produce the limiting approach temperature. The next step is to modify the flowsheet (done manually) in such a way that the particular streams involved in the pinch move apart on the temperature versus heat content diagram, while creating no pinch conditions elsewhere on the diagram. Changes include increasing a distillation column pressure so the condenser becomes a hotter source stream, reducing a column pressure so the reboiler becomes a colder heat sink, introducing a portion of the reboiler duty to a column part way up the column (see Peterson and Wells (1977)) and thus at a lower temperature, and running a reactor slightly hotter or colder. After making process alternations that move apart the temperatures which create the pinch, the two curves on the temperature versus heat content diagram can be moved together even more until another pinch point occurs. Again, this pinch point suggests the place where useful modifications might be made. This process continues until no further modifications exist, which hopefully occurs when the curves pinch in several places along their entire length. Each modification probably costs money which can be balanced against the savings in utilities. The engineer can stop when the modification costs more than the savings or when he or she does not like the process resulting from the heat integration.

There is no doubt that these insights lead to useful tools for the design engineer, and industry can no longer afford to ignore them. If no other results came from the synthesis literature, these alone would justify its existence.

Future synthesis questions to be answered in energy integration should include mechanical energy as well as heat energy. For example, the energy conservation schemes required for an ammonia plant require one to consider turbines as well as heat exchangers.

Separation System Synthesis. The first paper which deals with separation system synthesis appears to be that by Lockhart (1947). The next was by Harbert (1957) and third by Rod and Marek (1959). Then a flood of papers started around 1970. The majority of these papers deal with finding separation sequences comprising simple separators only; that is, each component in the feed to a separator leaves the separator in only one of its two product streams. A second assumption made is that each separator method, e.g. simple distillation, extractive distillation with extractive agent S1, or extractive distillation with extractive agent S2, imposes an ordering on the components in the feed mixture to the problem. That is, each separator method has its own component ordering, where the first component in the ordering is the most "volatile" and the last the least "volatile." Splits using a particular method can only occur between adjacent components on the ordered list for that method, and all those components above the split on the ordered list go to one product, all those below to the other.

The representation used for this problem is always some variation of the split list, originally described by Hendry and Hughes (1972). Figure 4 illustrates.

Thompson and King (1972) presented a computer program to invent separation system flowsheets based entirely on simple separators. The program is entirely heuristic, is very fast when it works, but has a problem with cycling and thus not always terminating with an answer.

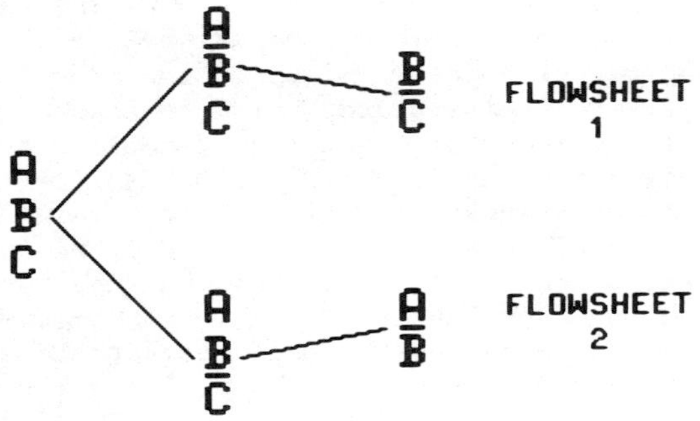

Figure 4. Split-list representation for simple separation sequence

Simultaneous with the above paper, Hendry and Hughes (1972) presented an algorithm for the same problem but based on dynamic programming. Following these were a number of papers for the same problem using branch and bound algorithms (Westerberg and Stephanopoulos (1975), Rodrigo and Seader (1975), Gomez and Seader (1976)) and evolutionary search (Stephanopoulos and Westerberg (1976)).

Of the algorithms for "list splitting" sequences, one which seems rather successful even when applied by hand combines heuristics and evolution (Seader and Westerberg (1977)). First a set of heuristics is given in a rank ordering, with the heuristic deemed most effective by the authors given first, the one which is deemed next most effective second, etc. To solve a separation synthesis problem, these heuristics are used to invent a good first solution. Often the heuristics are in conflict, but the higher ranked ones are used in preference to the lower ranked ones initially. Then, using a set of evolutionary rules, the designer invents a complete set of neighboring structures to his first solution. (A much improved set of rules to develop neighbors is given by Westerberg in the notes for the AIChE advanced seminar on Process Synthesis (Motard and Westerberg (1978)). The neighbors are placed onto an ordered list, where the ordering indicates the rank of the conflicting heuristic which suggests this neighbor might be better. Next this list of neighbors is searched in sequence until either a neighbor gives an improved structure or until no neighbor gives an improvement. In the first case, the neighbor giving an improvement becomes the next candidate solution; in the second case the search stops. A variation to reduce the effort is to drop all neighbors from the list not supported by any heuristic.

Nath and Motard (1978) present a closely related algorithm but also include an heuristically invented screening function to order the list of neighbors. They also invent a neighboring structure by destroying all "downstream" structure from the point where the change occurs and using the heuristics to invent the needed additional structure.

Another class of separation problem attacked has been that of designing the most effective thermally coupled distillation column arrangement to separate a multicomponent mixture. Sargent and Gaminibandara (1975) present a general column superstructure which they optimize. Imbedded in the superstructure are all the alternative thermally coupled and ordinary column sequences to be considered. The optimization eliminates those portions of the superstructure which are not economic leaving, hopefully, the optimal substructure.

Tedder and Rudd (1978) present the results of a study of the economics of separating a variety of three component mixtures using a variety of thermally coupled and ordinary columns. Their goal was to expose trends. They show which feed characteristics favor which column structure.

No one as yet has developed an algorithm suitable for a general separation problem which uses flash units, bleed streams, mixers, etc., as well as "list splitter" columns in the solution. Such an algorithm would have to handle partial splits. At best the algorithms for a total flowsheet might give an approach but not a solution to this much more general problem.

Synthesis of Separation Systems with Heat Integration. A variation on the synthesis of separation systems which leads to an even more complex synthesis problem is to synthesize such systems while also attempting to reduce the use of utilities in the reboilers and condensers. A reboiler is a heat sink or, in our earlier terminology, a cold stream which needs to be heated. A condenser is a heat source, that is a hot stream. Within a single column, the cold stream (reboiler) is at a higher temperature than the hot stream (condenser). To get a match between them requires the expenditure of work, which is of course the idea behind a vapor recompression cycle. On the other hand the condenser of one column can become the source of heat for a reboiler of another. Also the column temperature levels can be increased or decreased by increasing or decreasing column pressure. Increasing pressure will usually reduce the relative volatilities, thus increasing the number of stages and the reflux rates needed to obtain a given separation. We therefore have the intriguing problem of selecting the best internal flows, pressures, and column sequence to accomplish the desired separation task while saving to the extent possible on the use of utilities. The first efforts at handling this problem were by Rathore and Powers (1974a, 1974b). They developed a highly combinatorial approach based on dynamic programming. Sophos, Stephanopoulos and Morari (1978) have developed a related approach but advocate an branch and bound algorithm based on dual bounding through the use of Langrangian techniques.

Faith and Morari (1979) further develop the ideas of using dual bounding through the use of Lagrangian techniques for this problem. They describe refinements which allow one to make a good first estimate to the Lagrange multipliers (needed for the bounding) and to develop rather easily a "lower" lower bound. They still need to apply these ideas to several examples to demonstrate their effectiveness.

A subproblem to the above is to fix the separation sequence and attempt to improve the heat integration by adjusting pressures, using intermediate column reboilers (reboilers part way up the column) and intermediate condensers, using vapor recompression, etc. King, Gantz and Barnes (1972) very successfully developed by hand a sequence of evolutionary steps to improve heat integration within a demethanizer column. The ideas we discussed earlier of Umeda, Niida and Shiroko (1979) were first applied to heat integration of columns and provide a real tool for this subproblem.

Thermally coupled columns are also a means to heat integrate column sequences, and, while not mentioned yet in the literature specifically as part of a synthesis strategy, it is clear they should be. One serious difficulty associated with such column arrangements has been the lack of shortcut design procedures for them so one can quickly investigate among the alternatives. Cerda (1979) has produced a suite of interactive programs for this problem and will be reporting on them at the November AIChE meeting this year (Cerda and Westerberg (1979a)).

Reaction Path Synthesis. For the material in this section the author is relying heavily on material originally prepared by Motard (Motard and Westerberg (1978)). The usual synthesis problem here is to find a sequence of reactions which can be used to reach a given target molecule from a catalogue of available raw materials. Almost all the work on this problem has been done by organic chemists with the dominant names in the field being Corey (see Pensak and Corey (1977)), Wipke (see Wipke et al. (1977, 1978)), Gelertner (a computer scientist—see Gelertner et al. (1977)), Hendrickson (1976) and Ugi (see Brandt et al. (1977)). In chemical engineering, Govind, Blower and Powers (1976) and Agnihotri and Motard (1978) have been the most active researchers.

This synthesis problem has problems with every aspect: representation, evaluation and strategy, with evaluation probably being the most difficult. Basically three representations exist. The ones by Corey, Wipke and Gelertner are rather direct, using essentially linked lists and a variety of function codes to encode the molecule. Hendrickson has a more generalized representation in which he classifies each carbon atom in a molecule by the types of bonding it has with other carbon atoms and by the types of heteroatoms attached to it. His representation does not distinguish among heteroatoms, except to note if they are more or less electronegative than carbon. With this representation a number of molecules are treated as equivalent if they have the same structure and the same general functionality groups attached. Govind, Blower and Powers extend this representation and use it in their work. The last representation is that by Ugi where a molecule is represented by a bond electron connection matrix. The bond electron connection matrix (BECM) is a symmetric matrix whose rows and columns are the individual atoms in each of the molecules being considered. A particular molecule is represented by matrix elements a_{ij} where a_{ij} is the number of electrons in the bond between the two atoms i and j. A reaction is represented by a matrix which when added to the BECM for the reactants yields the BECM for the products. Ugi shows that these reaction matrices are restricted as to their structure. Agnihotri and Motard use Ugi's representation.

As indicated earlier, evaluation is probably the most difficut problem. How does one conjecture the value of an arbitrary reaction? Thermodynamics can tell one how far a reaction may proceed, but no one can yet predict how fast a reaction will proceed since the rate depends on such things as the effect of catalysts. Also assessing the processing costs is a complex exercise. Generally, evaluation is done very indirectly in these programs by allowing the use only of named chemistry and then, for those reactions which might work, other effects such as stereo hindrance are sometimes examined.

The synthesis strategy is almost always to start from the target molecule and work backwards to the source material chemicals. In this manner one selects the source material chemicals indirectly. To synthesize a target molecule by working forward from a prescribed set of source material chemicals may be possible if one has particularly good insight, but such an approach will generally be less efficient and may lead to no solution if one guesses incorrectly on which materials to use initially.

The one implicit heuristic encoded in almost all the programs produced so far is that each step backward from the target molecule must be from a source molecule which is structurally no more complex than the immediate target. Thus reaction paths are not found which pass through complex intermediate molecules.

The search strategies seem to be variations on two: develop the alternatives 1) by looking at functionality and 2) by looking at skeletal structure of the molecule. To look at functionality is to look for such things as double bonds, -COOH groups, -OH group and the like. To look at the skeletal structure on the other hand is to look for patterns in the molecules, like repeated rings. The first is a microscopic view and the second a macroscopic view. Hendrickson prefers the latter with most others the former. In the former, typically a computer routine or equivalent is written for each reaction type, with perhaps 300 to 500 such routines existing. Each routine has a check list at the beginning. The check list looks at the target molecule to see if it has the correct functional groups, to see if the carbon "β" to the reaction site has the right appendages, etc. If that list can be passed through successfully with a target molecule, the reaction becomes a candidate. For each step backward from a target molecule, each of the 300 to 500 reactions is rated and only those with the higher ratings are kept. In one set of programs the user is then presented with the preferred options, and he or she interactively selects those he or she wishes to pursue. Hendrickson prefers applying his method by hand. He has written a book on his method which contains a very long list of so-called half reactions and a general synthesis tree that one uses to develop his or her reaction paths.

Some of the computer programs developed to do reaction path synthesis apparently are very impressive. Subject to the above restriction forbidding complex intermediates, some of them may well out-perform many organic chemists for the class of problems they can handle.

May and Rudd (1976) present a very interesting paper on reaction path synthesis for a special class of inorganic reactions which they titled solvay clusters. A solvay cluster is a sequence of reactions equivalent to one overall net reaction which refuses to go at reasonable industrial conditions. To be a cluster the reaction sequence must make use of other species which are first used and then fully recovered later in the sequence, an example being

Cluster

$$2HCl + Mn_2O_3 + 2MnCl_3 = 2MnO + H_2O + 2MnCl_4$$
$$2MnCl_4 = 2MnCl_3 + Cl_2$$
$$2MnO + CO = Mn_2O_3 + C$$
$$C + H_2O = CO + H_2$$

Net Reaction

$$2HCl = H_2 + Cl_2$$

Each of the reactions in a successful cluster will proceed at reasonable industrial conditions.

May and Rudd show that clusters are very conveniently represented pictorially using what they term "nested polygons." Thus one can quickly generate alternative solvay clusters. They also describe a convenient method to represent pictorially the free energy of the reactions versus reaction conditions (T and P). Using this representation one can quickly select reactions sequences which form a thermodynamically viable solvay cluster. Included in the article are some clever ways to extend the list of potential reactions.

As May and Rudd point out, their approach only allows one to find a thermodynamically plausible cluster, but one will need further evaluation methods to see if the reactions rates are adequate and if the implied processing problems can be reasonably solved.

Total Flowsheet Synthesis. Several approaches to total flowsheet synthesis appear in the literature. The first was the AIDES system developed by Siirola, Powers and Rudd (1971). A second is the BALTAZAR system by Mahalec and Motard (1977a,b). The latest and one which seems to be very intriguing is by Johns and Romero (1979). In these efforts the problem solved is to develop an entire process flowsheet given the reactions to be allowed, desired product streams and available raw materials. The programs automatically invent the entire structure needed from a functional point of view including separators, reactors, splitters, mixers, pressure changers, temperature changers, and the routing of all flows including recycles where appropriate. It is probably fair to say that these programs do a reasonable job on many problems but that they cannot yet compete with a competent design engineer.

The representation for a total flowsheet is handled similarly by AIDES and BALTAZAR. Their view is to transform streams defined by temperature, pressure, total flow, state, chemical species and their mole fractions by applying a sequence of tasks such as temperature changing (heaters, coolers), pressure changing (valves, pumps, compressors), composition and flow changing (separators, mixters, splitters) and kind of species (reactors).

Johns and Romero view streams more crudely. They use binary (0 or 1) flags to note the presence or absence of a component or to note that a stream is at high or low pressure or high or low temperature or to note the presence or absence of a contaminant, etc. Their description of a process is to list all possible streams represented by their binary flag pattern and then to work out a structure which connects them "optimally" to meet the stated design goals.

In each case the evaluation of a flowsheet is done very crudely, with such guidelines as minimum mass flow into separation tasks being used. Of course such crude evaluation is all that can be done in the preliminary stages because one is not able to evaluate on an economic basis until the process is fully synthesized and the equipment selected; these programs stop short of selecting actual equipment.

For these programs the problem is to create a structure which can transform raw materials sources into desired product streams. BALTAZAR uses basically a depth first strategy followed by evolution to discover its solution. AIDES on the other hand uses a breadth first solution, selecting the better alternatives by doing a one step look ahead at each decision step.

Johns and Romero view the synthesis problem as one which enumerates all streams, and, whenever a transformation is needed between two streams, their approach is to look at all process options capable of doing that transform and picking the best. If one has three alternative simple separation methods (e.g. distillation, extractive distillation with extractive agent S1, and crystallation) to separate a 10 component mixture, the Thompson and King (1972) formula mentioned earlier predicts over 9.5×10^7 alternative configurations. However, only 1023 distinct process streams can ever occur in all these processes. Thus the Johns and Romero strategy is to represent a process by the total list of all the streams it can produce, viewing the streams crudely so only a small number of them exist. They estimate the problem size will grow by 3^n where n is the number of binary flags used to describe a stream. They conjecture about 12 such flags can readily be handled and that this will be sufficient for most preliminary synthesis problems.

BALTAZAR starts with a product stream and makes it the first goal to meet. It searches among the source streams (initially, the raw material sources) to see if they can supply this goal in terms of species and their amounts. Sources are rank ordered depending on how close they are to the goal—is it exactly what is needed except for amount, does it have only species needed in the goal and no others, does it have any species needed plus others which are not needed, etc.

The closest stream is selected as the source and appropriate splitters, mixers and separator tasks are invented to transform this source stream into the goal stream. While working on these transformations other goals and sources are likely created. Goals arise because the source may only provide part of the original product stream; the part not provided becomes a new goal. Sources arise because a source may have left over portions not needed for the goal. These goals and sources are listed, with the last goal listed being the next one to be satisfied. If a species in a goal is missing from all sources, a reaction must be selected which has the missing specie as a reaction product. The reactor feed becomes a new goal. In this manner an entire structure is created which can produce the first product before the second product is looked at, thus it can be viewed as a depth first strategy.

When developing the structure needed for the second product, all the streams of the existing structure can be used as sources thus the second product knows of the first, but not the reverse. BALTAZAR is therefore sensitive to the order in which it looks at the products. To remove this dependence, BALTAZAR goes into an evolutionary mode, where all structure for the first product is destroyed. Then the structure is reinvented, but this time the structure for the other products exists. The structure is repeatedly destroyed and reinvented for each product in sequence until the structure does not change. Few iterations seem to be required to get the structure to stabilize, and the result seems not to depend on the original order in which the products were listed.

AIDES has to be given the particular reactions to be used for the process. Its first step is to select which raw materials it will use and how much of each by selecting those which cost the least overall. It assumes all reactant species will be recycled and ultimately converted when selecting the amounts needed.

Next, it does a species allocation, deciding in one step where all species shall pass through the flowsheet. A scoring function is created for each species going from a source to each potential destination. This scoring function attempts to account for potential separation costs which might result if the match were made. Which separation tasks will occur depends on all matches made so this scoring step cannot be done exactly. The essential idea is that after working out all possible species source/destination matches, developing a score for each, the total species matching is done all at once (by solving a linear program). The matching is then fixed. Next the separation tasks which actually result are looked at and solved. These result from having different species in a source stream matched to different destination streams.

AIDES uses this look ahead, scoring, and then making of all match decisions at once as its way of developing the better flowsheets. It can be viewed as a breadth first algorithm therefore. From the evidence, it is not yet possible to say which approach is better, that used by AIDES or that by BALTAZAR. Of interest is that two such different approaches exist, and both seem to get reasonable structures.

Johns and Romero use a very different strategy to create a flowsheet. As stated earlier their approach is based on finding the best transformations among the small number of streams which can exist, regardless of the process structure. Obviously they need a crude definition for a stream to keep the number of streams small. For separation sequences, where each separator sharply splits the feed into two products, each of which have no species in common, the crude stream approximation is in fact accurate and was the first problem solved using this approach (Johns (1977)).

The synthesis starts with a single specified feed and a list of acceptable process output streams. It also starts with a list of possible unit transformations. The units are defined by conditions which must be met by the input streams and limitations which describe the outputs in terms of inputs. No more than two input or two output streams are allowed for a unit. A crude cost correlation is also required for each unit.

The algorithm is one which starts with the specified input stream and systematically works its way through a process to allowed output streams. Using a mixture of branch and bound and dynamic programming based arguments, the algorithm locates the least cost flowsheet structure. In essence, any time a pair of streams within the process can be connected by a unit, then all units are examined which can make that connection and the least cost one selected. The calculations associated with the enormous number of alternative structures are very significantly reduced by this two level approach.

Johns and Romero discuss how the program may be used to solve a variety of synthesis problems including reactors, multiple (more than two) input and/or output units, processes with recycles and/or bypass streams. Each involves using the program by being clever on the unit input/output restrictions and/or the allowed process products. In some instances one risks generating unrealizable flowsheets, particularly for the recycle case, but they give "gambits" for overcoming such difficulties.

Johns and Romero justify their very crude stream definition by arguing that the program is for very early process synthesis, to discover the general structure likely to be needed. The algorithm appears both interesting and restricted by the approach, but one might guess many of the restrictions may disappear when one finds a clever way around them, still within the "spirit" of the program. It seems unlikely the method can deal effectively with a heat exchanger network synthesis problem. It is clear that it can deal very well with synthesizing a sequence of sharp simple separators.

Friedler, Blickle, Gyenis and Tajan (1979) describe an algorithm for generating all alternative flowsheets which can be produced from given raw materials, products and single step transformations. The method uses an abstract algebraic method. They define as the basic "operand" in their algebra, a process stream at a given set of conditions so, if the temperature or flow rate or composition is altered, one has a different operand. Like Johns and Romero, it appears they must use a crude stream definition (i.e. they must judiciously discretize the continuous variables) to prevent an explosive growth in the number of operands and a resulting combinatorial growth in the number of flowsheets. While it is interesting to be able to generate all possible flowsheets, one must still remember that, for most synthesis problems, the problem is that there are far too many for total enumeration to be practical. It is difficult to assess the power of the approach until some example problems are presented.

Another class of total flowsheet synthesis algorithms is that of "embedding," where (as stated earlier) a superstructure is invented which contains all the desired flowsheet arrangements as substructures within it. Optimization is used to remove those portions of the superstructure which are not worthwhile, leaving the optimal flowsheet.

This approach still requires one to invent the superstructure before it is similar in capability to the three algorithms just described. However, as with the Romero and Johns view, a superstructure may be relatively small because of the limited number of streams possible.

Several papers on this approach convert the optimization problem to one using continuous "structural parameter" variables (Umeda, Hirai and Ichikawa (1972); Umeda, Shindo and Tazaki (1972); Osakada and Fan (1973); Ichikawa and Fan (1973); Mishra, Fan and Erickson (1973a,b); Himmelblau (1975); Sargent and Gaminibandara (1975)). Here the idea is to split the exit flow of each unit into part-flows to each of the potential target units for that output, with the split fractions being continuous variables ranging in value from zero to one and adding to unity. These split fractions are labeled structural parameters and, if optimized to zero, indicate the associated stream does not exist in the final flowsheet. The hope that one has converted the discrete problem of including or deleting portions of a structure in the final flowsheet to a continuous variable optimization problem is fraught with danger as shown by Shah and Westerberg (1977). Without considerable care inequality constraints may defeat the approach because they may cause discontinuities to occur at the boundaries of the feasible region. In essence the discrete decisions are often really still present. The method may also suffer from the presence of local optima (Westerberg and Shah (1978); Shah and Westerberg (1979)). Such optima are almost guaranteed if the same substructure can be created from the superstructure in more than one way.

Grossmann and Santibanez (1979) suggest that the approach of embedding can be very effectively and quite generally handled by reformulating the problem as a mixed integer linear program. This approach brings back memories of quite early work in the oil industry where the same tool was used for similar problems. One's first reaction is to reject the adequacy of casting the problem as a linear one, but, as Grossmann and Santibanez show, the use of discrete (zero/one) decisions allow one to include to a very good approximation many of the nonlinearities. For example, a zero/one variable can be associated with the existence or non-existence of a unit. In the cost function that discrete variable can cause one to add in a fixed charge for the unit only if it exists. Also one can define a continuous "flow" variable for the unit which can be forced to zero if the unit does not exist by the linear constraint:

$$\text{Flow} \leq \text{Discrete Variable} \times \text{Maximum Flow}$$

A variable cost for the unit can be assessed proportional to the continuous flow variable. Note that a fixed cost plus variable cost for a unit is not a convex function.

Mixed integer linear programming codes exist which can solve very large problems of that type (several thousand constraints). They generally use branch and bound algorithms. Grossmann and Santibanex solve two very different synthesis problems using embedding and a mixed integer linear programming formulation. The results are impressive.

<u>Control System Synthesis</u>. The synthesis problem for control systems is to select the controllers needed to meet specified control goals for a fixed plant configuration. The designer selects his goals by stating which variables he wishes controlled. The synthesis algorithm selects which variables to measure, which to manipulate and what type of controllers to use.

Govind and Powers (1976, 1978) have developed a synthesis algorithm which deliberately uses very crude process models. Their idea is to see how well one can automatically generate a control scheme using essentially no more information than control engineers currently appear to use. The controllers permitted are conventional PID controllers. The control schemes include simple, cascaded and split range structures, using both feed-forward and feed-back schemes. The crude process models are called cause-and-effect models. They allow only certain directions of information flow, reflecting that which the units being modeled permit. For example, the temperature of the inlet to a heat exchanger is allowed to affect the outlet temperature but not the reverse. The models include process gains to indicate if a small change in a "causal" variable will lead to a large or small, positive or negative change in each affected variable. Also, process delay times are estimated and are part of the models.

Using these models, alternative sets of variables are generated which can be measured to allow an estimate of each variable whose value is to be controlled. Similarly the system generates alternative sets of variables whose manipulation permits control of each variable to be controlled. Clearly if the controlled variable can be measured and manipulated inexpensively and accurately, it is. Otherwise alternative sets are generated and ranked. Long delay times or poor gains between a measured or manipulated variable and the desired control variable will give that choice a low rating. Govind and Powers also state a number of qualitative rules for liking or disliking various options for variables to be measured or manipulated, and these rules aid in selecting which ones to choose.

Cascading is tried if the cascaded loops have different expected response times so they will not fight each other.

Stephanopoulos and Morari (1976, 1977) developed ideas related to those of Govind and Powers. They developed a modified view to selecting the manipulated and measured variables, aided by the concepts of "structural controllability" and "structural observability." Structural controllability and structural observability are similar to controllability and observability except they are based only on the zero/nonzero pattern of the appropriate matrices. The question asked is if any numbers were to be allowed in these matrices, would the system be controllable or observable. Their other heuristic arguments for accepting or rejecting alternatives are very similar to those of Govind and Powers.

Umeda and Kuriyama (1978) describe a two level approach to control system synthesis. At the inner level control schemes are developed for each unit in the flowsheet. While these schemes would work for the unit in isolation, they likely will fight each other if used together. Thus an outer level coordinating step examines and modifies these schemes to eliminate undesirable interactions. The two levels of activity are repeated until they both give the same structure.

Brosilow and coworkers (Weber and Brosilow (1972); Easterday (1973); Joseph and Brosilow (1978a); Brosilow and Tong (1978); Joseph and Brosilow (1978b)) present another control synthesis technique. They term it linear inferential control, and its purpose is to select which variables to measure when developing a control scheme. This problem is obviously only part of that for the general control system synthesis problem. The proposed method ignores dynamics. They select the variables from among a given candidate set by examining the steady state sensitivity and the noise, and usually they pick more than one measurement for each control loop. Stephanopoulos and Morari (1977) add dynamic compensation considerations to this approach. Mellefont and Sargent (1978a) deal with selecting the measurement variables also, only they pose the problem as one to minimize a quadratic performance index for a linear stochastic system. The index included a quadratic term for cost of the controllers. They use a branch and bound algorithm to search for the best measurements to make.

Mellefont and Sargent (1978b) also treat a more general problem in which the measurement subsets change during the control interval.

Shinsky (1979) uses a relative gain matrix to select which variables to manipulate and measure. Again dynamics are ignored. The method allows one to find which variables influence which others the most if they were put into a feedback control loop.

Using embedding with a structural parameter formulation, Nishida, Liu and Ichikawa (1976) state the necessary conditions for optimality for both the structure and the control of a dynamic process system. They also permit some of the system parameters to take on uncertain values from within allowed ranges. The problem is stated as a minimax problem, where the maximum value of the performance index with respect to selection of the uncertain parameter values is minimized with respect to the control variables, the design decisions and the structural parameters.

This problem is obviously a large one in that it includes all the problems of optimal control with uncertain parameters as well as embedding in synthesis. Two example problems are given, with one illustrating that the minimax structure may well be different from the steady-state optimal structure.

Arkun, Stephanopoulos and Morari (1978) have added a new twist to control system synthesis. They developed the theory and then demonstrated on two example problems how to move from one control point to another for a chemical process. They note that the desirable control point is likely at the intersection of a number of inequality constraints, the particular set being those that give optimal steady-state performance for the plant. Due to process upsets or slow changes with time, the point may move at which one wishes to operate. Also, the inequality constraints themselves may shift relative to each other. Arkun, Stephanopoulos and Morari provide the theory to decide when to move, and then develop alternative paths along which to move to the new target. The alternative paths involve following different constraints. As new constraints become active and old ones are released along a path, new control structures may be needed; i.e. new loops must be structured dynamically. The synthesis problem is to select the original control structure, to select the path to travel and to restructure the controllers as one moves. This paper introduces what may become a very interesting topic in control.

Other Synthesis Problems. One recent synthesis publication by Nishitani and Kunugita (1979) is difficult to classify under the above headings. It deals with selecting the optimal vapor/liquid flow patterns to use for a multiple effect evaporator system. The two obvious patterns are cocurrent (the liquid and vapor proceed through the system together) and countercurrent. Other patterns are possible and often significantly improve the economics.

The authors pose the problem as a multiple objective function problem, and, in their example, consider the two competing objectives of total evaporator heat exchanger area and steam usage. The paper is really a clever way of eliminating a number of flow patterns as never being candidates for the optimal solution regardless of the relative importance of the cost for area versus steam costs. For a particular triple effect example problem, they show that, depending on the liquid feed temperature, 3, 4 and even 5 of the patterns out of 6 can be eliminated.

The paper simplifies the analysis required for each configuration by eliminating all continuous variable optimization problems. (This sounds familiar, does it not?) Most notably they require all evaporator areas to be equal and add other specifications sufficient in number to absorb all the degrees of freedom. One might view some of these added specifications as "heuristics."

Final Comments

Synthesis is an active research area. Significant results have emerged from this activity, particularly in heat exchanger network design. The ability to establish bounds on utility usage and number of exchangers required and the use of the temperature versus heat content diagram to find pinch points are very significant tools for design. Also very competent computer programs now exist to aid in organic reaction path synthesis. A number of the other synthesis strategies must be further developed to make them useful for a wide range of industrial problems. Separation system synthesis, for example, needs to handle a wider class of separators. Many of the control system synthesis ideas which establish the total control structure are as yet untried on real problems. The total flowsheet synthesis algorithms are as yet only a beginning.

Synthesis research has caused a number of people to organize their approach to process design. People now ask the right question about design even if the answers as yet are lacking. How does one locate the best structure for a process? How does one teach another person to do synthesis?

The past emphasis in teaching design has been on the analysis of a given process. The current emphasis is moving toward teaching how to invent the structure, and the synthesis tools give a framework from which to teach. The author has available a suite of interactive analysis programs including flash units, shortcut distillation and absorption. If students are not taught a design strategy, few will systematically examine all the alternatives for a separation scheme. On the other hand, if they are given the slightest insight into the synthesis problem and the established heuristics, they readily do a respectable job at inventing a flowsheet.

Abstract

Synthesis is the step in design where one conjectures the building blocks and their interconnection to create a structure which can meet stated design requirements. This review paper first defines chemical process synthesis and indicates the nature of the research problems—to find representations, evaluation functions and search strategies for a potentially nearly infinite problem. It then discusses synthesis research and the most significant results in each of six areas—heat exchanger networks, separation systems, separation systems with heat integration, reaction paths, total flowsheets and control systems.

The main conclusions of the review are that industrially significant synthesis results now exist in energy conservation and reaction path synthesis and that the area of synthesis is a valuable research area. As a research area, it will produce significant results, though many of those available now are of limited usefulness on real problems.

Literature Cited

Agnihotri, R.B., Computer-Aided Investigation of Reaction Path Synthesis, Ph.D. Dissertation, University of Houston, July 1978.

Amorese, L., Cena, V. and Mustacchi, C., "A Heuristic for the Compact Location of Process Components," Chemical Engineering Science, Vol. 32, pp 119-124, 1977.

Arkun, Y., Stephanopoulos, G. and Morari, M., "Design of Steady-State Optimizing Control Structures for Chemical Processes," AIChE 71st Annual Meeting, Miami Beach, FL., November 12-16, 1978.

Bannon, R.P. and Marple, Jr., S., "Heat Recovery in Hydrocarbon Distillation," Chemical Engineering Progress, pp 41-45, July 1978.

Bansal, K.M. and Powers, G.J., "A Study of Mass Balance and Separation Problems in Coal Liquefaction," Department of Chemical Engineering, Carnegie-Mellon University, Pittsburgh, PA., January 1977.

Barnes, F.J. and King, C.J., "Synthesis of Cascade Refrigeration and Liquefaction Systems," I&EC Process Design and Development, Vol. 13, No. 4, pp 421-433, 1974.

Boland, D. and Linnhoff, B., "The Preliminary Design of Networks for Heat Exchange by Systematic Methods," HTFS Research Symposium, Oxford, England, September 1978.

Brandt, J., Fredrich, J., Gasteiger, J., Jochum, C., Schubert, W., and Ugi, I., "Computer Assisted Organic Synthesis, ACS Symposium Series 61, p 33, 1977.

Brosilow, D.B. and Tong, M., "Inferential Control of Processes: Part II. The Structure and Dynamics of Inferential Control Systems," AIChE J. Vol. 24, No. 3, pp 492-500, 1978.

Bush, M.J. and Silveston, P.L., "Optimal Synthesis of Waste Treatment Plants," 82nd National AIChE Meeting, Atlantic City, N.J., August 1976.

Cena, V., Mustacchi, C. and Natali, F., "Synthesis of Heat Exchanger Networks: A Non-Iterative Approach," Chemical Engineering Science, Vol. 32, pp 1227-1231, 1977.

Cerda, J., A Shortcut Methodology for Conventional and Complex Columns, M.S. Thesis, Carnegie-Mellon University, Pittsburgh, PA., 1979.

Cerda, J. and Westerberg, A.W., "A Shortcut Methodology for Conventional and Complex Columns," 72nd Annual Meeting, AIChE, San Francisco, CA., November 1979a.

Donaldson, R.A.B., Paterson, W.R. and Ponton, J.W., "Design of Complex Heat Recovery Networks: Synthesis, Simulation and Uncertainty," 1976.

Douglas, J.M., "Quick Estimates of the Design of Plate-Type Gas Absorbers," I&EC Fundamentals, Vol. 16, No. 1, pp 131-138, 1977.

Douglas, J.M., Manuscript for Process Design Text, 1979.

Doukas, N. and Luyben, W.L., "Economics of Alternative Distillation Configurations for the Separation of Ternary Mixtures," I&EC Process Design and Development, Vol. 17, No. 3, pp 272-281, 1978.

Easterday, D.E., The Use of Secondary Measurements in Estimating Unmeasurable Process Outputs," Ph.D. Thesis, Case Western Reserve University, Cleveland, OH., 1973.

Elshout, R.V. and Hohmann, E.C., "The Heat Exchanger Network Simulator," Chemical Engineering Progress, pp 72-77, March 1979.

Faith, D.C. and Morari, M., "Synthesis of Distillation Schemes with Energy Integration," CACE '79, Montreux, Switzerland, April 8-11, 1979.

Fan, L.T. and Mishra, P.N., "A Random Sampling Apporach to Process System Design and Synthesis," 76th National AIChE Meeting, Tulsa, OK., March 10, 1974.

Flower, J.R. and Linnhoff, B., "Thermodynamic Analysis in the Design of Process Networks," CACE '79, Montreux, Switzerland, April 6-11, 1979a.

Flower, J.R. and Linnhoff, B., "A Thermodynamic-Combinatorial Approach to the Design of Optimum Heat Exchanger Networks," 85th National Meeting, AIChE, Philadelphia, PA., June 1978.

Freshwater, D.C. and Henry, B.D., "The Optimal Configuration of Multicomponent Distillation Trains," Chemical Engineer, pp 533-536, September 1975.

Freshwater, D.C. and Ziogou, E., "Reducing Energy Requirements in Unit Operations," Chemical Engineering Journal, Vol. 11, pp 215-222, 1976.

Friedler, F., Blickle, T., Gyonis, J. and Tarjan, K., "Computerized Generation of Possibilities for the Realization of Given Technological Aim," CACE '79, Montreux, Switzerland, April 8-11, 1979.

Gelertner, H.F., et al., "Empirical Explorations of SYNCHEM," Science 197, No. 4308, pp 1041-9, September 1977.

Gomez, M.A. and Seader, J.D., "Separation Sequence Synthesis by a Predictor Based Ordered Search," AIChE Journal Vol. 22, No. 6, pp 970-979, November 1976.

Govind, R., Blower, P. and Powers, G.J., "Reaction Path Synthesis," 1976.

Govind, R. and Powers, G.J., "Control System Synthesis Strategies," 82nd National Meeting, AIChE, Atlantic City, N.J., August 1976.

Govind, R. and Powers, G.J., "Synthesis of Process Control Systems," IEEE Trans. on Sys. Man and Cyber, Vol. FMC-8, No. 11, p 792, 1978.

Grossmann, I.E. and Santibanez, J., "Applications of Mixed Integer Linear Programming in Process Synthesis," 72nd Annual Meeting, AIChE, November 1979.

Grossmann, I.E. and Sargent, R.W.H., "Optimum Design of Multipurpose Chemical Plants," I&EC Process Design and Development 18 (2), pp 343-348, 1979.

Grossmann, I.E. and Sargent, R.W.H., "Optimum Design of Heat Exchanger Networks," Computers and Chem. Eng. 2 (1), p 1, 1978.

Harbert, V.D., "Which Tower Goes Where?" Petroleum Refiner, Vol. 36, No. 3, pp 169-174, March 1957.

Hartmann, K., "Experience in the Synthesis of Optimum Chemical Process Systems," CACE '79, Montreux, Switzerland, April 8-11, 1979.

Heaven, D.L., Optimum Sequencing of Distillation Columns in Multicomponent Fractionation, M.S. Thesis, University of California, Berkeley, CA., 1970.

Hendrickson, J.B., Topics in Current Chem. 61, p 1, 1976.

Hendry, J.E. and Hughes, R.R., List Processing Representation of Separation Operations, University of Wisconsin, August 20, 1975.

Hendry, J.E. and Hughes, R.R., "Generating Separation Process Flowsheets," Chemical Engineering Progress Vol. 68, No. 6, pp 71-76, June 1972.

Hendry, J.E., Rudd, D.F. and Seader, J.D., "Synthesis in the Design of Chemical Processes," AIChE J. Vol. 19, No. 1, pp 1-15, January 1973.

Himmelblau, D.M., "Optimal Design Via Structural Parameters and Nonlinear Programming." U.S. Japan Joint Seminar on Application of Process Systems Engineering to Chemical Technology Assessment, Kyoto, Japan, June 1975.

Hlavacek, V., "Synthesis in the Design of Chemical Processes," Computers and Chem. Eng. 2, pp 67-75, 1978.

Hohmann, E.C., Optimum Networks for Heat Exchange, Ph.D. Thesis, University of Southern California, 1971.

Hohmann, E.C. and Lockhart, F.J., "optimum Heat Exchanger Network Synthesis," 82nd AIChE National Meeting, Atlantic City, August-September 1976.

Hohmann, E.C. and Nash, D.B., "A Simplified Approach to Heat Exchanger Network Analysis," 85th National AIChE Meeting, Philadelphia, PA., June 1978.

Hwa, C.S., "Mathematical Formulation and Optimization of Exchanger Networks Using Separable Programming," AIChE-IChE Symp. Ser. No. 4, 101, 1965.

Hwang, F. and Elshout, R., "Optimizing the Heat Recovery of Crude Units," Chem. Eng. Prog. Vol. 72, No. 7, p 68, 1976.

Ichikawa, A. and Fan, L.T., "Optimal Synthesis of Process Systems—Necessary Condition for Optimal System and its Use in Synthesis of Systems," Chemical Engineering Science Vol. 28, pp 357-373, 1973.

Johns, W.R. and Romero, D., "The Automated Generation and Evaluation of Process Flowsheets," CACE '79, Montreux, Switzerland, April 8-11, 1979.

Joseph, B. and Brosilow, C.B., "Inferential Control of Processes: Part 1. Steady-Stated Analysis and Design, Part III. Construction of Optimal and Suboptimal Dynamic Estimations," AIChE J. Vol. 24, No. 3, pp 485-492, pp 500-509, 1978a,b.

Kesler, M.G. and Parker, R.D., "Optimal Networks of Heat Exchange," Chem. Eng. Progr. Symp. Ser. No. 92, 65, 111, 1969.

King, C.J., Separation Processes, McGraw Hill, New York, Chapters 13, 14, 1971.

King, C.J., Gantz, D.W. and Barnes, F.J., "Systematic Evolutionary Process Synthesis," I&EC Process Design and Development, Vol. 11, No. 2, pp 271-283, 1972.

Kobayashi, S., Umeda, T. and Ichikawa, A., "Synthesis of Optimal Heat Exchange Systems—An Approach by the Optimal Assignment Problem in Linear Programming," Chemical Engineering Science, Vol. 26, pp 1367-1380, 1971.

Krishna, R., "A Thermodynamic Approach to the Choice of Alternatives to Distillation," I. Chem. E. Symposium Series, No. 54, pp 185-214, 1978.

Lee, K.F., Masso, A.H. and Rudd, D.F., "Branch and Bound Synthesis of Integrated Process Design," I&EC Fundamentals, Vol. 9, No. 1, pp 48-58, February 1970.

Linnhoff, B. and Flower, J.R., "Synthesis of Heat Exchanger Networks II. Evolutionary Generation of Networks with Various Criteria of Optimality," AIChE J. Vol. 24, No. 4, pp 642-654, 1978a.

Linnhoff, B. and Flower, J.R., "Synthesis of Heat Exchanger Networks I. Systematic Generation of Energy Optimal Networks," AIChE J. Vol. 24, No. 4, pp 633-642, 1978b.

Linnhoff, B., Mason, D.R. and Wardle, I., "Understanding Heat Exchanger Networks," CACE '79, Montreux, Switzerland, April 8-11, 1979.

Mahalec, V. and Motard, R.L., "Procedures for the Initial Design of Chemical Processing Systems," Computers and Chem. Eng. Vol. 1, pp 57-68, 1977.

Mahalec, V. and Motard, R.L., "Evolutionary Search for an Optimal Limiting Process Flowsheet," Computers and Chem. Eng. Vol. 17, No. 1, pp 1-12, 1977.

Maikov, V.P., Vilkov, G.G. and Gallstov, A.V., "Optimum Design of Multicolumn Fractionating Plants from the Thermoeconomic Standpoint," International Chemical Engineering Vol. 12, No. 3, pp 426-432, July 1972.

Masso, A.H. and Rudd, D.F., "The Synthesis of System Designs: Heuristic Structuring," AIChE J. Vol. 15, No. 10, 1969.

May, D. and Rudd, D.F., "Development of Solvay Clusters of Chemical Reactions," Chem. Eng. Sci. 31, pp 39-69, 1976.

McGalliard, R.L. and Westerberg, A.W., "Structural Sensitivity Analysis in Design Synthesis," Chemical Engineering Journal, Vol. 4, pp 127-138, 1972.

Mellefont, D.J. and Sargent, R.W.H., "Selection of Measurements for Optimal Feedback Control," Ind. Eng. Chem. Process Design and Development 17 (4), pp 549-552, 1978a.

Mellefont, D.J. and Sargent, R.W.H., Proceedings of the 8th IFIP Conference on Optimization Techniques, Wurzburg J. Steer, Ed., Vol. I, pp 166-176, Springer-Verlag, Berlin, 1978b.

Menzies, M.A. and Johnson, A.I., "Synthesis of Optimal Energy Recovery Networks Using Discrete Methods," Can. J. Chem. Eng. 50, 290, 1972.

Mishra, P.N., Fan, L.T. and Erickson, L.E., "Biological Wastewater Treatment System Design—Part I. Optimal Synthesis," Canadian Journal of Chemical Engineering, Vol. 51, pp 694-701, December 1973.

Mishra, P.N., Fan, L.T. and Erickson, L.E., "Biological Wastewater Treatment System Design—Part II. Effect of Parameter Variations on Optimal Process System Structure and Design," Canadian Journal of Chemical Engineering, Vol. 51, pp 702-708, December 1973b.

Motard, R.L. and Westerberg, A.W., Notes for Process Synthesis AIChE Advanced Seminar, 1978.

Nash, D.B., Hohmann, E.C., Beckman, J. and Dobrzanski, L., "A Simplified Approach to Heat Exchanger Network Analysis," AIChE 85th National Meeting, Philadelphia, PA., June 1978.

Nath, R. and Motard, R.L., "Evolutionary Synthesis of Separation Processes" 85th National Meeting, AIChE, Philadelphia, PA., June 1978.

Nishida, N., Kobayashi, S. and Ichikawa, A., "Optimal Synthesis of Heat Exchange Systems. Necessary Conditions for Minimum Heat Transfer Area and their Application to Systems Synthesis," Chemical Engineering Science, Vol. 26, pp 1841-1856, 1971.

Nishida, N., Liu, Y.A. and Ichikawa, A., "Studies in Chemical Process Design and Synthesis II. Optimal Synthesis of Dynamic Process Systems with Uncertainty," AiChE J. Vol. 22, No. 3, pp 539-550, May 1976.

Nishida, N., Liu, Y.A. and Lapidus, L., "Studies in Chemical Process Design and Synthesis III. A Simple and Practical Approach to the Optimal Synthesis of Heat Exchanger Networks," AIChE J. Vol. 23, No. 1, pp 77-93, January 1977.

Nishimura, H., "A Theory for the Optimal Synthesis of Heat Exchanger Systems," Accepted for publication in JOTA.

Nishimura, H., "Intrinsic Approach in Process Optimization and Synthesis" (Manuscript of General Lecture at Chemical Engineering Department—MIT), October 3, 1978.

Nishimura, H. and Hiraizumi, Y., "Optimal System Pattern for Multicomponent Distillation Systems," International Chemical Engineering, Vol. 11, No. 1, pp 188-193, January 1971.

Nishitani, H. and Kunugita, E., "The Optimal Flow-Pattern of Multiple Effect Evaporator Systems," CACE '79, Montreux, Switzerland, April 8-11, 1979.

Osakada, K. and Fan, L.T., "Synthesis of an Optimal Large-Scale Interconnected System by Structural Parameter Method Coupled with Multilevel Technique," Canadian Journal of Chemical Engineering, Vol. 51, pp 94-101, February 1973.

Pensak, D.A., Corey, E.J., "LHASA" in Computer-Assisted Organic Synthesis," ACS Symp. Ser. 61, p 1, 1977.

Petlyuk, F.B., Platonov, V.M. and Slavinski, D.M., "Thermodynamically Optimal Method for Separating Multicomponent Mixtures," International Chemical Engineering, Vol. 5, No. 3, pp 555-561, July 1965.

Petterson, W.C. and Wells, T.A., "Energy-Saving Schemes in Distillation," Chemical Engineering, pp 78-86, Sept. 26, 1977.

Pho, T.K. and Lapidus, L., "Topics in Computer-Aided Design: Part II. Synthesis of Optimal Heat Exchanger Networks by Tree Search Algorithms," AIChE Journal, Vol. 19, No. 6, pp 1182-1189, November 1973.

Ponton, J.W. and Donaldson, R.A.B., "A Fast Method for the Synthesis of Optimal Heat Exchanger Networks," Chemical Engineering Science, Vol. 29, pp 2375-2377, 1974.

Powers, G.J., "Heuristic Synthesis in Process Development," Chem. Eng. Prog., Vol. 68, No. 8, p 88, 1972.

Powers, G.J., "Non-Numerical Problem Solving Methods in Computer Aided-Design," IFIPS Conference on Computer-Aided Design, Eindhoven, The Netherlands, 1972.

Powers, G.J. and Jones, R.L., "Reaction Path Synthesis Strategies," AIChe J. Vol. 19, p 1204, 1973.

Rathore, R.N.S. and Powers, G.J., "A Forward Branching Scheme for the Synthesis of Energy Recovery Systems," I&EC Process Design and Development, Vol. 14, No. 2, pp 175-181, 1975.

Rathore, R.N.S., Wormer, K.A.V. and Powers, G.J., "Synthesis Strategies for Multicomponent Separation Systems with Energy Integration," AIChE J. 20, No. 3, p 491, 1974a.

Rathore, R.N.S., Wormer, K.A.V. and Powers, G.J., "Synthesis of Distillation Systems with Energy Integration," AIChE J. 20, No. 5, p 940, 1974b.

Rodrigo, F.R. and Seader, J.D., "Synthesis of Separation Sequences by Ordered Branch Search," AIChE J. Vol. 21, No. 5, pp 885-894, October 1975.

Rod, V. and Marek, J., "Separation Sequences in Multicomponent Rectification," Coll. Czech. Chem. Comm., Vol. 24, pp 3240-3248, 1959.

Rudd, D.F., Powers, G.J. and Siirola, J.J., Process Synthesis, Prentice Hall, New York, Chapter 5, 1973.

Sargent, R.W.H. and Gaminibandara, K., "Optimum Design of Plate Distillation Columns," IMA Conference on Optimization in Action, University of Bristol, Bristol, England, January 1975.

Seader, J.D. and Westerberg, A.W., "A Combined Heuristic and Evolutionary Strategy for Synthesis of Simple Separation Sequences," AIChE J. Vol. 23, No. 6, pp 951-954, Nov. 1977.

Shah, J.V. and Westerberg, A.W., "Process Synthesis Using Structural Parameters: A Problem with Inequality Constraints," AIChE J. Vol. 23, No. 3, pp 378-380, May 1977.

Shah, J.V. and Westerberg, A.W., "EROS—A Program for Quick Evaluation of Energy Recovery Systems," to appear, Computers and Chem. Eng., 1979.

Shinsky, F.G., Process Control Systems, 2nd Ed., McGraw Hill, New York, N.Y., pp 197-202, 1979.

Siirola, J.J., "Status of Heat Exchanger Network Synthesis," 76th National AIChE Meeting, Tulsa, OK., March 1974.

Siirola, J.J., "Progress Toward the Synthesis of Heat-Integrated Distillation Schemes," 85th National AIChE Meeting, Philadelphia, PA., June 1978.

Siirola, J.J., Powers, G.J. and Rudd, D.F., "Synthesis of System Designs: III. Toward a Process Concept Generator," AIChE J. Vol. 17, No. 3, pp 677-682, May 1971.

Siirola, J.J. and Rudd, D.F., "Computer-Aided Synthesis of Chemical Process Designs from Reaction Path Data to the Process Task Network," I&EC Fundamentals, Vol. 10, No. 3, pp 353-362, 1971.

Simon, H.A., The Sciences of the Artificial, MIT Press, Cambridge, MA., 1969.

Sophos, A., Stephanopoulos, G. and Morari, M., "Synthesis of Optimum Distillation Sequences with Heat Integration Schemes," National AIChE Meeting, Miami, FL., November 1978.

Stephanopoulos, G., Synthesis of Networks of Heat Exchangers— A Self-Study Block Module, Project PROCEED, Massachusetts Institute of Technology, Cambridge, MA., 1977.

Stephanopoulos, G. and Morari, M., "Synthesis of Control Structures for Chemical Processes," Conference on Computers in Chemical Engineering, Czechoslovakia, October 1977.

Stephanopoulos, G. and Morari, M., "Toward the Synthesis of Control Structures—A Progress Report," 82nd National AIChE Meeting, Atlantic City, N.J., August 1976.

Stephanopoulos, G. and Westerberg, A.W., "Studies in Process Synthesis—II. Evolutionary Synthesis of Optimal Process Flowsheets," Chemical Engineering Science, Vol. 31, pp 195-204, 1976.

Stephanopoulos, G. and Westerberg, A.W., "Synthesis of Optimal Process Flowsheets by an Infeasible Decomposition Technique in the Presence of Functional Non-Convexities," Canadian Journal of Chemical Engineering, Vol. 53, pp 551-555, 1975.

Takamatsu, T., Naka, Y., Tomita, S. and Ochi, H., "Determination of Process Structures of Azeotropic Distillation System," International Congress on Contribution of Computers...., Paris, France, March 7-10.

Tedder, D.W. and Rudd, D.F., "Parametric Studies in Industrial Distillation: Part I. Design Comparison Part II. Heuristic Optimization and Part III. Design Methods and their Evaluation," AIChE J. Vol. 24, No. 2, pp 303-334, 1978.

Thompson, R.W. and King, C.J., "Systematic Synthesis of Separation Schemes," AIChE J. Vol 18, p 941, 1972.

Umeda, T., Harada, T. and Shiroko, K., "A Thermodynamic Approach to the Synthesis of Heat Integration Systems in Chemical Processes," CACE '79, Montreux, Switzerland, April 8-11, 1979.

Umeda, T., Hirai, A. AND Ichikawa, A., "Synthesis of Optimal Processing System by an Integrated Approach," Chemical Engineering Science, Vol. 27, pp 795-804, 1972.

Umeda, T. and Ichikawa, A., "A Rational Approach to Process Synthesis—An Extensive Use of Task Assignment Concept," Chem. Eng. Sci., Vol. 30, p 699, 1975.

Umeda, T., Itoh, J. and Shiroko, K., "Heat Exchange System Synthesis," Chemical Engineering Progress, pp 70-76, July 1978.

Umeda, T. and Kuriyama, T., "A Logical Structure for Process Control System Synthesis," Proc. IFAC Congress, Helsinki, Finland, 1978.

Umeda, T., Niida, K. and Shiroko, K., "A Thermodynamic Approach to Heat Integration in Distillation Systems," AIChE J. Vol. 25, No. 3, pp 423-429, May 1979.

Umeda, T., Shindo, A. and Ichikawa, A., "Process Synthesis by Task Assignment," Chemical Engineering Science, Vol. 29, pp 2033-2040, 1974.

Umeda, T., Shindo, A. and Tazaki, E., "Optimal Design of Chemical Process by Feasible Decomposition Method," I&EC Proc. Design and Development, Vol. 11, p 1, 1972.

Weber, R. and Brosilow, C.B., "The Use of Secondary Measurements to Improve Control," AIChE J. Vol. 18, p 614, 1972.

Westerberg, A.W., Process Design Strategy, Notes for Design Course Presented at INTEC, Santa Fe, Argentina, 1978.

Westerberg, A.W. and Stephanopoulos, G., "Studies in Process Synthesis—I. Branch and Bound Strategy with List Techniques for the Synthesis of Separation Schemes," Chemical Engineering Science, Vol. 30, pp 963-972, 1975.

Westerberg, A.W., Stephanopoulos, G. and Shah, J.V., "The Synthesis Problem with some Thoughts on Evolutionary Synthesis in the Design of Engineering Systems," Symp. on Basic Questions in Design Theory, Columbia University, 1974.

Wipke, W.T. and Howe, W.J., "Computer Assisted Organic Synthesis," ACS Symposium Series 61, 1977.

Wipke, W.T., Ouchi, G.I. and Krishnan, S., "Simulation of Evaluation of Chemical Synthesis—SECS: An Application of Artificial Intelligence Techniques," Artificial Intelligence, Vol. 11, pp 173-193, 1978.

Zhavoronkov, N.M., Kafarov, V.V., Perov, V.L. and Meshalkin, V.P., "New Principles for the Analysis and Synthesis of Chemical Engineering Systems," Teoreticheskie Osnovy Khimicheskoi Tekhnologii, Vol. 4, No. 2, pp 152-167, 1970.

RECEIVED November 5, 1979.

4

Advanced Control Strategies for Chemical Processes: A Review

THOMAS F. EDGAR

Department of Chemical Engineering, University of Texas, Austin, TX 78712

The decade 1960 - 1970 was a very active period in university control research, stimulated by parallel developments in the space and communication fields. However, there was very little diffusion of this technology into the industrial sector, simply because the implementation of this so-called modern control theory could not be justified economically. Industrial control technology remained fairly static during this period, mainly relying on analog hardware and the PID controller.

Since about 1970 a quantum change in industrial activity in control technology has taken place, spurred on by major cost and scientific breakthroughs in computer hardware technology, namely the minicomputer and microprocessor. These developments have radically changed the prospects of applying advanced control strategies. University researchers in the 1970's appear to be more oriented towards industrial control applications than their earlier counterparts and more eclectic in developing theory tied to those problems. Evidence of this tilt toward applications include the Asilomar Conference on Chemical Process Control (1) and the fact that many process control investigators today employ experimental testing of computer control and estimation algorithms as an integral part of their research program.

The rapidly changing world energy picture has created a further impetus for new advances in control technology. Increases in the cost of energy have already caused major changes in process operation and design for the process industries. In this milieu the tuning of individual control loops is superseded by the need to coordinate both plant control and optimization strategies, the latter often referred to as "supervisory control". Established algorithms such as linear programming or one-dimensional search are used in most supervisory control or set point optimization applications. Simplicity is a key feature of these methods, but the results are impressive.

The impact of process dynamics and control on energy utilization should not be interpreted narrowly as affecting only operat-

0-8412-0549-3/80/47-124-089$05.75/0

ting costs. As mentioned above, design changes implemented to minimize energy consumption can have a profound influence on how efficiently a process can be controlled. The introduction of these design changes, which often requires integration of heat producing and heat consuming components of the plant, usually makes such a process more interacting and less susceptible to simple control strategies. In the future industry will probably be forced to operate against constraints that tax the operability of the system; self-regulating systems may not be so omnipresent.

Because of the heavy dependence on the profit incentive, complicated control strategies have not received wide acceptance in the process industries. It has been difficult to justify the costs of engineering manpower and control hardware and instrumentation balanced against potential benefits of advanced control techniques. As discussed by Edgar (4), most of the recent activity in energy conservation has been of the simple, low investment type, not necessarily even requiring a control computer. Such projects include adding insulation, monitoring steam consumption and leaks more closely, controlling combustion efficiency more tightly, and more frequent cleaning of heat transfer surfaces. Future energy savings will probably be tied to improvements in computer control and monitoring of energy intensive processes.

It is indicative of the topicality of industrial energy conservation that in the March, 1979 issue of Chemical Engineering Progress, engineers from two major energy companies (DuPont and Exxon) discussed innovative engineering design changes to energy-intensive processes (5), (6). Such design modifications are motivated principally by steady state decreases in energy consumption, while the use of on-line process control and optimization is secondary. However, it is clear that the success of the conservation strategies discussed in these articles is dependent upon the development and implementation of suitable computer control/optimization algorithms. One particularly interesting example cited by Robnett (5) involves several refrigeration machines running in parallel to satisfy a demand for chilled water. The chilling system load is controlled by steam flow. One refrigeration system is electrically driven; a second is operated by a condensing turbine. High pressure steam from a power house is pressure-reduced for use by the plant as it passes through one air compressor turbine, one refrigeration machine turbine, and the high pressure section of the turbogenerator. The steam turbines require one-third the energy of an electric motor. Each refrigeration unit has a different horsepower/ton characteristic, which also depends upon ambient conditions. There are hard constraints on compressor loads and cost penalties (soft constraints) on electrical load. Steam, refrigeration, compressed air, and electrical loads to the plant vary continually. While the author does not suggest that the optimum operating and control strategy is known, he does imply that a computer control system is the only way to operate the plant in an optimum manner.

A second example, discussed by Kenney (6), involved energy integration of a train of distillation columns in a refinery/aromatics complex. The integration was performed in order to eliminate furnace reboilers; this design change reduced Btu consumption by 50 per cent in the distillation train but made pressure control more difficult. As reported by the author, the towers also became more sensitive to variations in throughput and feed composition, which are common disturbances in distillation systems. These results are generally expected when energy integration is implemented.

It is important to note some of the salient characteristics of the above systems.

(1) They are multivariable in nature, and there are interactions which exist among the various control and output variables, making a simplistic single input-output (SISO) strategy somewhat hazardous if not inefficient.

(2) Set point optimization and dynamic optimization may be coupled, thus confounding a priori specification of controller parameters. Changing set points may change model parameters, although some parameters may be inherently time-varying due to system nonlinearities. This suggests that on line identification and/or adaptive control approach might be fruitful.

(3) Not all variables of interest are directly measurable, i.e., it may not be technically or economically feasible to obtain needed data in real-time. It is important to select the minimum number of measurements necessary for effective control.

(4) The trade-offs among process design, optimization and control must be considered. The hierarchical or distributed nature of the plant or process may need to be exploited in an advanced control scheme. The operation of energy-integrated plants requires design of control systems which are decentralized (such as with microprocessors) but which respond to overall plant objectives via a communication link to a larger computer.

In this review the new thrusts of advanced control theory into the four areas mentioned above will be discussed. The current state of control theory and practice for these areas will be mentioned, and new contributions which hold promise for future application will be outlined. Most of the literature cited in this review falls under the general classification of linear control theory, which in the author's opinion holds the most near-term promise for applications. Hence such topics as nonlinear estimation and control are not covered in any depth here.

Control and Modeling Philosophies

The output or state feedback approach is still the most accepted controller structure, whether for analog or digital models or for single input-single output (SISO) or multiple input-multiple output (MIMO) processes. Feedback control design techniques can be classified under the following general headings:

(1) frequency domain (open or closed loop frequency response)
(2) root locus, where closed loop eigenvalues and sometimes eigenvectors can be specified
(3) optimal control, where a performance function is optimized (e.g., minimum variance)

Special considerations, such as dead time compensation, can be discussed under the three areas.

The goal of the above techniques is to achieve the following desirable controller characteristics:

(1) adequate disturbance rejection
(2) quick response to set point changes
(3) insensitive to model and measurement errors
(4) avoids controller saturation or excessive control action
(5) requires minimum process information
(6) stable in the face of instrument failure
(7) suitable over a wide range of operating conditions

While it may be impossible to achieve all of these goals simultaneously, it is clear that a "super-controller" which is inexpensive to implement and always achieves superior performance is the paragon against which all controllers must be measured. This has led to a design approach which is often based on interactive graphics, allowing successive evaluation of many alternatives.

If one is dealing with an SISO system with constant coefficients, there are no advantages to employing so-called modern techniques vs. the well-established classical methods based on frequency response. The equivalence of optimal control, pole placement, and frequency response methods can be demonstrated under certain conditions, especially for low order systems. Therefore, the individual preference of the designer can be followed in most cases. However, if appreciable dead time is present in the process, then simple feedback techniques may be disadvantageous. For high performance special methods, such as feedforward control or a Smith predictor (7), should be superimposed on the feedback controller. It is interesting to note that while the well-established tuning rules for analog controllers rarely incorporate any special considerations for dead time compensation, the use of a dead time predictor for digital controllers is a well-established procedure (8). Presumably this dichotomy has arisen because analog hardware could not easily simulate pure time delays, as required by the Smith predictor, while time delays are readily treated by sampled data (computer) controllers. Dead time compensation naturally arises out of optimal control methods, as will be discussed later.

One preliminary consideration in controller design is how the selection of the model affects the nature of the controller. Use of modern (multivariable) control theory usually requires better models, often more detailed than first or second order plus dead time. These models may be based on a fairly rigorous interpretation of physical and chemical principles. Using distributed physical or chemical measurements for state feedback is certainly appealing. However, for large scale systems (such as a distillation column or a multi-unit system) the use of physical models is less attractive, due to the large amounts of manpower and time required in model development (for the example of modeling a distillation column, see refs. (9) and (10)). Therefore for these systems the so-called black box models are favored for most "on-line" applications. That is not to say that physical models do not lend insight into synthesis of the control structure as well as provide a tool for simulation; these are perhaps the most valuable roles for physical models.

In the literature there have been significant efforts in physical modeling and controller design for the following multivariable unit operations [Rijnsdorp and Seborg (11)]:

(1) distillation columns - Buckley (12), Edgar and Schwanke (9).
(2) reactors-Padmanabhanan and Lapidus (13), Wallman and Foss (14).
(3) fluid catalytic crackers - Kurihara (15), Schuldt and Smith (16).
(4) Wastewater treatment - Olsson (17).
(5) paper processing - Church (18).
(6) furnaces - Clelland (19).
(7) double effect evaporator - Fisher and Seborg (20).

All design methods to be discussed in this section are for linear dynamic systems. Such systems can be described in the time domain by the state space regulation equation

$$\dot{x} = \underline{A}x + \underline{B}u + \underline{F}d \qquad (1)$$

$$y = \underline{C}x \qquad (2)$$

x, u, and d are vectors which represent deviations from a selected steady state operating point, x represents the state vector of dimension nx1; the definition of n state variables is necessary for complete specification of the dynamic system. The inputs in equation (1) are the control vector u, of dimension r, and the disturbance vector, of dimension p.

The output vector, y, is of dimension m, which represents the linear combination of the states which are directly measurable. With state variable notation, one can achieve dynamic compensation through linear feedback (proportional) control and by using augmentation of the state vector. This is a necessary step in allowing for integral control with time domain design methods. Feedback control can be implemented either in terms of $\underline{x}$ (state variable feedback) or $\underline{y}$ (output feedback).

An equivalent description of equations (1) and (2) in terms of multivariable transfer functions (Laplace domain) can also be given:

$$y(s) = \underline{G}(s)u(s) + \underline{G}^*(s)d(s) \tag{3}$$

$$\underline{G}(s) = \underline{C}(s\underline{I} - \underline{A})^{-1}\underline{B} \qquad \underline{G}^*(s) = \underline{C}(s\underline{I} - \underline{A})^{-1}\,\underline{F} \tag{4}$$

The poles for $\underline{G}(s)$ are equivalent to the eigenvalues of $\underline{A}$ in equation (1). The output feedback controller in the s domain becomes $u(s) = \underline{K}(s)\ y(s)$.

Controllers can be designed on a regulation basis ($d\neq 0$, $x(0) = 0$) or on a servomechanism basis ($d\neq 0$, $x(0)=0$). When $d\neq 0$, the controller must counteract the disturbance, although the addition of integral control can force the error of some outputs to zero, depending upon the nature of $\underline{A}$. The closed loop transfer function for set point changes (servomechanism) is

$$y(s) = \{\underline{I} + \underline{G}(s)\underline{K}(s)\}^{-1}\ \underline{G}(s)\ \underline{K}(s)\ r(s) \tag{5}$$

For a fairly complete review of linear multivariable controller design techniques, the reader is referred to articles by MacFarlane (21) and Edgar (22).

Frequency Domain Design Techniques. The design of SISO systems based on frequency response characteristics includes such well-established methods as the Bode plot and the Nyquist and inverse Nyquist diagrams. During the past ten years there has been a serious effort to extend these methods to treat MIMO systems. In fact, there are several graphics-based software packages developed in Great Britain which are now being marketed commercially. (23), (24). These are generally based on Laplace transform representation of process dynamics and control. The two pioneers which must be credited with leading these developments are H. H. Rosenbrock and A. G. J. MacFarlane.

Decoupling, or non-interacting control, is the oldest multivariable control technique (21). The general philosophy of non-interacting control is to cancel the interactions by choosing a controller of appropriate structure. If $(\underline{G}(s) \cdot \underline{K}(s))$ in equation (5) can be made a diagonal matrix by properly selecting $\underline{K}(s)$, then the product matrix has no interactions (off-diagonal terms are zero). Thus the controller synthesis problem reduces to treating each diagonal element separately as in the single loop problem. In other works, in the decoupled system, r_i only affects y_i but r_i does not affect y_j $(i\neq j)$. In order to obtain a straightforward design problem, controller performance is sacrificed. The loop decoupling approach also can suffer from extreme sensitivity to model errors (if a parameter changes, the design is no longer non-interacting). Pathological cases arise in decoupling when deadtime or positive zeroes occur in the transfer function matrix. In the former case, a controller with a "prediction" element ($e^{\theta s}$) may arise, while in the latter case the controller will contain an unstable element (21). Another disadvantage of exact decoupling is controller complexity. Experimental applications of decoupling have been popular for distillation columns, as reviewed by Edgar and Schwanke (9).

There are several less binding options available for decoupling. One is to use approximations (sometimes ad hoc) to the required decoupling controllers, often simplifying the controller forms. A second approach is to use partial ("one-way") decoupling. This approach recognizes that one loop may be more sensitive to input-output interactions than another; partial decoupling is implemented by setting one cross-controller (K_{ij}, $i \neq j$) equal to zero [Shinskey, (25)]. Partial decoupling is more tolerant of model errors. Recent studies by McAvoy and coworkers (26), (27) and Toijala and Fagervik (28) have examined the effects of model errors in decoupling in distillation column models, explaining some difficulties which have been reported for experimental application of decoupling. A still simpler approach is static decoupling, where the dynamics of each element in the transfer function matrix are neglected in the cross-controllers; only the steady state gains are utilized.

The inverse Nyquist array technique, as presented by Rosenbrock (29), is a set of diagrams corresponding to the elements of the inverse of the open loop transfer function $\hat{\underline{Q}}(s) = (\underline{G}(s)\underline{K}(s))^{-1}$ of the system. Central to Rosenbrock's technique is the principle of diagonal dominance, where the designer attempts to select $\underline{K}(s)$ so that $\hat{\underline{Q}}(s)$ is diagonally dominant. This requirement is less restrictive than exact decoupling, where the off-diagonal elements are forced to zero. Rosenbrock's stability theorem requires that for all s each diagonal element of $(\underline{G}(s)\underline{K}(s))^{-1}$ must be larger in magnitude than the sum of all of the off-diagonal elements in that same row. This insures that interactive effects are "small". Once $\hat{\underline{Q}}(s)$ has this property, only the inverse Nyquist diagram for components of the diagonal needs to be constructed. Thus a set of approximate single loop control problems, corresponding to the diagonal elements, is solved. However, the inverse Nyquist diagram, which for SISO systems is a locus of points on the complex plane, is in the multivariable case a locus of circles for each loop. These circles account for the uncertainty incurred by ignoring the system interactions (referred to as Gershgorin bands), and are constructed by adding the magnitudes of the off-diagonal elements in a row or column for a given frequency. The validity of this approach is supported theoretically. As in single loop theory, the critical point for determination of closed loop system stability is (-1, 0) on the complex plane.

To make $\hat{\underline{Q}}(s)$ diagonally dominant, it is necessary to select a specific type of controller. Rosenbrock has proposed that $\underline{K}(s) = \underline{K}_a\underline{K}_b(s)\underline{K}_c(s)$, i.e., a product of three controller matrices. $\underline{K}_a$ is a permutation matrix, which scales the elements in $\underline{G}(s)\underline{K}(s)$ and makes some preliminary assignment of single loop connections, usually to assure integrity. This step can be used to make $(\underline{G}(s)\underline{K}(s))^{-1}$ diagonally dominant as $s \to 0$. $\underline{K}_b(s)$ can be chosen to meet stability criteria. Finally the elements of $\underline{K}_c(s)$, a diagonal matrix, can be selected to improve performance of the system. The proper selection of $\underline{K}_a$ and $\underline{K}_b(s)$ are the most difficult parts of the design process, and this step should be considered iterative, especially for an inexperienced user.

Another design technique, the characteristic loci controller, is based on the approach of MacFarlane and Belletrutti (30). This method has some similarities to the strategy employed in the inverse Nyquist array. However, rather than approximate stability calculations performed using a locus of circles (inverse Nyquist diagram = $((G(s)K(s))^{-1})$, a vector of exact characteristic loci is computed (Nyquist diagram=G(s)K(s)). As with Rosenbrock's method, a controller formed by a cascade of several transfer function matrices is utilized. MacFarlane and Belletrutti have identified seven different types of controller transfer functions, which when combined attempt to satisfy conditions of closed loop stability and high integrity (closed loop stability when a transducer fails), minimize interaction due to off-diagonal elements at high and low frequency, and achieve good performance. Diagonal dominance is not required in this design method, although it is helpful in obtaining a system of high integrity. Application of this design method does yield a feasible multivariable controller, but it is difficult to obtain a suitable design without a great deal of experience and intuition.

Other frequency domain techniques which have been proposed include the commutative controller (31), sequential return difference (32), and the direct Nyquist array (33). In chemical process control, a number of recent applications of multivariable frequency response methods include distillation columns (34), (35), and reactors (36).

Dead time compensation for multivariable systems can be accomplished using a matrix version of the Smith predictor, as recently presented by Ogunnaike and Ray (37). This multivariable block diagram for the Smith predictor is structurally the same as for the SISO case (7). The resulting compensator allows one to tune the controller as if there were no dead time in the characteristic equation, thus allowing higher controller gains and more effective regulation. The question of which multivariable technique is best to use with the Smith predictor is still open to question, as is the problem with sensitivity to changes in the time delays.

Root Locus Techniques. The discussion of root locus methods is almost a standard feature in most undergraduate textbooks, although this procedure is generally acknowledged to be inferior to frequency response methods. It is normally expected that if the closed loop eigenvalues are shifted further to the left of the complex plane, the system will be faster responding. However, this is not always the case. The key problem with root locus or pole placement methods is that they ignore the effects of control on the system eigenvectors.

The primary interest in the pole placement literature recently has been in finding an analytical solution for the feedback matrix so that the closed loop system has a set of prescribed eigenvalues. In this context pole placement is often regarded as a simpler alternative than optimal control or frequency response methods. For a single control (r=1), the pole placement problem yields an analytical solution for full state feedback (e.g., (38), (39)). The more difficult case of output feedback pole placement for MIMO systems has not yet been fully solved(40).

In the past few years, a number of workable pole-placement algorithms have been published. However, their application to MIMO systems with incomplete state variable feedback are often unsatisfactory in that:

(1) Only a limited number of poles can be placed arbitrarily
(2) Nothing can be said about the remaining unassigned eigenvalues, i.e., their stability is not guaranteed.
(3) For complete pole placement, it is usually required that $r+m \geq n+1$, thus the total number of inputs and outputs are considerably larger than the minimum condition $r \times m \geq n$. Here the minimum condition means that when $r \times m \geq n$, it is likely that a solution exists for the resulting set of nonlinear equations.
(4) Usually the algorithm returns a feedback matrix with very large components. This may be unacceptable for a control system with constrained inputs. Finding a feedback matrix with smaller entries by trial and error can be very tedious.
(5) The close loop response depends not only on the closed loop eigenvalues but also on eigenvectors. Intuitive specification of closed loop eigenvalues may be difficult.
(6) Time delays are not readily treated.

It is well known that there are closed loop locations which can not be reached by constant proportional control using less than full state feedback. The common approach in the case where proportional output feedback cannot yield a satisfactory design is to add an observer to the system. A similar but somewhat different approach is to use a dynamic controller. As an example, consider the control of a second order SISO plant by an ideal PID controller cascaded with a first order filter, which is basically a second order dynamic controller of restricted structure.

$$K(s) = \frac{\beta_o + \beta_1 s + \beta_2 s^2}{\alpha_1 s + s^2} \qquad (6)$$

Let the second order plant be

$$G(s) = \frac{b_1 s + b_o}{s^2 + a_1 s + a_o} \qquad (7)$$

Then the closed loop transfer function becomes

$$L(s) = \frac{G(s)}{1+K(s)\ G(s)} \qquad (8)$$

$$= \frac{\alpha_2 b_1 s^3 + (\alpha_2 b_o + \alpha_1 b_1)s^2 + \alpha_1 b_o s}{s^4+(\alpha_1+a_1+b_1\beta_2)s^3+(\alpha_1 a_1+a_o+b_1\beta_1+b_o\beta_2)s^2+(a_o\alpha_1+\beta_o b_1+\beta_1 b_o)s+\beta_o b_o}$$

The closed loop poles are specified to be p_1, p_2, p_3 and p_4. This implies

$$1 + K(s)G(s) = \prod_{i=1}^{4} (s-p_i)$$

$$= s^4+C_3 s^2+C_2 s^2+C_1 s+C_o \qquad (9)$$

Comparing equation (9) with the denominator of equation (8), the controller parameters β_o, β_1, β_2, α_1 are determined from solving the simultaneous equations:

$$\begin{bmatrix} b_o & 0 & 0 & 0 \\ b_1 & b_o & 0 & a_o \\ 0 & b_1 & b_o & a_1 \\ 0 & 0 & b_1 & 1 \end{bmatrix} \begin{bmatrix} \beta_o \\ \beta_1 \\ \beta_2 \\ \alpha_1 \end{bmatrix} = \begin{bmatrix} C_o \\ C_1 \\ C_2-a_o \\ C_3-a_1 \end{bmatrix} \qquad (10)$$

Note that a set of linear algebraic equations results. More details on pole placement with a dynamic controller have been reported by Brasch and Pearson (41).

Pole and zero placement using a dynamic compensator for an SISO system can be accomplished by specifying analytically the closed loop servo response (e.g., first or second order with deadtime). Suppose that the specified response is defined by P(s); solving the closed loop equation (5) yields an analytical form for the feedback controller:

$$K(s) = \frac{P(s)}{G(s)-P(s)G(s)} \qquad (11)$$

This design may yield controllers which are quite sensitive to model errors and require high order derivative action. If the dead time in P(s) is the same as the dead time in G(s), the controller contains dead time compensation, as in the Smith predictor. Bristol (42) has extended this idea to apply to multivariable systems, although he treats the controller in a more general form, allowing a pre-compensation block before $\underline{G}(s)$ and a post-compensation block after $\underline{G}(s)$ in the direct path between r(s) and y(s).

A related approach which has been used successfully in industrial applications occurs in discrete-time control. Both Dahlin (43) and Higham (44) have developed a digital control algorithm which in essence specifies the closed loop response to be first order plus dead time. The effective time constant of the closed loop response is a tuning parameter. If z-transforms are used in place of s-transforms in equation (11), we arrive at a digital feedback controller which includes dead time compensation. This dead time predictor, however, is sensitive to errors in the assumed dead time. Note that in the digital approach the closed loop response is explicitly specified, which removes some of the uncertainties occurring in the traditional root locus technique.

Optimal Control. Optimal control is extension of the principles of parameter optimization to dynamic systems. In this case one wishes to optimize a scalar objective function, which may be a definite integral of some function of the state and control variables, subject to a constraint, namely a dynamic equation, such as Equation (1). The solution to this problem requires the use of time-varying Lagrange multipliers; for a general objective function and state equation, an analytical solution is rarely forthcoming. However, a specific case of the optimal control problem does lend itself to analytical solution, namely a state equation described by Equation (1) and a quadratic objective function given by

$$J(u) = \int_o^{tf} (x^T \underline{Q} x + u^T \underline{R} u)dt \qquad (12)$$

where t_f is the final time. This problem is known as the linear-quadratic problem (LQP). J(u) is to be minimized when $\underline{Q}$ is positive semi-definite and $\underline{R}$ is positive definite. As demonstrated in standard texts on optimal control, e.g., Athans and Falb (45) and Lapidus and Luus (46), the optimal control for the servo problem (d = o) is $u(t) = \underline{K}(t)x(t)$, or linear proportional feedback. This multivariable design method has been the subject of intensive research efforts over the last fifteen years. For all feedback matrices designed by the LQP, the closed loop system is asymptotically stable.

The LQP is the only general optimal control problem for which there exists an analytical representation for the optimal control in closed-loop or feedback form. For the LQP, the optimal controller gain matrix $\underline{K}$ becomes a constant matrix for $t_f \to \infty$. $\underline{K}$ is independent of the initial conditions, so it can be used for any initial condition displacement, except those which, due to model nonlinearities, invalidate the computed state matrices. There is a substantial amount of material developed for the LQP and its extensions: the December, 1971, issue of IEEE Transactions on Automatic Control and a review article by Edgar et al. (47) provide a fairly complete exposition on the effects of non-linearities in the state equations, parameter variations, constraints, disturbances, incomplete measurement, and system noise. There are several general purpose computer programs for solving the LQP, as discussed by Edgar et al (47).

Edgar (22) has also discussed the quadratic objective function issue, i.e., whether it incorporates economic realities, and he concluded that it is not wholly satisfactory in this regard. At the present time the LQP appears to be a method which can usually yield results more or less equivalent to other design techniques, although possibly with larger effort. Other features of the LQP controller synthesis approach are as follows:

(1) Physical state space models are more attractive for use with the LQP (especially when state variables are directly measurable), while multivariable black box models are probably better treated by frequency response methods (22) or minimum variance control (discussed later in this section).

(2) The equivalence of tuned PID controllers and optimal controllers can be demonstrated by augmentation of the state vector and judicious selection of the objective function (47), (48); ordinarily an optimal feedback controller contains higher order derivative terms, yielding significant phase advance (which can cause noise amplification and controller saturation).

(3) Optimal feedforward control strategies can be treated within the framework of the LQP (49).

(4) Time delays can also be handled with the LQP, although the discrete-time formulation (46) of the LQP is better suited to the time delay problem (especially when there are only a few such elements in the differential equations).

(5) The optimal filtering problem (the Kalman-Bucy filter) can be solved independent of the optimal control for the LQP and provides a means for estimating unmeasured state variables which may be corrupted by process and instrument noise.

(6) The major deficiencies of the LQP feedback control

technique lie in on-line tuning (or de-tuning), sensitivity calculations, noise effects and interpreting the frequency response, although Safonov and Athans (50) have obtained some results on the gain and phase margin characteristics of the LQP. These authors claim that the optimal controller is surprisingly robust, and they have derived conditions for the effects of model errors and nonlinearities on asymptotic stability.

Recently there has been great interest in discrete-time optimal control based on a one-step ahead optimization criterion, also known as minimum variance control. A number of different approaches for minimum variance control has been developed in the last decade. MacGregor (51) and Palmor and Shinnar (52) have provided overviews of these minimum variance controller design techniques.

A generalized linear discrete model (SISO) is

$$y_t = \frac{z^{-b} \sum_{i=0}^{s} w_i z^{-i}}{1 - \sum_{i=1}^{r} \delta_i z^{-i}} \cdot u_t + \frac{z^{-d} \sum_{i=0}^{p} \theta_i z^{-i}}{1 - \sum_{i=1}^{q} \Phi_i z^{-i}} \cdot \varepsilon_t \qquad (13)$$

Another form used in the literature is

$$y_t = G_p(B)\, u_t + G_n(B)\, a_t \qquad (14)$$

In equation (13) b and d represent the time delays in the system, (p, q) is the order of the noise model, (r, s) is the order of the deterministic model, and z and b are shift operators defined as $z^{-i}y_t = B^i y_t = y_{t-i}$. The component ε_t or a_t represents uncorrelated white noise which is passed through a transfer function to describe the noise or disturbance model. u_t is the input and y_t is the output. G_p and G_n are discrete transfer functions. Both Astrom (53) and Box and Jenkins (54) have developed modeling approaches for equation (13), which involve obtaining maximum likelihood estimates of the parameters in the postulated model followed by diagnostic checking of the sum of the residuals. The Box and Jenkins method also develops a detailed model for the process disturbance. Both of the above references include derivations of the minimum variance control.

The minimum variance control for an SISO system finds the unrestricted minimum of the expected value of a quadratic objective function:

$$\min_{u_t} \quad E\,[y_t^2 + \lambda(\nabla^d u_t)^2] \qquad (15)$$

Note that λ is a tuning parameter which can be selected to minimize controller saturation. ∇^d represents a differencing operator (d can be zero). In some cases $\lambda = 0$ is allowable if the sampling time is large enough (spreading out the control effort).

As shown in the above works, an optimal feedback/feedforward controller can be derived as an analytical function of the numerator and denominator polynomials of $G_p(B)$ and $G_n(B)$. No iteration or integration is required to generate the feedback law, as a consequence of the one step ahead criterion. Shinnar and Palmor (52) have also clearly demonstrated how dead time compensation (discrete time Smith predictor) arises naturally out of the minimum variance controller. These minimum variance techniques can also be extended to multi-variable systems, as shown by MacGregor (51).

Finally, it should be noted that many existing digital controller design techniques (Kuo (55)) yield an optimal or "perfect" control; in deadbeat or minimal prototype controller designs, the exact form of the response is pre-specified, rather than minimizing the integral of the squared error.

On-Line Optimization and Control

As discussed in previous sections, the design of optimization and control schemes for systems described by linear differential equations with constant coefficients has evolved to a satisfactory level for reasonable size models. Many techniques are available, giving a control engineer much flexibility in the choice of techniques. However, the chief failing in this type of control/optimization structure is the assumption that the parameters of the process remain constant. In most actual processes, the parameters are either poorly known (usually due to measurement and/or modeling deficiencies) or are time-varying in nature. One solution to this problem is to design a worst case controller; however, this solution is definitely inferior to an adaptive controller, where on-line state and parameter identification of the process is incorporated into the controller action. A "gain-adaptive" controller is presently commercially available, but this is only a first step towards more powerful adaptive control methods which could be implemented in industry.

An adaptive controller normally will incorporate the highly successful feedback structure. In the field of adaptive control, three general approaches have been developed (56):

(1) design an "insensitive" or robust controller
(2) adjust the controller parameters in response to output performance characteristics
(3) measure on-line the plant parameters and adjust the control law based on prior analysis

The first two approaches appear to be the most suitable for chemical process applications; the robust controller is particularly attractive for microprocessor-based control. The second approach is usually superior to the third because parameter measurement delays can negate the adaptive control advantages.

The development of an insensitive controller can of course be accomplished by repetitive simulations, but this by itself is an inefficient and usually impractical approach. The design of such a controller using standard linear optimal control methods has not proven to be fruitful as yet, since inclusion of sensitivity measures in the performance index does not yield to a closed form solution (57), (58). There is a need for improved methods which can realize desired sensitivity characteristics as well as high performance without resorting to extensive interactive calculations; Davison (59) has recently suggested one such approach.

Many of the successful adaptive control applications have been based on feedback control, e.g., minimum variance control or the self-tuning regulator (60), (61), model reference adaptive control (57), and hyperstability methods (62). The minimum variance controller developed by Astrom has been extended for the purposes of adaptive control. This self-tuning regulator requires that the model parameters in equation (14) be estimated using a least squares method. These estimates are then used in the minimum variance control law. Astrom and Wittenmark (61) have shown that if the parameter estimates converge, the regulator will converge to the minimum variance regulator that could be obtained if the order and the parameters of the system were known, although this proof is based on weak optimality conditions. Model reference adaptive control is based on a Lyapunov stability approach, while the hyperstability method uses Popov stability analysis. All of the above methods have been tested on experimental systems, both SISO and MIMO (53), (54), (55). The self-tuning regulator is now available as a commercial software package, although this method is not satisfactory for variable time delays, an important industrial problem.

Other recent developments in the field of adaptive control of interest to the processing industries include the use of pattern recognition in lieu of explicit models (Bristol (66)), parameter estimation with closed-loop operating data (67), model algorithmic control (68), and dynamic matrix control (69). It is clear that discrete-time adaptive control (vs. continuous time systems) offers many exciting possibilities for new theoretical and practical contributions to system identification and control.

Control with Limited Measurements

One of the major questions in control system design is the selection of process measurements. An important deficiency of state variable control is that measurements or estimates of all state variables are required. Usually only a few of the states can be monitored instantaneously, because of sensor cost or time delays caused by the need for chemical analysis. Distillation columns with many components and large numbers of trays would create special difficulties. The multivariable frequency domain methods require output information only; linear optimal control, on the other hand, does require complete state measurement or state estimates. Observer theory or filtering theory can be used to provide estimates of the unmeasured state variables from input/output data. These estimates can then be used with the computed optimal control law; the combination of the Kalman-Bucy filter with the optimal feedback matric is optimal for the stochastic LQP. The filter approach reduces the phase advance and reduces the system sensitivity to high frequency noise, but at the expense of extra on-line computation and system performance. An observer has the opposite effect, increasing phase advance of the system even more.

An approach called inferential control has been developed by Brosilow and coworkers (70), (71) to address the measurement limitation problem, especially when unmeasured disturbances are present. These disturbances, when persistent, are problematic for the Kalman filter approach. Weber and Brosilow (70), in their research with distillation columns, have developed a static estimator which predicts the product quality based on readily available measurements; measurements can be selected so that the estimator is relatively insensitive to modeling errors and measurement noise. Their approach also avoids the need for observers or dynamic state estimators. This inferential control approach has an extra advantage in that composition measurement loop and sampling delays can be eliminated. The net result is a tremendous reduction in number of state variables and measurements (although not necessarily yielding a single input-single output coupling). The number of measurements is selected so the control system is insensitive to modeling errors. The control system uses the inferred measurements to adjust the control effort and counteract the unmeasurable disturbances. A dynamic compensation scheme for the static estimator/controller based on simple lead-lag elements has been developed by Brosilow and Tong (71).

A related idea in process control which has received much interest recently is the analysis of interactions among states, outputs, and controls. The analytical technique used in many commercial applications is the relative gain array (Bristol, (72)). Rather than being explicitly based on system dynamics, it yields a measure of the steady state gain between a given input/output pairing. By using the most sensitive SISO connections, control magnitudes can be minimized. The relative gain array can be obtained analytically, computationally, or experimentally, and the basis for computing the relative gain matrix, of dimension mxm (m = number of outputs and the number of controls) is

$$\phi_{ij} = \left.\frac{dy_i}{du_j}\right|_{u_{i \neq j}} \tag{16}$$

ϕ_{ij} is a measure of the sensitivity of output i to controller j; it is computed by varying the jth controller output while holding all other controller outputs constant. Interaction is quantitatively measured by

$$\mu_{ij} = \underline{\phi} \cdot (\underline{\phi}^t)^{-1} \tag{17}$$

As shown by Bristol (72) for controllers with heavy reset action, this measure has very interesting properties. Input/output pairs are selected for those μ_{ij} approaching 1. A negative element in μ_{ij} indicates instability or non-minimum phase behavior, so a glimpse of the dynamics can be obtained from Bristol's method.

McAvoy (73) has explored the use of this index and a dynamic version of the index to analyze two-point composition control in distillation columns. Input-output pairing using equation (16) can often lead to poor control, while the opposite pairings can actually yield better results. This is especially true for time delay and non-minimum phase processes. Tung and Edgar (74) have developed a comprehensive theory of control-output dynamic interactions for linear systems which includes the steady state relative gain index as a special case. They have applied this dynamic interaction index to analysis of a distillation column and a fluid catalytic cracker. Gagnepain and Seborg (75) have also proposed an interaction measure based on open loop step responses and have provided some interesting comparisons with McAvoy's results. The subject of the interaction index employed as a process design tool is also addressed in the following section.

Integration of System Design and Control Considerations

In the practice of engineering the synthesis of control systems is normally performed after the system design, i.e., after selection of steady state parameters is completed. Thus a system which may appear attractive based on steady state analysis may have very undesirable dynamic properties, making successful control system design a difficult task. When energy costs were low, the decoupling of the design and control steps usually did not lead to uncontrollable systems; however, with increasing fuel prices energy integration has been introduced to greater degrees in order to reduce energy requirements, and the design and control steps have become more strongly related.

The problem of satisfying both steady state and dynamic objectives transcends the problem of control of individual processes. Design procedures for single pieces of equipment are well-established, although for reasonably complicated processes (such as reactors and distillation columns), there are still some questions to be resolved. The more challenging research problems fall under the heading of plant control, where several units are integrated, for example, to conserve energy. Simple decomposition of the overall process into discrete blocks is usually very difficult.

In heat recovery applications there can be a large number of feasible plant configurations. After the configuration is optimized based on steady state considerations (which may not be an easy problem), the evaluation of the effectiveness of various control schemes can be performed. The dynamic plant operation must be evaluated in terms of economics, regulation, reliability, and safety over a broad range of operating regimes.

On the other hand, the control evaluation could be performed in tandem with the design study, thus ruling out candidate design configurations rather early because of control difficulties. It would be advantageous to have quick and uncomplicated screening methods to evaluate potential control structures in the design phase; alternatively, the control structure optimization could be incorporated as part of the steady state design optimization. This would avoid the necessity of actual controller synthesis, which is obviously unattractive and could be quite time-consuming. One approach for control evaluation discussed earlier is the relative gain array (72), (74). No actual synthesis of the controller is required in these algorithms. The development of such screening tools is still in its infancy but appears to be quite promising for concurrent design/control evaluation. Such techniques, if simple to use, would be immediately acceptable for use by major engineering firms and the process industries.

The overall plant control concept and incorporation of more detailed control design with plant design requires the selection of the following elements:

(1) control objectives
(2) output variables
(3) measurements
(4) manipulated variables
(5) control structure

There are a number of available techniques for evaluating the control system, which can be classified as follows:

(1) analysis of control constraints (rather than dynamics) (76)
(2) generating alternative control structures for each unit and minimizing conflicts among the various structures, using a multilevel analytical approach (77)
(3) satisfying product quality and controlling the material balance as primary objectives (steady state control),

followed by dynamic analysis (78), (79)
(4) aggregation of units that have common functional goals in terms of control and economics (80)

While these techniques have been applied to energy-related processes such as heat-integrated distillation columns and fluid catalytic cracking reactors, there is still extensive research required before the concept of plant design/control is reduced to practice.

The operation of energy-integrated plants will make it necessary to design control systems which are decentralized but which also respond to overall plant objectives. Existing modern control theory is really not adequate for these large scale problems, since there may be over 50 state variables. These systems are often made up of interconnected and often similar elements and must be controlled by an hierarchy of computers - micro, mini, and macro. The questions of system structure, representation, and modeling and control, measurement, and optimization strategies are fertile ones. A recent issue of the IEEE Transactions on Automatic Control (April, 1978) was devoted entirely to this subject. There have been only a few applications of multi-level or distributed control reported in the literature (81), (82), (83), but during the next decade this area promises to be one of great activity.

Literature Cited

1. Foss, A. S.; and Denn, M.M., Ed., "Chemical Process Control," AIChE Symp. Ser., 1976, 72, No. 159.
2. Seborg, D. E., "A Survey of Process Control Education in the U.S. and Canada," Annual AIChE Meeting, Miami, Florida, November, 1978.
3. Coughanowr, D. R.; and Koppel, L. B., Process Systems Analysis and Control, McGraw-Hill, New York, 1965.
4. Edgar, T. F., "Assessment of the Use of Process Dynamics and Control for Influencing Energy Utilization in Industrial Processes," paper presented at Department of Energy Workshop on Process Dynamics and Control, Denver, Colorado, June, 1979.
5. Robnett, J.D., "Engineering Approaches to Energy Conservation," Chem. Engr. Prog., 1979, 75 (3), 59.
6. Kenney, W. F., "Reducing the Energy Demand of Separation Processes," Chem. Engr. Prog., 1979, 75 (3), 68.
7. Smith, O.J.M., "Closer Control of Loops with Dead Time," Chem. Engr. Prog., 1957, 53, 217.
8. Smith, C. L., Digital Computer Process Control, In text, Scranton, Pa., 1972.
9. Edgar, T. F.; and Schwanke, C. O., "A Review of the Application of Modern Control Theory to Distillation Columns," Proc. Joint Automatic Control Conference, 1977, 1370.
10. Tung, L.S.; and Edgar, T. F., "Development and Reduction of a Multivariable Distillation Column Model with Tray Hydraulics,"

AIChe National Meeting, Houston, Texas, April, 1979.

11. Rijnsdorp, J. E.; and Seborg, D. E., "A Survey of Experimental Applications of Multivariable Control to Process Control Problems," AIChE Symp. Ser., 1976, 72 (159), 112.
12. Buckley, P. B., "Distillation Column Design Using Multivariable Control," Inst. Tech., 1978, 25(3), 115.
13. Padmanabhan, L.; and Lapidus, L., "Control of Chemical Reactors" in Chemical Reactor Theory: A Review, Ed. by L. Lapidus and N. R. Amundson, Prentice-Hall, Englewood Cliffs, N. J., 1977.
14. Wallman, P. H.; and Foss, A. S., "Multivariable Integral Controls for Fixed Bed Reactors," AIChE Annual Meeting, Miami, Florida, November, 1978.
15. Kurihara, H., Optimal Control of Fluid Catalytic Cracker, Report ESL-R-309, M.I.T., June, 1967.
16. Schuldt, S. B.; and Smith, F. B., "An Application of Quadratic Performance Synthesis Techniques to a Fluid Catalytic Cracker," Proc. Joint Automatic Control Conference, 1971, 270.
17. Olsson, G., "State of the Art in Sewage Treatment Plant Control, "AIChE Symp. Ser., 1976, 72 (159), 52.
18. Church, D. F., "Cement and Projected Pulp and Paper Industry Problems in Process Control and Process Modeling," AIChE Symp. Ser., 1976, 72 (159), 19.
19. Clelland, P. J., "Recent Furnace Draft Control Considerations," Proc. IEEE Conf. on Dec. and Cont., 1977, 1101.
20. Fisher, D. G.; and Seborg, D. E., Multivariable Computer Control-A Case Study, North-Holland Publishing Co., Amsterdam, 1976.
21. MacFarlane, A. G. J., "A Survey of Some Recent Results in Linear Multivariable Feedback Theory," Automatica, 1972, 8, 455.
22. Edgar, T. F., "Status of Design Methods for Multivariable Control," AIChE Symp. Ser., 1976, 72, (159) 99.
23. Munro, N.; Broland, B. J.; and Brown, L. S., "The UMIST Computer-Aided Control System Design Suite," National Research Development Corporation, Great Britain, 1973.
24. Rosenbrock, H. H., Computer-Aided Control System Design, Academic Press, New York, 1974.
25. Shinskey, F. G., "The Stability of Interacting Control Loops with and without Decoupling," Proc. Fourth IFAC Multivariable Technological Systems Conference, 1977, 21.
26. Jafarey, A.; and McAvoy, T. J., "Degeneracy of Decoupling in Distillation Columns," IECPDD, 1978, 17, 485.
27. McAvoy, T. J., "Steady State Decoupling Sensitivity with Application to Distillation Control," AIChE National Meeting Houston, Texas, April, 1979.
28. Toijala (Waller), K. V.; and Fagervik, K. C., "A Digital Simulation Study of Two Point Feedback Control of Distillation Columns," Kem. Teollisuus, 1972, 29, 5.

29. Rosenbrock, H. H., "Design of Multivariable Control Systems Using the Inverse Nyquist Array," Proc. IEEE, 1969, 116, 1929.
30. MacFarlane, A. G. J.; and Belletrutti, J. J., "The Characteristic Locus Method," Automatica, 1973, 9, 575.
31. MacFarlane, A. G. J., "Commutative Controller: A New Technique for the Design of Multivariable Control Systems," Electron Letters, 1970, 6, 121.
32. Mayne, D. Q., "The Design of Linear Multivariable Systems," Automatica, 1973, 9, 201.
33. Kuon, J. F., "Multivariable Frequency Domain Design Techniques," Ph.D. dissertation, University of Alberta, Edmonton, Canada, 1975.
34. Bilec, R.; and Wood, R. K., "Multivariable Frequency Domain Controller Design for a Binary Distillation Column," AIChE National Meeting, Houston, Texas, April, 1979.
35. Schwanke, C. O.; Edgar, T. F.; and Hougen, J. O., "Development of Multivariable Control Strategies for Distillation Columns," ISA Trans., 1977, 16 (4), 69.
36. Foss, A. S.; Edmunds, J. M.; and Kouvaritakis, B., "Multivariable Control System for Two-Bed Reactors by the Characteristic Locus Method," IEC Fund., 1980, in press.
37. Ogunnaike, B. A.; and Ray, W. H., "Multivariable Controller Design for Linear Systems Having Multiple Time Delays," AIChE J., 1979, in press.
38. Simon, J. D., and Mitter, S. K., "A Theory of Model Control," Inf. Control, 1968, 13, 316.
39. Davison, E., "On Pole Placement in Linear Systems with Incomplete State Feedback," IEEE Trans. Auto. Cont., 1970, AC-15, 348.
40. Vardulakis, A. I., "Generalized Root Locus Assignment of All Poles of a Multivariable System by Output Feedback," Int. J. Cont., 1976, 23, 39.
41. Brasch, F. M.; and Pearson, J. B., "Pole Placement Using Dynamic Compensators," IEEE Trans. Auto. Cont., 1970, AC-15, 34.
42. Bristol, E. H., "Recent Results in Interaction in Multivariable Process Control," AIChE Annual Meeting, Miami, Florida, November, 1978.
43. Dahlin, E. B., "Designing and Tuning Digital Controllers," Inst. Cont. Syst., 1968, 41 (6), 77.
44. Higham, J. D., "Single-Term Control of First and Second Order Processes with Dead Time," Control, 1968, February, p. 136.
45. Athans, M.; and Falb, P. L., Optimal Control, McGraw-Hill, New York, 1966
46. Lapidus, L; and Luus, R., Optimal Control of Engineering Processes, Blaisdell, Waltham, Mass., 1967.
47. Edgar, T. F.; Vermeychuk, J. G.; and Lapidus, L., "The Linear-quadratic Control Problem: A Review of Theory and

Practice," Chem. Engr. Comm., 1973, 1, 57.
48. O'Connor, G. E.; and Denn, M. M., "Three Mode Control as an Optimal Control," Chem. Engr. Sci., 1972, 27, 121.
49. Balchen, J. G.; Endresen, T.; Fjeld, M.; and Olsen, T. O., "Multivariable PID Estimation and Control in Systems with Biased Disturbances," Automatica, 1973, 9, 295.
50. Safonov, M. G.; and Athans, M., "Gain and Phase Margin for Multiloop LQP Regulators," IEEE Trans. Auto. Cont., 1977, AC-22, 173.
51. MacGregor, J. F., "Optimal Discrete Stochastic Control Theory for Process Application," Can. J. Chem. Engr., 1973, 51, 468.
52. Palmor, Z. J.; and Shinnar, R., "Design of Sampled Data Controllers," IEC PDD, 1979, 18 (1), 8.
53. Astrom, K. J., Introduction Stochastic Control Theory, Academic Press, New York, 1970.
54. Box, G. E. P.; and Jenkins, G. M., Time Series Analysis, Forecasting, and Control, Holden-Day, San Francisco, 1970.
55. Kuo, B. C., Digital Control Systems, SRI Publishing Company, Champaign, IL., 1977.
56. Landau, Y. D., Adaptive Control, Marcel Dekker, New York, 1978.
57. Newman, M. M., "On Attempts to Reduce the Sensitivity of the Optimal Linear Regulator to a Parameter Change,", Int. J. Cont., 1970, 11, 1079.
58. Sannuti, P.; and Cruz, J. B., "A Note on Trajectory Sensitivity of Optimal Control Systems," IEEE Trans. Auto. Cont., 1968, AC-13, 111.
59. Davison, E. J., "Decentralized Robust Control of Unknown Systems Using Tuning Regulators," IEEE Trans. Auto. Cont., 1978, AC-23, 276.
60. Wittenmark, B., "Stochastic Adaptive Control Methods-a Survey," Int. J. Cont., 1975, 21, 705.
61. Astrom, K. J.; and Wittenmark, B., "On Self-Tuning Regulators," Automatica, 1973, 9, 185.
62. Popov, V. M., "The Solution of a New Stability Problem for Controlled-Subsystems," Automation and Rem. Cont., 1963, 24, 1.
63. Borisson, U.; and Wittenmark, B., "An Industrial Application of a Self-Tuning Regulator," 4th IFAC Conference on Digital Computer Applications to Process Control, Zurich, Switzerland, 1974.
64. Shah, S. L., Fisher, D. G.; and Karim, N. M., "Hyperstability Adaptive Control-A Direct Input-Output Approach without Explicit Model Identification," Proc. Joint Automatic Control Conference, 1979, 481.
65. Oliver, W. K.; Seborg, D. E.; and Fisher, D.G., "Model Reference Adaptive Control Based on Lyapunov's Direct Method," Chem. Engr. Comm., 1973, 1, 125.

66. Bristol, E. H., "Pattern Recognition: an Alternative to Parameter Identification in Adaptive Control," Automatica, 1977, 13, 197.
67. Box, G. E. P.; and MacGregor, J. F., "Parameter Estimation with Closed Loop Operating Data," Technometrics, 1976, 18, 371.
68. Richalet, J.; Rault, A.; Testud, J. L.; and Papon, J., "Model Predictive Heuristic Control: Applications to Industrial Processes," Automatica, 1978, 14, 413.
69. Cutler, C. R.; and Ramaker, B. L., "Dynamic Matrix Control-A Computer Control Algorithm," AIChE National Meeting, Houston, Texas, April, 1979.
70. Weber, R.; and Brosilow, C.B., "The Use of Secondary Measurements to Improve Control," AICHE J., 1972, 18, 614.
71. Brosilow, C. B.; and Tong, M., " The Structure and Dynamics of Inferential Control Systems," AIChE J., 1978, 24, 492.
72. Bristol, E. H., "On a New Measure of Interaction for Multivariable Process Control," IEEE Trans. Auto. Cont., 1966, AC-11, 133.
73. McAvoy, T. J., "Interacting Control Systems: Steady State and Dynamic Measurement of Interaction," ISA Trans., 1978, 16 (3), 35.
74. Tung, L. S.; and Edgar, T. F., "Analysis of Control-Output Interactions in Dynamic Systems," AIChE Meeting, Miami, Florida, November, 1978.
75. Gagnepain, J. P.; and Seborg, D. E., "An Analysis of Process Interactions with Applications to Multi-Loop Control System Design," AIChE Annual Meeting, San Francisco, Nov., 1979.
76. Govind, R.; and Powers, G. J., "Synthesis of Process Control Systems," IEEE Trans. Sys. Man. Cyb., 1978, FMC-8, 792.
77. Umeda, T.; and Kuriyama, T., "A Logical Structure for Process Control System Synthesis," Proc. IFAC Congress, Helsinki, Finland, 1978.
78. Buckley, P. S., Techniques of Process Control, Wiley, New York, 1964.
79. Douglas, J. M., "Quick Estimates of Control Economics," AIChE Annual Meeting, New York, 1977.
80. Morari, M.; Arkun, Y.; and Stephanopoulos, G.,"An Integrated Approach to the Synthesis of Process Control Structures," Proc. Joint Automatic Control Conference, 1977, 243.
81. Ray, W. H.; and Lainiotis, D. G., Distributed Parameter Systems, Marcel Dekker, New York, 1978.
82. Kociuba, T.; and Ponstingl, J. C., "Energy Management Systems for Industry," Proc. Fourth Annual Advanced Control Conference, Purdue University, April, 1977.
83. Al-Shaikh, A.; and Leffler, N., "Application of Multi-Level Control of Calenders in the Tire Industry," Automatica, 1978, 14, 1.

RECEIVED November 14, 1979.

SPECIFIC REVIEWS

5

Computation of Phase and Chemical Equilibrium: A Review

WARREN D. SEIDER, RAJEEV GAUTAM[1], and CHARLES W. WHITE, III

Department of Chemical and Biochemical Engineering, University of Pennsylvania, Philadelphia, PA 19104

Figure 1 illustrates the physical situation in which a mixture of chemicals is allowed to reach equilibrium. The number of phases, P, and their compositions are unknown at equilibrium.

Usually the state of the feed and temperature and pressure of the products are known, and the Gibbs free energy is minimized

$$G = \sum_{j=1}^{S} G_j^o n_j^c + \sum_{j=S+1}^{C} \sum_{\ell=1}^{P} G_{j\ell} n_{j\ell} \quad (1)$$

subject to mass balance constraints, where C is the number of chemical species, P is the number of mixed phases (vapor, liquid, solid), S is the number of condensed species (which appear in only one pure phase, are normally solid, and do not distribute amongst other phases), $n_{j\ell}$ is the number of moles of compound j in phase ℓ, $G_{j\ell}$ is the chemical potential of compound j in phase ℓ, given by

$$G_{j\ell} = G_{j\ell}^o + RT \ln \frac{\bar{f}_{j\ell}}{f_{j\ell}^o}$$

and often, $\bar{f}_{j\ell} = y_{j\ell}\phi_{j\ell}P$ for vapor and $= x_{j\ell}\gamma_{j\ell}f_{j\ell}$ for liquid phases.

When the number of independent chemical reactions equals C - ρ, where ρ is the rank of the atom matrix (m_{jk}), Gibbs free energy is minimized subject to atom balance constraints:

[1] Current address: Union Carbide Corporation, South Charleston, West Virginia

0-8412-0549-3/80/47-124-115$05.00/0

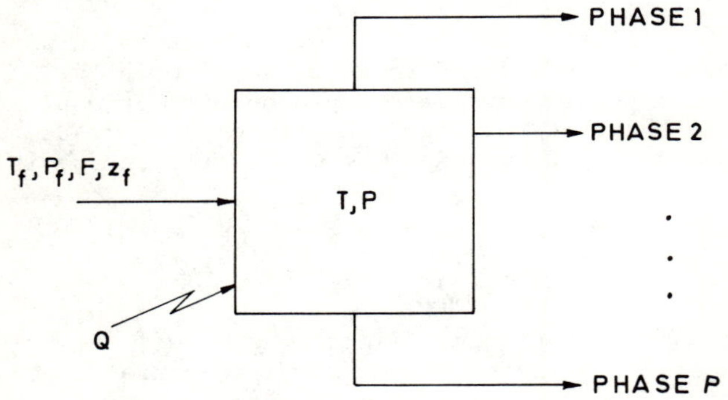

Figure 1. Physical situation. Mixture reaches equilibrium at T and P. The number of phases, P, *and their compositions are unknown.*

$$b_k = \sum_{j=1}^{S} m_{jk} n_j^c + \sum_{j=S+1}^{C} \sum_{\ell=1}^{P} m_{jk} n_{j\ell} \qquad k=1,\ldots,E \qquad (2)$$

where b_k is the number of gram-atoms of element k, m_{jk} is the number of atoms of element k in compound j, and E is the number of elements.

After computing product compositions, the energy balance gives heat duty.

Other specifications for the product, such as pressure and heat duty (e.g., adiabatic processes) or pressure and entropy (e.g., isentropic processes) also involve finding the extremum of a thermodynamic function. Given pressure and heat duty, the entropy is maximized

$$\mathcal{S} = -\frac{1}{T}\left[\sum_{j=1}^{S} G_j^o n_j^c + \sum_{j=S+1}^{C} \sum_{\ell=1}^{P} G_{j\ell} n_{j\ell}\right] \qquad (3)$$

subject to mass balance constraints and the energy balance

$$Q + H_F = \sum_{j=1}^{S} h_j^c n_j^c + \sum_{\ell=1}^{P} h_\ell n_\ell \qquad (4)$$

Alternatively, Gibbs free energy (1) is minimized subject to these constraints. Given pressure and entropy, the enthalpy is minimized

$$H = \sum_{j=1}^{S} G_j^o n_j^c + \sum_{j=S+1}^{C} \sum_{\ell=1}^{P} G_{j\ell} n_{j\ell} \qquad (5)$$

subject to mass balance constraints and the entropy balance, which for isentropic systems is

$$\mathcal{S}_F = \sum_{j=1}^{S} s_j^c n_j^c + \sum_{\ell=1}^{P} s_\ell n_\ell \qquad (6)$$

This is equivalent to minimizing Gibbs free energy. Less common specifications such as temperature and heat duty or temperature and entropy are treated similarly.

Solution of the optimization problem requires algorithms that locate the phase distribution corresponding to the global optimum. Castillo and Grossmann (1) term this a mixed-integer programming problem and suggest a tree to enumerate the possible combinations of phases at equilibrium.

Some computational problems are specific to chemical equilibrium, while others are specific to phase equilibrium. Hence, we consider techniques for chemical equilibrium in a single phase and for phase equilibrium without reaction prior to those for chemical and phase equilibrium.

Before proceeding, two reviews are worthy of mention. Zeleznik and Gordon (2) and VanZeggeren and Storey (3) concentrate on thermodynamic fundamentals and numerical methods. Their conclusions should be reassessed in view of recent developments in numerical algorithms.

Chemical Equilibrium

Until the early 1940's, with temperature and pressure given, compositions in chemical equilibrium were computed manually by solving a set of nonlinear equations:

$$\sum_{j=1}^{C} \nu_{ij} G_j = 0 \qquad i=1,\ldots,R \tag{7}$$

subject to the atom balances in Equation (2). Equation (7) was written for each of R independent reactions and was obtained by differentiating Equation (1):

$$\frac{dG}{d\xi_i} = 0 \qquad i=1,\ldots,R$$

where ξ_i is the extent of reaction i and:

$$n_j = n_j^o + \sum_{i=1}^{R} \nu_{ij} \xi_i \qquad j=1,\ldots,C \tag{8}$$

n_j^o is the moles of compound j in the feed mixture and ν_{ij} are the stoichiometric coefficients of compound j in reaction i given by the atom balances $\underline{\underline{M}}^T \underline{\underline{\nu}}^T = 0$.

Brinkley (4,5) postulated C species at equilibrium. ρ species, referred to as "components," were selected to have linearly independent formula vectors, where ρ is the rank of the atom matrix, (m_{jk}), and Y_j is the formula vector for the jth species, $[m_{j1}, m_{j2}, \ldots m_{jE}]$. Given the choice of components, the stoichiometric coefficients for an independent set of chemical reactions are computed:

$$\sum_{j=1}^{\rho} \nu_{ij} Y_j = Y_i \qquad i=\rho+1,\ldots,C \tag{9}$$

where the number of independent reactions, R = C - ρ, and the species on the right-hand-side are referred to as "derived species." Compositions at equilibrium are calculated using Equation (7) and conservation of mass for each component

$$G_i - \sum_{j=1}^{\rho} \nu_{ij} G_j = 0 \qquad i=\rho+1,\ldots,C \tag{10}$$

$$n_j + \sum_{i=\rho+1}^{C} \nu_{ij} n_i = q_j \qquad j=1,\ldots,\rho \tag{11}$$

where the feed mixture is distributed amongst components, $j = 1, \ldots, \rho$, and $q_j = n_j^o$. Kandiner and Brinkley (6) assumed values for the derived species, n_i^*, and solved Equation (11) for the moles of components, n_j. Equations (10) were solved for n_i and convergence criteria checked. For convergence, components need be selected as those species present in the highest concentration at equilibrium. When this is not the case, Equation (11) frequently gives negative values for n_j.

Browne, et al. (7) suggested the "optimum component" procedure in which components are selected as those species expected to be in the highest concentration at equilibrium. Cruise (8) altered the components during iterative calculation to reflect changes in composition.

Cruise (8) solved Equations (10) and (11) by adjusting extents of reaction. To prevent divergence, Smith and Missen (9) also improved initial estimates by using a linear programming procedure which neglects the logarithmic terms in Equation (1).

Because of these problems, most recent methods avoid the distinction between components and derived species and take the moles of all species as iteration variables.

Naphtali (10) solved Equations (7) by obtaining corrections to extents of reaction from:

$$d\xi_i = -\Delta G_i d\lambda \qquad i=1,\ldots,R \tag{12}$$

where $d\lambda$ is the step-size and

$$\Delta G_i = \sum_{j=1}^{C} \nu_{ij} G_j$$

He showed that positive $d\lambda$ give $d\xi_i$ that reduce Gibbs free energy. This method is analogous to that of steepest descent, a first-order method for minimization of Gibbs free energy. Ma and Shipman (11) used Naphtali's method to estimate compositions at equilibrium and the Newton-Raphson method to achieve convergence.

Several nonlinear programming methods have been applied to

minimize Gibbs free energy. But these are normally formulated to determine compositions in chemical and phase equilibrium and are presented in the section "Chemical and Phase Equilibrium."

Heidemann (12) observes that "the chemical reaction equilibrium problem in a homogeneous phase is known to have an unique solution except when the thermodynamic model of the phase can exhibit diffusional instability." Hence, for chemical equilibrium in a single phase, local minima in Gibbs free energy do not occur and the search is simplified.

Phase Equilibrium

Algorithms for computation of compositions in vapor and liquid phases at equilibrium usually solve the nonlinear algebraic equations

$$G_{jV} = G_{jL} \qquad j=1,\ldots,C \tag{13}$$

in the form

$$y_j = k_j\{T,P,\underline{x},\underline{y}\}x_j \qquad j=1,\ldots,C \tag{14}$$

together with mass balances and the energy balance. Equations (13) are obtained by differentiating Equation (1) for $S = 0$ and $\ell = 2$ with $dn_{jV} = -dn_{jL}$.

With temperature and pressure given, the flash equation:

$$\sum_{j=1}^{C} \frac{z_j(1 - k_j)}{1 + \alpha(k_j - 1)} = 0 \tag{15}$$

can be solved for α (moles of vapor/moles of feed) using simple root-finding procedures.

For non-ideal solutions k_j is a function of liquid composition and α. Hence, Equation (15) must be solved iteratively with material balance equations and computation times increase.

Boston and Britt (13) present an algorithm, given the state of the feed and six practical sets of specifications shown in Figure 2. They define volatility parameters

$$u_j = \ln \frac{k_j}{k_b} \qquad j=1,\ldots,C \tag{16}$$

where k_b is the "reference equilibrium ratio" defined by:

$$\ln k_b = \sum_{j=1}^{C} w_j \ln k_j \tag{17}$$

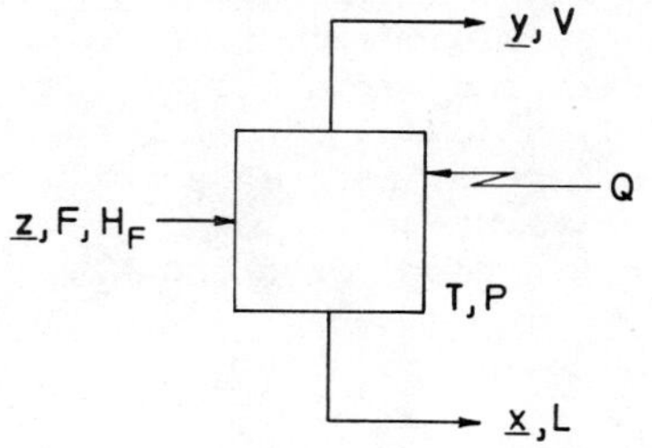

	Specified	Calculated
I	Q, P	T, V
II	Q, T	V, P
III	Q, V	T, P
IV	V, T	P, Q
V	V, P	T, Q
VI	T, P	V, Q

Figure 2. Specification by Boston and Britt (13) for vapor–liquid equilibrium

w_j are weighting functions computed as a function of V/L, $\underline{y}$, and $\underline{k}$, as described in their paper. Furthermore, a second relation for the reference equilibrium ratio is defined as:

$$\ln k_b = A + B\left(\frac{1}{T} - \frac{1}{T^*}\right) \tag{18}$$

A and B are parameters and T* is a reference temperature. $\underline{u}$, A, and B are calculated using accurate models for $\underline{k}$ at guesses for $\underline{x}$, $\underline{y}$, T, and P (when P is unknown, Eqn. (18) includes the term, $B_p \ln(P/P^*)$). Then, the material and energy balance and equilibrium (MESH) equations are rewritten using $\underline{u}$ and k_b and solved for new values of $\underline{x}$, $\underline{y}$, V, L, T, and P holding $\underline{u}$, A, and B constant. Given these new values, $\underline{u}$, A, and B are recalculated using accurate models for $\underline{k}$. $\underline{u}$, A, and B, along with similar parameters for enthalpy, are taken as iteration variables because they have weak dependence on $\underline{x}$, $\underline{y}$, and T, and each other. Broyden's method, beginning with $\underline{\underline{I}}$ as the approximation to the Jacobian matrix, adjusts these parameters to convergence in few iterations. Furthermore, the MESH equations, in terms of $\underline{u}$ and k_b, are solved for the iteration variable

$$R = \frac{k_b V}{k_b V + k_b^o L} \tag{19}$$

R is principally a function of temperature and composition (i.e., k_b) for wide-boiling systems and V/L for narrow-boiling systems. Hence, few iterations are needed to achieve convergence for wide- and narrow-boiling systems. The FORTRAN program XFLASH (14) implements this algorithm.

For highly non-ideal solutions, involving two or less liquid phases, Henley and Rosen (15) developed an algorithm that attempts to solve for compositions at equilibrium among a vapor and two liquid phases using k-values. When a solution cannot be obtained attempts are made to determine compositions at equilibrium between two liquid or vapor and liquid phases. Boston and Fournier (16) first solve for compositions at equilibrium assuming a vapor and one liquid phase using the algorithm by Boston and Britt (13). They split the liquid phase into two liquid phases using an approximate solution model, followed by rigorous solution, using k-values, for compositions of a vapor and two liquid phases at equilibrium. The FORTRAN program YFLASH (14) implements this algorithm.

Nonlinear programming methods have been applied to minimize Gibbs free energy with mass balances:

$$n_j = \sum_{\ell=1}^{P} n_{j\ell} \qquad j=1,\ldots,C$$

These are described in the next section. Note that when atom balances are used, Dluzniewski and Adler (17) show that "fictitious elements" prevent reaction. Consider a reactor that produces ethylbenzene by reaction of benzene and ethylchloride in the presence of $AlCl_3$ catalyst. For calculation of phase equilibrium, downstream of the reactor, fictitious element A replaces a hydrogen atom in benzene (C_6H_5A) and the moles of each species remain unchanged.

Chemical and Phase Equilibrium

Sanderson and Chien (18) solve Equations (7), (8), and (13) to determine compositions of vapor and liquid phases in chemical and phase equilibrium given temperature and pressure. A set of independent chemical reactions is selected with guesses for extent of reaction. Solution of Equation (13) leads to compositions in phase equilibrium, but applies only for a vapor and liquid in equilibrium. Residuals of Equations (7) and (8) are computed and extents of reaction, ξ_i, and moles of species j, n_j^t, are adjusted using Marquardt's method (15).

White (19) prepared the FORTRAN program EQUILK using the Sanderson and Chien algorithm and XFLASH program. It permits any combination of specifications in Figure 2.

It should be possible to solve Equations (4), (7), (8), and (15) simultaneously using an algorithm like that of Boston and Britt (13). But, we are unaware of such an algorithm.

In 1958, White, Johnson, and Dantzig (20), working at RAND Corporation, computed compositions at equilibrium using Newton's method to minimize the Gibbs free energy of an ideal gas mixture with temperature and pressure given. Their algorithm was extended to include pure solid phases by Boynton (21), Oliver, et al. (22), Balzhiser, et al. (23), one or more mixed liquid or solid phases by Dluzniewski and Adler (17), and electrolytic phases containing ionic species by Gautam and Seider (24).

A quadratic Taylor Series is used to approximate the Gibbs free energy as a function of composition (Equation (1)). Then, an unconstrained objective function, using Lagrange multipliers and mass balance constraints, is minimized to give improved compositions.

When solid phases are incorrectly postulated to occur at equilibrium, values of n_j^c become negative. Oliver, et al. (22) recommend that such species be removed and the algorithm restarted. They showed that the criteria for introducing a solid species is:

$$\frac{G_j^o}{RT} - \sum_{k=1}^{E} \pi_k m_{jk} < 0 \qquad j \leq S \qquad (20)$$

where the Lagrange multipliers, π_k, are computed at equilibrium without the solid species. Such a check should be made, especially after a solid species has been removed.

When a mixed phase ℓ is postulated that does not exist at equilibrium, the RAND Method decreases its concentrations to zero. This causes the row and column associated with extraneous phase ℓ in the coefficient matrix to approach zero and singularities to develop. Hence, it is necessary to remove these rows and columns from the matrix during matrix inversion when the maximum pivot element is less than an arbitrarily small value. This can occur with poor guesses for composition and lead to a "constrained" minimum in Gibbs free energy. Figure 3 is an attempt to show this phenomenum in three dimensions. Gibbs free energy is plotted as a function of the moles of component 1 in phases L_1 and L_2. If $n_1^{L_2}$ approaches zero and is removed from the search space, the search for minimum Gibbs free energy is constrained to the boundary, along the curve RST and a constrained minimum at T is found.

To avoid constrained minima, Gautam and Seider (25, 26) introduce an algorithm to split unstable phases between iterations of an optimization method and test the algorithm with the RAND Method. If the Gibbs free energy is reduced, composition variables for two new phases replace composition variables for the parent phase, and the minimization algorithm is continued. A FORTRAN program GIBBS (14) was created for this purpose.

Gordon and McBride (27), working at the NASA Lewis Research Center, developed an algorithm similar to the RAND Method. They minimized:

$$F\{\underline{n}\} = G\{\underline{n}\} + RT \sum_{k=1}^{E} \pi_k (b_k - \sum_{j=1}^{S} m_{jk} n_j^c - \sum_{j=S+1}^{C} m_{jk} n_j) \quad (21)$$

by setting

$$\frac{\partial F\{\underline{n}\}}{\partial n_j^c} = \frac{\partial F\{\underline{n}\}}{\partial n_j} = \frac{\partial F\{\underline{n}\}}{\partial \pi_k} = 0 \quad (22)$$

The resulting C + E equations are nonlinear in unknowns n_j^c, n_j, and π_k; but ln n_j are iteration variables since n_j occur in logarithmic terms. These equations are linearized using first-order Taylor Series (Newton-Raphson method), in the variables Δn_j^c, $\Delta(\ln n_j)$, and π_k, and with $n = \sum_j n_j$ are reduced to S + 1 + E linear equations in unknowns ΔN_j^c, $\Delta(\ln N)$, and π_k. When extended to include P mixed phases, we have shown that they are nearly identical to the equations of the RAND Method and have the same coefficient matrix.

George, et al. (28) eliminate constraints, $n_j \geq 0$, by introducing an allocation function

$$A\{u_j\} = \frac{1}{1 + e^{-u_j}} \quad (23)$$

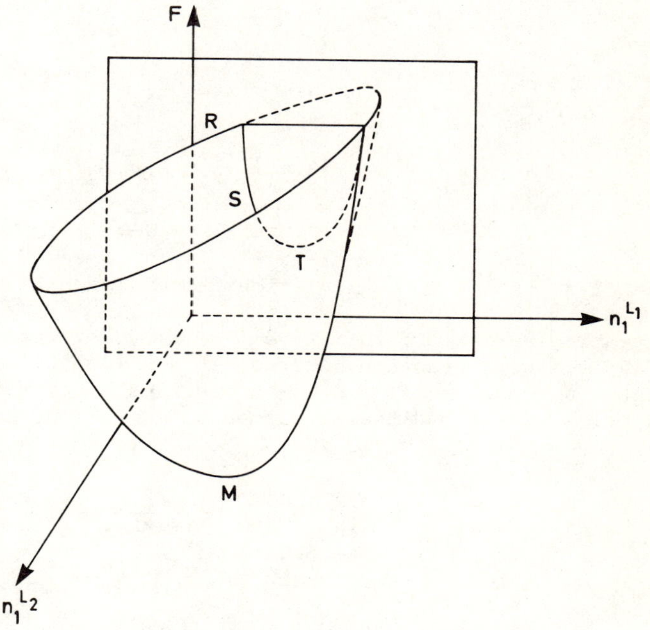

Figure 3. Hypothetical surface of Gibbs free energy shows $G\{n_1^{L_1}, n_1^{L_2}\}$. *Additional axes cannot be shown for* $n_j^{L_1}$ *and* $n_j^{L_2}$ *in multicomponent mixtures.*

where $-\infty < u_j < \infty$ and $0 \leq A\{u_j\} \leq 1$. For two phases in equilibrium, without reaction,

$$\left.\begin{aligned} n_{j1} &= A\{u_j\}n_j^t \\ n_{j2} &= (1 - A\{u_j\})n_j^t \end{aligned}\right\} \quad j=1,\ldots,C \qquad (24)$$

and for three phases, without reaction,

$$\begin{aligned} n_{j1} &= A\{u_j\}n_j^t \\ n_{j2} &= (1 - A\{u_j\})A\{u_{j+C}\}\, n_j^t \qquad (25) \\ n_{j3} &= (1 - A\{u_j\})(1 - A\{u_{j+C}\})n_j^t \\ & \qquad j=1,\ldots,C \end{aligned}$$

where $n_j^t = \sum_{\ell=1}^{P} n_{j\ell} = n_j^o$. In calculations to minimize G, u_j replace $n_{j\ell}$ as iteration variables; there are $(P - 1)C$ elements in $\underline{u}$.

For systems with reaction, George, et al. (28), like Brinkley (4), select ρ "components" and C - ρ "derived species." For the derived species, C - ρ additional elements, u_j, are defined such that

$$n_j^t = A\{u_{j+(P-1)C}\}n_{j_{max}} \qquad j=\rho+1,\ldots,C \qquad (26)$$

and

$$n_{j_{max}} = \frac{1}{\max\limits_{k=1,\ldots,E}\left[\dfrac{m_{jk}}{b_k}\right]}$$

where $n_{j_{max}}$ is the maximum amount of species j at equilibrium. These elements, u_j, are also adjusted to minimize G. As recognized by Kandiner and Brinkley (6), care must be taken to select "components" from amongst those species present in highest concentration at equilibrium. Otherwise, during calculations to minimize G, large values of n_j^t for the derived species ($j = \rho+1, \ldots, C$) can give negative values of n_j^t for trace components when solving Equations (11).

George, et al. (28) implemented Powell's method (29), a quasi-Newton method, that uses approximations to the Hessian matrix and its inverse, $H\{\underline{u}\}$ and $H\{\underline{u}\}^{-1}$, to calculate new compositions. A FORTRAN program is available from the authors. For each system, u_j are assigned for species in phases expected at equilibrium. As u_j become small for all species in a phase, variables are retained and new values computed using Powell's method.

Gautam and Seider (25) implemented the quadratic programming algorithm of Wolfe (30, 31) because compositions satisfy inequality constraints ($\underline{n} \geq 0$) without transformations. Moles of solid species approach zero, rather than large negative values, when solid species are removed during minimization calculations. Hence, their variables can be retained during subsequent calculations. Similarly, for solution phases, when the moles of all species approach zero, variables may be retained. Gautam and Seider eliminate these variables to avoid unnecessary operations, introducing the possibility of obtaining constrained minima in Gibbs free energy, which they prevent with a phase-splitting algorithm. For 7 systems, beginning with correct phase distributions and the same guessed compositions, they found quadratic programming and the RAND Method to have comparable speeds, whereas the implementation of Powell's method by George, et al. (28) was 2-8 times slower.

Another technique for handling the mass balances was introduced by Castillo and Grossmann (1). Rather than convert the objective function into an unconstrained form, they implemented the Variable Metric Projection method of Sargent and Murtagh (32) to minimize Gibbs's free energy. This is a quasi-Newton method which uses a rank-one update to the approximation of H^{-1}, with the search direction "projected" onto the intersection of hyperplanes defined by linear mass balances.

Since $\partial G/\partial n_{j\ell}$ approaches $-\infty$ as $n_{j\ell}$ approaches zero, the term $\ln \bar{f}_{j\ell}/f^o_{j\ell}$ is approximated with a quadratic function that remains finite when $\bar{f}_{j\ell}/f^o_{j\ell} < \varepsilon$. Like quadratic programming, it is not necessary to remove variables when the moles of a phase approach zero. Unlike Gautam and Seider (25, 26), Castillo and Grossmann do not eliminate these variables. For the system ethanol, acetic acid, ethyl acetate, and water, at 358.2°K and 1.013 bar (28), Castillo and Grossmann report 17 sec of UNIVAC 1106 CPU time to compute compositions of a vapor phase at equilibrium. When only the vapor phase was assumed at equilibrium, only 1 sec was used. Gautam and Seider assumed all liquid, split the liquid into vapor and liquid phases, and removed the liquid phase in 2.14 sec of UNIVAC 90/70 CPU time. They did not perform calculations assuming a vapor phase. As the 90/70 is judged to be 20-50 percent slower, for this case the method of Castillo and Grossmann is approximately an order of magnitude slower. Additional cases should be run, preferably using the same computer and examining computation times for minimization and phase-splitting before conclusions are reached.

For mixtures containing isomers, the number of independent

chemical reactions R often is not given by C - ρ, where ρ is the rank of the atom matrix. Whitwell and Dartt (33) describe methods for determining R and mass balance constraints for minimization of Gibbs free energy.

Given specifications other than temperature and pressure, in principle, nonlinear programming algorithms can optimize the appropriate thermodynamic function. When H and P are specified, entropy is maximized. But with compositions as iteration variables, the relation $\mathcal{S} = \mathcal{S}\{H,P,\underline{n}\}$ is needed. Similarly, given $\mathcal{S}$ and P, enthalpy is minimized; but with compositions as iteration variables, $H = H\{\mathcal{S},P,\underline{n}\}$ is needed. Since these relations are usually not available, entropy can be maximized using Equation (3) and enthalpy minimized using Equation (5), with temperature and compositions as iteration variables.

In one approach, T is assumed and Gibbs free energy minimized to give compositions. Then T is adjusted to satisfy the specification; e.g., H or $\mathcal{S}$. Naphtali (10) suggests this procedure, but does not present an algorithm for adjusting T.

Alternatively all iteration variables are adjusted simultaneously, as demonstrated by Gordon and McBride (27) in the NASA method. They add the specification equation (e.g., (4) or (6)) to the equations that give minimum Gibbs free energy (22). Linearization leads to S + 2 + E equations in the unknowns ΔN_j^c, $\Delta(\ln N)$, π_k, and $\Delta(\ln T)$. A formulation to permit multiple mixed phases is presented by Zeleznik and Gordon (2).

Restricted Chemical Equilibrium

For some mixtures, experimental measurements show that fewer than C - ρ reactions describe the mixtures at equilibrium. This occurs when some of C - ρ reactions do not occur due to the formation and depletion of free radicals in the reaction mechanism (34, 35) or because "certain reaction paths are not available for chemical change (36)." The latter occurs for certain reactions, in the absence of catalyst, that are too slow to occur.

For these systems, mass balances must account for these restrictions and replace the atom balances (2) as constraints for minimization of Gibbs free energy. These mass balances are given by Schott (37):

$$q'_j = \sum_{j=1}^{S} m'_{jj'} n_j^c + \sum_{j=S+1}^{C} \sum_{\ell=1}^{P} m'_{jj'} n_{j\ell} \qquad j'=1,2,\ldots,\rho' \tag{27}$$

where ρ' = C - R and R is the number of reactions observed experimentally. Schott solves for unknown coefficients in the $\underline{\underline{M}}'$ matrix, where

$$\underline{\underline{M}}'\underline{\underline{\nu}}^T = 0 \tag{28}$$

$\underline{\underline{M}}'$ is a C x ρ' matrix given by $[\underline{\underline{I}}|\underline{\underline{X}}]^T$ and $\underline{\underline{X}}$ is a submatrix of unknown coefficients. Bjornbom (34) shows that extents of R independent reactions must vary independently.

In many systems, slow reactions occur to an appreciable extent in the presence of fast, reversible reactions. Given extents of reaction measured experimentally, or computed by integration of a kinetic model, it is desired to compute compositions of a system. This has been accomplished using EQUILK (19) given specifications for extents or temperature approaches for less than R reactions, or the moles of some species in the product. The RAND Method has been extended to permit these specifications, as well (38). In principle, these specifications are easily added to the mass balances for any nonlinear programming algorithm.

Conclusions

For calculation of compositions in chemical equilibrium, we do not recommend the use of derived species as iteration variables, as recommended by Brinkley (4,5), because solution of the mass balances can lead to negative moles of components. This problem does not arise when the moles of all species are taken as iteration variables.

For calculation of compositions in vapor and liquid phases at equilibrium without reaction, when solving nonlinear algebraic equations involving K-values, we recommend the use of volatility and enthalpy parameters, as defined by Boston and Britt (13). This method handles narrow-boiling, wide-boiling, and highly nonideal systems efficiently.

When the phase distribution is uncertain at equilibrium, we recommend the minimization of Gibbs free energy using algorithms for nonlinear programming and phase-splitting (25,26). A single phase is assumed at equilibrium and new phases are added along the trajectory in search of the minimum Gibbs free energy. A less efficient procedure involves determining compositions of a vapor and single liquid phase only at minimum Gibbs free energy, followed by phase-splitting to give guesses for an algorithm to determine compositions of vapor and two liquid phases at equilibrium (16).

It is also possible to guess compositions for the most phases expected at equilibrium and to use a nonlinear programming algorithm, such as a gradient projection method, to minimize Gibbs free energy. When the moles of species in a phase approach zero, the zeroes can be retained, as in the work of Castillo and Grossmann (1). But our results suggest that it is more efficient to eliminate these variables and to use algorithms for phase-splitting to avoid constrained minima in Gibbs free energy. In future work, a more complete comparison should be undertaken. Furthermore, the gradient projection methods with zeroes eliminated and algorithms for phase-splitting should be tested.

Given accurate methods for computing the chemical potential

of species in the phases in which they occur, calculation of compositions at phase and chemical equilibrium can be accomplished reliably.

In nonlinear programming algorithms that evaluate derivatives, such as Newton's method, quadratic programming, and gradient projection methods, properties that depend on composition (activity and fugacity coefficients) are evaluated at the guesses for composition. Hence, simple analytical expressions give partial molal Gibbs free energies. These derivatives give good search directions, even for highly nonideal systems.

One unresolved question concerns whether it is possible to use volatility parameters as iteration variables in a nonlinear programming algorithm, with an approach similar to that introduced by Boston and Britt (13) for solution of nonlinear algebraic equations involving K-values. Our conclusion is that volatility parameters apply where K-values are used, and would be awkward to use in minimization of Gibbs free energy.

Nomenclature

A	Allocation function; Parameter of Eqn. (18)
B	Parameter of Eqn. (18)
b_k	Gram atoms of element k
C	Number of chemical species
E	Number of elements
$f_{j\ell}$	Fugacity of pure compound j in phase ℓ at T and P of system
$\bar{f}_{j\ell}$	Mixture fugacity of compound j in phase ℓ at T and P of system
$f^o_{j\ell}$	Fugacity of compound j in standard state for phase ℓ
F	Unconstrained objective function (see Eqn. (21))
G	Gibbs free energy
$G^o_{j\ell}$	Free energy of formation of species j in standard state for phase ℓ
$G_{j\ell}$	Partial molal free energy of compound j in phase ℓ (chemical potential)
H	Enthalpy
h	Specific enthalpy
$\underline{\underline{H}}$	Approximation to Hessian matrix
$\underline{\underline{H}}^{-1}$	Approximation to inverse of Hessian matrix
$\underline{\underline{I}}$	Identity matrix
k_b	Reference equilibrium constant

k_j	Vapor-liquid equilibrium constant for compound j
L	Flow rate of liquid product
m_{jk}	Number of atoms of element k in compound j
$\underline{\underline{M}}$	Atom matrix-elements are m_{jk}
$\underline{\underline{M}}'$	Mass balance matrix
$n_{j\ell}$	Moles of compound j in phase ℓ
n_j^c	Moles of condensed species j
n_j^o	Moles of compound j in feed mixture
n_j^t	Moles of compound j at equilibrium
$n_{j_{max}}$	Maximum moles of species j at equilibrium
$\underline{n}$	Guess vector for moles of compound j in phase ℓ
P	Number of mixed phases (vapor, liquid, or solid); Pressure
q_j	Equals n_j^o, where feed mixture is distributed amongst ρ "components"
q_j'	Equals n_j^o, where feed mixture is distributed amongst ρ' "components"
R	Number of independent reactions; Universal gas constant; Defined in Eqn. (14)
S	Number of condensed species - appear in only one pure phase - normally solid - do not distribute among other phases
$\mathcal{S}$	Entropy
s	Specific entropy
T	Absolute temperature
$\underline{u}$	Transformed variables; Volatility parameters
V	Flow rate of vapor product
$\underline{w}$	Weighting factors
$x_{j\ell}$	Mole fraction of species j in phase ℓ
$\underline{\underline{X}}$	Submatrix of $\underline{\underline{M}}'$
$y_{j\ell}$	Mole fraction of species j in vapor phase ℓ
Y_j	Formula vector for compound j $[m_{j1}, m_{j2}, \ldots, m_{jE}]$
z_j	Mole fraction of compound j in feed

Greek Symbols

α	Moles of vapor per mole of feed
$\gamma_{j\ell}$	Activity coefficient of compound j in phase ℓ
$d\lambda$	Step-size in Naphtali method
ν_{ij}	Stoichiometric coefficient of compound j in reaction i
$\underline{\underline{\nu}}$	Stoichiometric coefficient matrix
$\phi_{j\ell}$	Fugacity coefficient of compound j in phase ℓ
π_k	Lagrange multipliers
ρ	Rank of atom matrix (m_{jk})
ξ_i	Extent of reaction i

Superscripts

c	Condensed phase (comprised of a pure species that occurs in condensed phase only)

Subscripts

f	Feed

Acknowledgment

Financial support for this work has been provided by the Department of Energy under Contract No. E(49-18)-2295, Task Order No. 9, the ASPEN Project, and is appreciated.

Abstract

Algorithms are reviewed for determining composition in multiphase, reacting mixtures at equilibrium. Approaches are examined for initiating and carrying-out the search for global minimum in Gibbs free energy when the number of phases is uncertain. Some lead to constrained minima given poor guesses in compositions. Specifications include temperature and pressure or enthalpy and pressure, arising in process units where heat duty is known.

Literature Cited

1. Castillo, J. and I. E. Grossmann, "Computation of Phase and Chemical Equilibria," presented at 86th National Meeting of AIChE, Houston, April, 1979.
2. Zeleznik, F.J. and S. Gordon, "Calculation of Complex Chemical Equilibria," IEC, 60, 6 (1968).

3. VanZeggeren, F. and S.H. Storey, The Computation of Chemical Equilibria, Cambridge Univ. Press, 1970.
4. Brinkley, S.R., "Note on the Conditions of Equilibrium for Systems of Many Constituents," J. Chem. Phys., 14, 563-4, 686 (1946).
5. Brinkley, S.R., "Calculation of the Equilibrium Composition of Systems of Many Constituents," J. Chem. Phys., 15, 107-10 (1947).
6. Kandiner, H.J. and S.R. Brinkley, "Calculation of Complex Equilibrium Relations," Ind. Eng. Chem., 42, 850 (1950).
7. Browne, H.N., Williams, M.M., and D.R. Cruise, "The Theoretical Computation of Equilibrium Compositions, Thermodynamic Properties and Performance Characteristics of Propellant Systems," NAVWEPS Rept. 7043, NOTS - Tech. Publ. 2434 (June, 1960).
8. Cruise, D.R., "Notes on the Rapid Computation of Chemical Equilibria," J. Phys. Chem., 68, 3797-3802 (1964).
9. Smith, W.R. and R.W. Missen, "Calculating Complex Chemical Equilibria by an Improved Reaction - Adjustment Method," Can. J. Chem. Eng., 46, 269 (1968).
10. Naphtali, L.M., "Computing Complex Chemical Equilibria by Minimizing Free Energy," First Conference on Kinetics, Equilibria and Performance of High Temperature Systems, Butterworth, London, 1959.
11. Ma, Y.H., and C.W. Shipman, "On the Computation of Complex Equilibria," AIChE Journal, 18, No. 2, 299-303 (1972).
12. Heidemann, R.A., "Non-Uniqueness in Phase and Reaction Equilibrium Computations," Chem. Eng. Sci., 33, 1517 (1978).
13. Boston, J.F. and H.I. Britt, "A Radically Different Formulation and Solution of the Single-Stage Flash Problem," Comp. and Chem. Eng., 2, 109 (1978).
14. ASPEN Technical Documentation, ASPEN Project, Dept. of Chem. Eng., M.I.T., 1979.
15. Henley, E.J. and E.M. Rosen, Material and Energy Balance Computations, Wiley (1969).
16. Boston, J.F. and R.L. Fournier, "A Quasi-Newton Algorithm for Solving Multiphase Equilibrium Flash Problems," presented at 71st Annual Meeting of AIChE, Miami, Nov., 1978.
17. Dluzniewski, J.H. and S.B. Adler, "Calculation of Complex Reaction and/or Phase Equilibria Problems," I. Chem. E. Symp. Ser. No. 5, Intn. Chem. Engrs., London, 4:21 (1972).
18. Sanderson, R.V. and H.H.Y. Chien, "Simultaneous Chemical and Phase Equilibrium Calculation," Ind. Eng. Chem. Process. Des. Develop., 12, No. 1 (1973).
19. White III, C.W., ASPEN Technical Documentation for EQUILK, ASPEN Project, Dept. of Chem. Eng., M.I.T., 1979.
20. White, W.B., Johnson, S.M., and G.B. Dantzig, "Chemical Equilibrium in Complex Mixtures," J. Chem. Phys., 28, 751 (1958).
21. Boynton, F.P., "Chemical Equilibrium in Multicomponent Polyphase Systems," J. Chem. Phys., 32, 1880 (1960).

22. Oliver, R.C., Stephanou, S.E. and R.W. Baier, "Calculating Free Energy Minimization," Chem. Eng., 69, 4, 121 (1962).
23. Balzhiser, R.E., Samuels, M.R. and J. Eliassen, Chemical Engineering Thermodynamics, Prentice-Hall, Englewood Cliffs, New Jersey, 1972.
24. Gautam, R. and W.D. Seider, "Calculation of Phase and Chemical Equilibrium. Part III: Electrolytic Solutions," Accepted for publication in AIChE Journal, 1979c.
25. Ibid., "Calculation of Phase and Chemical Equilibrium. Part I: Local and Constrained Minima in Gibbs Free Energy," Accepted for publication in AIChE Journal, 1979a.
26. Ibid., "Calculation of Phase and Chemical Equilibrium. Part II: Phase-Splitting," Accepted for publication in AIChE Journal, 1979b.
27. Gordon, S. and B.J. McBride, "Computer Program for Calculation of Complex Chemical Equilibrium Compositions, Rocket Performance, Incident and Reflected Shocks, and Chapman-Jouguet Detonations," NASA SP-273 (1971).
28. George, B., Brown, L.P., Farmer, C.H., Buthod, P. and F.S. Manning, "Computation of Multicomponent Multiphase Equilibrium," Ind. Eng. Chem. Process Des. Dev., 15, No. 3 (1976).
29. Powell, M.J.D., "A New Algorithm for Unconstrained Optimization," in Nonlinear Programming, Academic Press, Inc., New York (1970).
30. Hadley, G., Non-Linear and Dynamic Programming, Addison-Wesley, Massachusetts, 1964.
31. Kunzi, H.P., Krelle, W. and W. Oettli, Non-Linear Programming, Blaisdell, Waltham, Massachusetts (1966).
32. Sargent, R.W.H. and B.A. Murtagh, "Projection Methods for Non-linear Programming," Math. Prog., 4, 245 (1973).
33. Whitwell, J.C. and S.R. Dartt, "Independent Reactions in the Presence of Isomers," AIChE Journal, 19, No. 6 (1973).
34. Bjornbom, P.H., "The Independent Reactions in Calculations of Complex Chemical Equilibria," Ind. Eng. Chem. Fund., 14, No. 2 (1975).
35. Bjornbom, P.H., "The Relation Between the Reaction Mechanism and the Stoichiometric Behavior of Chemical Reactions," AIChE Journal, 23, No. 3 (1977).
36. Prigogine, I. and R. Defay, "On the Number of Independent Constituents and the Phase Rule," J. Chem. Phys., 15, 614 (1947).
37. Schott, G.L., "Computation of Restricted Equilibria by General Methods," J. Chem. Phys., 40, 2065 (1964).
38. White III, C.W. and W.D. Seider, "Analysis of Chemical Reactors in Process Design," to be presented at 72nd Annual Meeting of AIChE, San Francisco, Nov., 1979.

RECEIVED November 5, 1979.

6

Inside-Out Algorithms for Multicomponent Separation Process Calculations

J. F. BOSTON

Massachusetts Institute of Technology, 20A-015, Cambridge, MA 02139

There are several characteristics common to the describing equations of all types of multicomponent, vapor-liquid separation processes, both single- and multi-stage, that make it possible to exploit the inside-out concept in similar ways to solve them efficiently and reliably. The equations have as common members component and total mass balance, enthalpy balance, constitutive and phase equilibrium equations. In addition, all such processes require K-value or fugacity coefficient and vapor and liquid enthalpy models. In all cases the describing equations have similar forms, and depend on the primitive variables (temperature, pressure, phase rate and composition) in essentially the same ways. Before presenting the inside-out concept, it will be useful to identify two classes of conventional methods and discuss their main characteristics.

Conventional Methods

Conventional methods for solving the coupled set of describing equations and thermo-physical property models are characterized by taking the primitive variables, or some subset of them, as the main iteration variables, and by working with the equations in essentially their "primitive" forms. Many methods have been proposed which may be regarded as conventional methods in this sense. For purposes of this paper, it is convenient to consider all conventional methods as members of one of two classes based on two fundamentally different approaches.

Class I Methods. The methods of the first class are based on tearing and partitioning the system so that subsets of the primitive variables are paired with subsets of the equations through which they typically show their greatest effect. Perhaps the best-known example of this approach for multicomponent distillation is the method of Wang and Henke (1).

0-8412-0549-3/80/47-124-135$05.00/0

Here the main iteration variables are the stage temperatures and inter-stage phase rates. The temperatures are paired with the combined constitutive and phase equilibrium equations, and the phase rates with the enthalpy and total mass balances. Unfortunately, this pairing is effective only for relatively narrow-boiling systems, hence the method frequently fails for wide-boiling systems. Further, the computational procedure involves a lag of the K-value composition dependence from iteration to iteration, which makes the method unsuitable when the composition dependence is strong.

The "sum-rates" method of Sujata (2) also uses the temperatures and phase rates as iteration variables, but reverses the pairing. The temperatures are paired with the enthalpy balances, and the phase rates are corrected by summing the component flow rates resulting from solution of the combined component mass balance and phase equilibrium equations. This method is especially effective for wide-boiling systems, such as absorbers, but is not suitable for narrow-boiling systems.

Tomich (3) presented a Class I method in which the pairing problem is avoided by adjusting the temperatures and phase rates simultaneously at each iteration. The adjustments are determined by considering simultaneously the system of equations consisting of the combined enthalpy and total mass balances, and the combined constitutive and phase equilibrium equations. The Jacobian of this system with respect to both temperatures and phase rates is initially calculated by finite difference approximations, and its inverse is updated thereafter using the quasi-Newton method of Broyden (4). The excessive computational effort associated with initial evaluation and inversion of the Jacobian is a major disadvantage of this method. In addition, there is a composition lag like that of the Wang and Henke method which makes it unsuitable for highly nonideal systems.

Class I methods for single-stage processes have been discussed by several authors (5,6,7). While a detailed discussion will not be given here, it is worth noting that the issues of composition lag and pairing of equations with their dominant variables arise in essentially the same ways, and are as important, as for multi-stage processes.

Class II Methods. The methods of Class II are those that use the simultaneous Newton-Raphson approach, in which all the equations are linearized by a first order Taylor series expansion about some estimate of the primitive variables. In its most general form, this expansion includes terms arising from the dependence of the thermo-physical property models on the primitive variables. The resulting system of linear equations is solved for a set of iteration variable corrections, which are then applied to obtain a new estimate. This procedure is repeated until the magnitudes of the corrections are sufficiently small.

The system Jacobian is full or nearly full for single-stage processes, but has a stage-wise sparse structure for multi-stage processes. In the latter case advantage can be taken of the fact that the sparsity pattern is known a priori to develop efficient solution procedures. In most cases, the Jacobian has a block-tridiagonal structure which can be exploited as first shown by Naphtali and Sandholm (8). Hofeling and Seader (9) have recently presented an efficient sparse algorithm to handle cases in which there are side-strippers or pump-arounds which destroy the block-tridiagonal structure, causing additional "fill-in" during the elimination process. Many other variations of the Newton-Raphson approach have been reported over the past few years (10-20) including applications to three phase distillation (21) and to single-stage processes (22).

One of the most important characteristics of the Newton-Raphson approach is that the convergence rate accelerates, becoming second order, as the solution is approached. This is at the same time an advantage and a disadvantage. If the initial estimates of the iteration variables are close to the solution, convergence may be achieved in relatively few iterations compared with other methods. On the other hand, if they are far from the solution, the convergence rate may be exceedingly slow, and unstable behavior may result unless the vector of corrections is appropriately truncated. Perhaps the most undesirable feature of Newton-Raphson methods is that, when it is not possible to provide sufficiently good initial guesses, there is little recourse but to apply some other type of method.

The Inside-Out Concept

The development of the inside-out concept was motivated by the hope of overcoming the major difficulties of both the Class I and Class II methods without introducing significant new difficulties. The goals were to achieve:

(1) Above all, robustness in the face of exceedingly poor initial guesses.
(2) Efficiency in terms of properly distributing and balancing the computational loads.
(3) Generality in terms of applicability to all types of systems, i.e., narrow-boiling, wide-boiling, highly nonideal.
(4) At least superlinear convergence rates as the solution is approached.

The challenges presented by these goals are to exploit ones knowledge of how the system behaves and how it is structured to select well-behaved iteration variables, and to recognize ways in which the equations can be beneficically rearranged and transformed to accommodate these variables.

The concept was first presented and implemented by Boston and Sullivan (23), who demonstrated the exceptional stability and efficiency of the resulting algorithm for multicomponent distillation problems. The algorithm has since been extended by this author to handle absorption and reboiled absorption, and to make it applicable to highly nonideal systems as well as narrow-boiling and extremely wide-boiling ones. It has also been modified to accommodate water-hydrocarbon systems in which water phases occur on internal stages, and has been successfully adapted to provide a rigorous treatment of both multi-stage (24) and single-stage (25) three-phase systems. Boston and Britt (7) discussed the application of the inside-out concept to the single-stage two-phase flash. This algorithm has recently been modified in two ways. One modification enables it to handle systems operating at near-critical conditions. The other is designed to solve the problem of simultaneous phase and chemical equilibrium. Finally, Boston (26) has shown how the inside-out concept can be adapted to solve multi-stage problems where it is desired to determine a set of design variables within specified bounds to satisfy arbitrary design constraints, or to minimize an arbitrary function subject to equality and inequality constraints.

In all of these implementations of the inside-out concept, there are six important common features. The first three of these will be listed and discussed in detail before considering the last three. They may be stated as follows:

(1) Complex K-value and enthalpy models are used only to generate parameters for simple models. These parameters are unique for each stage of a multi-stage system.

(2) These simple model parameters become the main (or "outer loop") iteration variables, the role played by the primitive variables temperature, pressure, vapor and liquid composition and phase rates in Class I and Class II methods.

(3) The new outer loop iteration variables are relatively free of interaction with each other, and are relatively independent of the primitive variables, hence precise initialization is not critical to good algorithm performance.

<u>Simple K-Value and Enthalpy Models</u>. In general, K-values and enthalpies depend on the primitive variables through a set of models, referred to here as the "actual" thermo-physical property models, as follows:

$$K_i = K_i(T, P, x, y) \quad (1)$$

$$H_L = H_L(T, P, x) \quad (2)$$

$$H_V = H_V(T, P, y) \quad (3)$$

While no assumptions are made in this paper about the specific forms of the actual K-value and enthalpy mdoels, it is assumed that they exhibit realistic behavior.

In developing a simple K-value model, it is first recognized that the dependence of K-values on temperature is represented very well by a model of the form:

$$\ln K_i = A_i + B_i(1/T-1/T^*)$$

where T* is some reference temperature, and B_i is largely independent of T. Because the magnitude of B_i does not vary widely from component to component, even in a wide-boiling system, it is useful to define a reference K-value, K_b, as a weighted average of the form:

$$\ln K_b \equiv \sum_i w_i \ln K_i \qquad (4)$$

The w's are weighting factors, expressions for which have been derived elsewhere (7) to make $d\ln K_b/dT$ an appropriate weighted average of the individual $d\ln K_i/dT$ values.

The dependence of K_b on T is then represented by the simple model:

$$\ln K_b = A + B(1/T-1/T^*) \qquad (5)$$

where T* is a reference temperature. Just as the K's are strongly dependent on T, and for a highly nonideal system, on x and y as well, K_b will also exhibit a strong dependence on these primitive variables. However, the coefficients A and B will be only weakly dependent on T. Furthermore, since the T dependence of the K's is usually not strongly affected by composition, B will also be only weakly dependent on x and y. As a result of these weak dependencies, A and B are excellent iteration variables. The values of A and B may be determined by evaluating the K's at two temperature levels, while holding x and y constant.

The simple K-value model is completed by defining a set of relative volatilities:

$$\alpha_i \equiv K_i/K_b \qquad (6)$$

which are much less sensitive to T than the individual K's because a weighted average temperature dependence is included in K_b. The α's, therefore, are also good iteration variables.

Simple vapor and liquid enthalpy models may be developed by recognizing that, on a mass basis, vaporization enthalpies and vapor and liquid heat capacities do not vary widely from component to component, and the latter are relatively independent of temperature. Enthalpy departure functions in mass units are first defined as follows:

$$\Delta \underline{H}_V \equiv \underline{H}_V - \underline{H}_V^o \quad (7)$$

$$\Delta \underline{H}_L \equiv \underline{H}_L - \underline{H}_L^o \quad (8)$$

where $\underline{H}_V^o$ and $\underline{H}_L^o$ are ideal gas enthalpies:

$$\underline{H}_V^o \equiv (\sum_i y_i \; H_i^{IG})/M_V \quad (9)$$

$$\underline{H}_L^o \equiv (\sum_i x_i \; H_i^{IG})/M_L \quad (10)$$

Note that $\Delta \underline{H}_V$ and $\Delta \underline{H}_L$ both have units of energy per unit mass rather than per mole. The temperature dependence of these functions is then represented by the simple models:

$$\Delta \underline{H}_V = C + D(T-T^*) \quad (11)$$

$$\Delta \underline{H}_L = E + F(T-T^*) \quad (12)$$

where T* is a reference temperature. The parameters D and F represent mean residual heat capacities for the vapor and liquid mixtures, respectively, over the temperature range from T* to T. C represents the vapor enthalpy departure at T*, and D the liquid mixture vaporization enthalpy (relative to the ideal gas) at T*. Since all four parameters are in mass units, they are relatively insensitive to composition as well as temperature. Hence they are very well-behaved iteration variables.

Modification of Describing Equations. In the simple K-value and enthalpy models just described, there are N_c+6 parameters, α, A, B, C, D, E and F, which have characteristics that make them excellent choices for the iteration variables of an outer iteration loop. This is in fact the essence of the inside-out concept. Its success, however, rests on the ability to transform and rearrange the describing equations to properly accommodate these variables. The result should be an efficient and well-behaved inner iteration loop in which values of the primitive variables - now regarded as dependent variables - are calculated. When the inside loop is converged, the actual K-value and enthalpy models can be employed to calculate new values of the simple model parameters.

These ideas are embodied in the last three common features of the inside-out concept, which may be stated as follows:

(4) The describing equations are expressed in terms of the simple models, and are rearranged in a novel way so that a complete solution for the primitive variables is possible. In most cases this is achieved by converging a N-dimensional inner iteration loop, where N is the number of stages. In multi-stage applications, this inner loop is particularly amenable to solution by

Broyden's quasi-Newton method, using the identity matrix as the initial Jacobian.

(5) The inner loop iteration variable for each stage is a unique combination of temperature and phase ratio which eliminates the need to make a distinction between wide- and narrow-boiling systems. In certain cases there are additional inner loop variables.

(6) The primitive variables resulting from the solution of the inner loop for a given set of values of the outer loop variables are used to calculate a new set of the latter through the actual K-value and enthalpy models. The entire problem is solved when these calculated values match the corresponding assumed values. In single-stage applications, convergence of the outer loop is accelerated using the Broyden quasi-Newton method with the identity matrix as the initial Jacobian. In multi-stage applications, it is converged either by direct substitution or by the bounded Wegstein method (27).

It is most convenient to illustrate these features using an adiabatic, single-stage flash with specified pressure. The describing equations in their primitive forms are as follows:

Phase equilibrium: $$y_i = K_i x_i \quad (i=1,2,\ldots,N_c) \tag{13}$$

Component mass balance: $$x_i L + y_i V = z_i F \quad (i=1,2,\ldots,N_c) \tag{14}$$

Total mass balance: $$L + V = F \tag{15}$$

Enthalpy balance: $$H_L L + H_V V = H_F F + Q \tag{16}$$

Constitutive: $$\sum_i x_i - \sum_i y_i = 0 \tag{17}$$

When z, F, H_F, Q and P are known, there are $2N_c+3$ of these equations, and $2N_c+3$ primitive variables: x, y, T, L and V. K, H_L and H_V are directly related to the primitive variables through the models, eqs. (1), (2) and (3), and therefore do not constitute additional independent variables. The basis for the choice of the form of eq. (17) is discussed elsewhere (7).

To facilitate rearrangement of these equations, the following new variables are introduced:

$$R \equiv K_b V/(K_b V+L) \tag{18}$$

$$p_i \equiv l_i/(1-R) \tag{19}$$

Then eqs. (6) and (13)-(19) can be combined and rearranged to give:

$$p_i = f_i/(1-R+R\alpha_i) \tag{20}$$

$$K_b = \sum_i p_i / \sum_i \alpha_i p_i \tag{21}$$

$$x_i = p_i/\Sigma_j p_j \tag{22}$$

$$y_i = \alpha_i p_i/\Sigma_j \alpha_j p_j \tag{23}$$

$$\underline{L} = M_L(1-R)\Sigma_i p_i \tag{24}$$

$$\psi \equiv (\underline{L}/\underline{F})(\underline{H}_V-\underline{H}_L) + \underline{H}_F - \underline{H}_V + Q/\underline{F} = 0 \tag{25}$$

Single-Stage Inside-Out Algorithm. Table I shows how the transformed and rearranged equations can accommodate the new iteration variables in an integrated algorithm having all six of the features discussed above.

The basis for holding B, D, and F constant after the first outer loop iteration, as indicated in steps (9) and (10), may be explained by noting that the correctness of the final solution requires only that α and one each of the pairs (A,B), (C,D), (E,F), be updated, and that the calculated values of these variables match the assumed values. This is not required of the remaining three variables, whose values may be fixed. The significance of this observation may be explained by considering the variable B. When B is updated, two evaluations of K are required, whereas if only α and A are updated, a single evaluation is sufficient. Since B is much less temperature and composition dependent than α and A, little is lost by not updating B, and much is gained in reduction of the computational load associated with K-value calculation. Similar remarks may be made with regard to the updating of D and F, with corresponding implications in the reduction of the required number of vapor and liquid enthalpy evaluations.

It is notable that the variable R is a unique combination of temperature (through K_b) and phase ratio which both reduces the inner loop to a single variable convergence problem, and precludes the necessity of switching roles depending on whether the system is wide- or narrow-boiling. As R changes for a wide-boiling system, the calculated L and V will remain nearly constant, but the calculated T will be relatively sensitive. In effect, then, the enthalpy balance error will be driven to zero by the change in temperature. For a narrow-boiling system, on the other hand, the calculated T will be relatively constant, but the calculated L and V will respond almost linearly to changes in R. In this case, the phase ratio will dominate in driving the enthalpy balance error to zero. For systems that are neither wide- nor narrow-boiling, the relative dominance of the temperature and phase ratio will change automatically depending on the nature of the system. This behavior was demonstrated by Boston and Britt (7).

Expressing the enthalpy balance function ψ, eq. (25), in mass rather than mole units precludes the necessity of using Holland's (28) "constant composition" formulation. While $\underline{H}_V$ and $\underline{H}_L$ depend on the compositions of different phases, the fact that they are in mass units means that their difference will not be close to zero, even though the molar enthalpy difference H_V-H_L could be. The loss of sensitivity of ψ to changes in R is thereby avoided.

The performance of the algorithm outlined in Table I has been illustrated using a variety of example problems, including highly nonideal ones, by Boston and Britt (7) and Boston and Fournier (25). In addition, Boston and Britt have shown how to modify the algorithm to handle combinations of specified quantities other than pressure and heat duty.

Multi-Stage Inside-Out Algorithms. The multicomponent distillation algorithm reported by Boston and Sullivan (23) is conceptually the same as the single-stage algorithm just described. While there are several differences in detail, the main differences are those indicated in the six common features of the inside-out concept outlined above. Boston and Sullivan illustrated the performance of the algorithm using a variety of example problems, including several with extremely wide-boiling feeds. The performance was compared with that of Tomich's (3) conventional algorithm and found to be superior.

Only a few modifications of the algorithm were required to make it applicable to absorption and reboiled absorption. The changes were mainly in the handling of the enthalpy and total mass balance equations to accommodate different specification combinations involving the reflux, heat duties, and top and bottom product flow rates. The results of two example problems, one each for absorption and reboiled absorption, are shown in Table II.

Application to Highly Nonideal Systems. When the inside-out algorithms discussed so far are applied to systems with strong liquid phase nonidealities, they might be expected to behave poorly. The absence of any form of composition dependence in the simple K-value model would seem to be a violation of the inside-out concept that would seriously impair peformance for highly nonideal systems. That this is not so for the single-stage algorithm discussed above may be rationalized by considering the characteristics of the outer loop convergence problem.

Because the calculated values of the parameters (step (9) of Table I) are relatively insensitive to changes in the assumed values of these parameters, the Jacobian of the outer loop tends to be diagonally dominant with diagonal elements not greatly different from -1. The main effect of system nonideality is to increase the interaction among the parameters, thereby increasing the magnitudes of the off-diagonal Jacobian elements. However, these effects are seldom so great as to eliminate diagonal dominance. Further, the Broyden quasi-Newton algorithm is capable of building up information to approximate the off-diagonal

TABLE I. Inside-Out Adiabatic Single-Stage Flash Algorithm

(1) Assume α, A, B, C, D, E, F

→ (2) Assume R

→ (3) Calculate p, K_b, T, x, y using eqs. (20), (21), (5), (22), (23), respectively.

(4) Calculate M_L, M_V, H^{IG}

(5) Calculate $\underline{H}_V^o$, $\underline{H}_L^o$, $\Delta\underline{H}_V$, $\Delta\underline{H}_L$, $\underline{H}_V$, $\underline{H}_L$ using eqs. (9), (10), (11), (12), (7), (8), respectively.

(6) Calculate L and ψ using eqs. (24), (25), respectively.

(7) Assume new R and return to (3) until $|\psi| < \varepsilon$

(8) Calculate K, H_L, H_V using actual models.

(9) Calculate $\hat{\alpha}$, $\hat{A}$, $\hat{C}$, $\hat{E}$ (and $\hat{B}$, $\hat{D}$, $\hat{F}$ first iteration only).

(10) Assume new values of α, A, C, E (and B, D, F first iteration only) and return to (2) until calculated values match assumed values. Use Broyden quasi-Newton method after first iteration.

TABLE II. Multi-Stage Algorithm Example Problems

Example type	Components	K-value model	Enthalpy models	No. stages[1]	No. outside loop iterations	Avg. no. inside loop iterations[2]	CPU time[3] (sec.)
Absorption	hydrogen sulfide nitrogen hydrogen methane-pentane (8 components) 442 NBP solvent	Activity coefficients by Scatchard-Hildebrand equation. Liquid fugacity coefficents by Chao Seader equation. Vapor fugacity coefficents by Redlich-Kwong equation.	Vapor enthalpy departure by Redlich-Kwong equation. Liquid enthalpy by corresponding states correlation.	7	7	2.7	2.5
Reboiled absorption	methane-butane (5 components) octane solvent			25	8	3.5	3.6
Nonideal distillation	acetone water	Activity coefficients by Wilson equation. Vapor pressures by Antoine equation. Vapor phase assumed ideal.	Vapor phase assumed ideal. Liquid enthalpy by polynomial in temperature.	20	12	3.8	5.4
Nonideal distillation	acetone chloroform methanol			15	10	2.0	1.7
Nonideal (extractive) distillation	6 C_4-hydrocarbons water furfural			15	9	5.3	12.9

(1) Including condenser and reboiler when present.
(2) Per outside loop iteration.
(3) IBM 360/168.

elements well enough in relatively few iterations. Therefore, not including composition effects in the simple K-value model actually represents a trade-off between the effects that the simple models ought to account for and the effects that the convergence procedure ought to account for.

In multi-stage systems, on the other hand, strong nonideality is a more serious problem. In the first place, it is not feasible to use Broyden's quasi-Newton method for convergence of the outer loop because there are too many parameters ($N(N_c+3)$), and the Jacobian is in general not sparse. The outer loop is therefore converged by either the bounded Wegstein method (27), or by direct substitution with moderate damping. Secondly, because interactions among the parameters tend to propagate over several stages, strong nonidealities frequently lead to poor convergence behavior or even failure to converge.

The difficulties associated with highly nonideal multi-stage systems have been overcome by introducing a simple model for the composition dependence of K-values. In keeping with the spirit of the inside-out concept, the parameters of the simple model become outside-loop iteration variables, and are determined by applying the actual models only in the outer loop. Further, they are as independent as possible of the primitive variables.

The basis of this approach is the representation of the K-value in terms of three factors as follows:

$$K_i = K_b \alpha_i \gamma_i^* \quad (26)$$

where γ_i^* is a new quantity introduced to account for the composition dependence. For simplicity, the most frequent case is considered, where most of the nonideality arises from the liquid phase, and is represented by an activity coefficient model. In this case, γ_i^* is a pseudo-activity coefficient which is obtained from the actual activity coefficent model at a reference temperature T*:

$$\gamma_i^* = \gamma_i(T^*,x)$$

The objective is to substantially reduce the composition dependence of the α's by using a simple model for γ_i^* in terms of parameters that will be well-behaved iteration variables. The treatment of K_b is the same as before, therefore α_i, which is now defined as follows:

$$\alpha_i \equiv K_i/K_b\gamma_i^* \quad (27)$$

should depend weakly on both temperature and composition.

The composition dependence of γ_i^* is based on the treatment of each component in turn as a pseudo-binary with all the other components. For each component, then, a simple model of the form:

$$\ln \gamma_i^* = a_i + b_i x_i \tag{28}$$

is developed. The parameters a_i and b_i are determined from the actual activity coefficient model at T^* with the distribution of all components other than i held constant. A unique set of these parameters is generated for each stage in a multi-stage system. The b's are handled in much the same way as the parameters B, D and F, being updated only once or until the change from iteration to iteration is within some relatively large tolerance. The a's are treated in the same way as the α's, A, C and E.

To show how the inside loop can be revised to accommodate the simple composition dependence model, the single-stage case is considered for simplicity.
The revised form of eq. (20) is:

$$(1-R)p_i + R\alpha_i \gamma_i^* p_i = f_i \tag{29}$$

and the simple model for γ_i^* can be written as:

$$\ln\gamma_i^* = a_i + b_i p_i / p^\sigma \tag{30}$$

$$p^\sigma \equiv \sum_i p_i \tag{31}$$

For a given value of R, eqs. (29), (30), and (31) can be solved readily for p_i and p^σ using the Newton-Raphson method. It is easy to arrange the linearized equations so that the coefficient matrix is upper triangular with only one non-null column above the diagonal. The solution therefore does not add appreciably to the computational load of the inside loop.

The multi-stage algorithm may be modified to incorporate the approach just discussed in much the same ways as the single-stage algorithm. The modifications were found to provide a dramatic improvement in performance, enabling solutions to be obtained in cases in which the previous algorithm failed to converge. Results are given for three highly nonideal example problems in Table II.

Application to Simultaneous Phase and Chemical Equilibrium. The single-stage process with simultaneous phase and chemical equilibrium is another application of the inside-out concept where the Newton-Raphson method has been employed in a judicious way in the inside loop. There would appear to be no reaction parameter having characteristics that make it suitable as an outside loop iteration variable in the spirit of the inside-out concept. On the other hand, the chemical equilibrium relationships are simple in form, and do not introduce new thermophysical properties that depend in a complicated way on other variables. Thus it makes sense to include them in the inside loop, and to introduce the reaction extents as a new set of inside loop variables.

Since chemical equilibrium expressions involve fugacity coefficients, rather than K-values, it is necessary to introduce a new set of outside loop variables. The K-value is factored as follows:

$$K_i = K_b \beta_i \theta_i \tag{32}$$

where

$$\beta_i \equiv \phi_i^L / K_b \tag{33}$$

$$\theta_i \equiv 1/\phi_i^V \tag{34}$$

and the parameters β and θ now replace α as outside loop iteration variables. Because ϕ_i^V is ordinarily not strongly dependent on T, P or y, and β_i is the same as $\alpha_i \phi_i$, both β and θ should be well-behaved.

To show how the inside loop may be modified to handle chemical equilibrium, consider for simplicity the case of specified pressure, and reaction occurring only in the liquid phase. For the jth reaction the equilibrium condition may be expressed as:

$$\sum_i \nu_{ij} \ln(\phi_i^L x_i) = g_j(T,P)$$

$$g_j \equiv -\frac{1}{RT} \sum_i \nu_{ij} G_i^{IG} - (\sum_i \nu_{ij}) \ln P/P_{ref}$$

Substituting eqs. (22), (31) and (32) we have:

$$\sum_i \nu_{ij} \ln(K_b \beta_i p_i / p^\sigma) = g_j(T,P) \tag{35}$$

The revised form of the component mass balance, eq. (20), is:

$$(1 - R + R\beta_i \theta_i) p_i = f_i + \sum_j \nu_{ij} \eta_j \tag{36}$$

and eq. (21) becomes:

$$K_b \sum_i \beta_i \theta_i p_i = p^\sigma \tag{37}$$

where p^σ is as defined by eq. (31).

For a given value of R, eqs. (5), (31), (35), (36) and (37) may be solved simultaneously by the Newton-Raphson method for the η's, p's, p^σ, K_b and T. Note that eq. (36) relates p_i explicitly to a subset of the η's, hence the number of variables and equations can easily be reduced to N_R+3, where N_R is the number of reactions.

The performance of an algorithm using the approach just described is illustrated in Table III for an example involving an esterification reaction. The performance is compared with that of the conventional algorithm of Sanderson and Chien (29).

TABLE III. Single-Stage Simultaneous Phase and Chemical Equilibrium Example Problem

Chemical reaction:
ethanol + acetic acid ⇌ ethyl acetate + water

Thermo-physical property models taken from Sanderson and Chien (29).

Feed composition:	50 mole per cent each of ethanol and acetic acid.
Specifications:	$P = 1.013 \times 10^5$ N/m^2, V/F = 0.5
Initial guesses:	T = 290 K, η = 0
Final values:	T = 333.7 K, η = 0.4206

Number outside loop iterations: 6

Note: Sanderson and Chien (29) reported 10 outside loop iterations using their conventional algorithm for the same system at the same pressure, but with T specified as 358 K. The number of inside loop iterations was not given. It is notable that each inside loop iteration of their algorithm requires that properties be calculated using the actual thermo-physical property models.

List of Symbols

A,B	K-value or K_b temperature dependence parameters
a,b	Activity coefficient composition dependence parameters
C,D	Vapor enthalpy temperature dependence parameters
E,F	Liquid enthalpy temperature dependence parameters
F	Total feed molar flow rate
$\underline{F}$	Total feed mass flow rate
f	Component feed molar flow rate
G	Gibbs free energy
H	Enthalpy, energy/mole
$\underline{H}$	Enthalpy, energy/mass
K	K-value
K_b	Weighted average K-value
L	Total liquid molar flow rate

$\underline{L}$	Total liquid mass flow rate
l	Component liquid molar flow rate
M	Mixture molecular weight
N_C	Number of components
N_R	Number of reactions
P	Pressure
p	Defined by eq. (19)
Q	Heat duty
R	Defined by eq. (18)
T	Temperature
V	Total vapor molar flow rate
w	K_b weighting factor
x	Liquid mole fraction
y	Vapor mole fraction
z	Feed mole fraction

Subscripts

i	component index
j	component index; reaction index
L	liquid phase
V	vapor phase
F	feed
ref	ideal gas reference state

Superscripts

L	liquid phase
V	vapor phase
σ	summation over components
o	ideal gas mixture
IG	ideal gas
*	evaluation at reference temperature, T*

Greek letters

α	Relative volatility
β	Defined by eq. (33)
γ	Activity coefficient
ε	Tolerance
η	Reaction extent
θ	Defined by eq. (34)
ν	Stoichiometric coefficient
ϕ	Fugacity coefficient
ψ	Defined by eq. (25)

Literature Cited

1. Wang, J.C.; Henke, G.E. Hydr. Proc., 1966, 45 (8), 155.
2. Sujata, A.D. Hydr. Proc. Petr. Ref., 1961, 40 (12), 137.
3. Tomich, J.R. AIChE J., 1970, 16 (2), 229.
4. Broyden, C.G. Math. Comp., 1965, 19, 577.
5. Henley, E.J.; Rosen, E.M. "Material and Energy Balance Computations"; John Wiley & Sons: New York, 1969.
6. King, C.J. "Separation Processes"; McGraw-Hill: New York, 1971.
7. Boston, J.F.; Britt, H.I. Comp. and Chem. Eng., 1979, 2, 109.
8. Naphtali, L.M.; Sandholm, D.P. AIChE J., 1971, 17, 148.
9. Hofeling, B.S.; Seader, J.D. AIChE J., 1978, 24, 1131.
10. Gentry, J.W. Can. J. Chem. Eng., 1970, 48, 451.
11. Goldstein, R.P.; Stanfield, R.B. Ind. Eng. Chem. Proc. Des. Dev., 1970, 9, 78.
12. Roche, E.C., Jr. Brit. Chem. Eng., 1971, 16, 821.
13. Harclerode, H.; Gentry, J.W. Can. J. Chem. Eng., 1972, 50, 253.
14. Ishii, Y.; Otto, F.D. Can. J. Chem. Eng., 1973, 51, 601.
15. Hutchison, H.P.; Shewchuk, C.F. Trans. Instn. Chem. Eng., 1974, 52, 325.
16. Kubicek, M.; Hlavacek, V.; Prochaska, F. Chem. Eng. Sci., 1976, 31, 277.
17. Brannock, N.F.; Verneuil, V.S.; Wang, Y.L. Chem. Eng. Prog., 1977, 73, 83.
18. Browne, D.W.; Ishii, Y.; Otto, F.D. Can. J. Chem. Eng., 1977, 55, 307.
19. Shewchuck, C.F. Trans. Instn. Chem. Eng., 1977, 55, 130.
20. Magnussen, T.; Michelsen, M.L.; Fredenslund, A. Paper presented at Distillation Symposium '79, The Institution of Chemical Engineers (London), Symp. Ser. 56, 1979.
21. Block, U.; Hegner, B. AIChE J., 1976, 22 (3), 582.
22. Robinson, D.B.; Peng, D.-Y. Paper presented at AIChE 70th Annual Mtg., New York, Nov. 13-17, 1977.
23. Boston, J.F.; Sullivan, S.L., Jr. Can. J. Chem. Eng., 1974, 52, 52.
24. Boston, J.F.; Shah, V.S. Paper presented at AIChE 86th National Mtg., Houston, April 1-5, 1979.
25. Boston, J.F.; Fournier, R.L. Paper presented at AIChE 71st Annual Mtg., Miami Beach, Nov. 12-16, 1978.
26. Boston, J.F. Paper presented at AIChE 86th National Mtg., Houston, April 1-5, 1979.
27. Kliesch, H.C. Ph.D. Dissertation, Tulane University: New Orleans, Louisiana, 1967.
28. Holland, C.D. "Multicomponent Distillation"; Prentice-Hall: Englewood Cliffs, N.J., 1963.
29. Sanderson, R.V.; Chien, H.H.Y. Ind. Eng. Chem. Proc. Des. Dev., 1973, 12, 81.

RECEIVED November 5, 1979.

7

A General Approach to Classify Operational Parameters and Rectify Measurement Errors for Complex Chemical Processes

G. STEPHANOPOULOS and J. A. ROMAGNOLI

Department of Chemical Engineering and Materials Science,
University of Minnesota, Minneapolis, MN 55455

The reliability of the process measurements' data is extremely important for good monitoring, control and optimization of chemical process. On-line rectification of a measurement error is possible be it a random error or a gross bias, if additional information is available. Such information is supplied by the extent to which the material and energy balances are satisfied by the recorded data. These balances are simple, involve parameters usually well known, and they should be satisfied independently of the measurements accuracy.

Previous works have attempted to resolve the above problem through the solution of a least squares problem, using linear balance equations. Swenker [1] was among the first who employed this idea, which was further developed by Hoffman [2] and Vaclaveck [3]. Almasy and his coworkers [4] outlined an iterative solution for the second order balance equations. They also applied the algorithm to linear dynamic system (Gertler, Almasy, [5]) and they extended the deterministic methods of balance error smoothing to stochastic system (Sztano, Almasy, [6]). Furthermore, Umeda et al. [7] presented a method for the plant data analysis and parameter estimation which relies on a general criterion for fitting multiresponse data suggested by Box and Draper [8]. All these algorithms yield to the Markov estimation of the system variables which correspond to the maximum likelihood estimation for errors with normal distribution and zero mean. The large scale of the correction problem involved in a chemical process make the above approaches cumbersome. This last feature motivated Vaclavek [3,9] to attempt to reduce the size of the least square problem through the classification of the measured and unmeasured process variables. Such classification allow the size reduction of the initial problem and its easier solution. A similar approach was undertaken by Mah et al[12] in their attempt to organize the analysis of the process data and systematize the estimation and measurement correction problems.

Of particular interest is the identification of biases and gross errors in the measurements. Vaclaveck and Vosolsobe [11],

0-8412-0549-3/80/47-124-153$05.50/0

Almasy and Sztano [6] and Mah and his coworkers [12] have dealt with this problem and developed structural or probabilistic rules that will determine the location of the gross error. A throuth review of the related problems and the proposed solutions can be found in [15]. When all the measurements are corrected, then they can be used to estimate the value of the variables which are not directly measurable. Such situation entails the solution of a nonlinear estimation problem, in general.

In the present work we will deal with all the above problems and provide a unified framework to deal with the error correction for static or dynamic systems using multicomponent mass and energy balances. The topological character of the complex process is exploited for an easy classification of the measured and unmeasured variable independently of the linearity or nonlinearity of the balance equations.

Statement of the Problem

Consider a chemical process containing K units. The multicomponent steady-state mass and energy balances are represented by

$$\sum_{j=1}^{J} a_{ik}\, u_j = 0 \qquad \text{for } k = 1, 2, \ldots K \tag{1}$$

where u_j is the quantity that is balanced. a_{jk} is the element of the process incidence matrix which denotes the topology of units and streams. $a_{jk} = 1$ if stream j is an input to unit k, $a_{jk} = -1$ if j is an output stream, and $a_{jk} = 0$ if stream j is not associated with unit k. J is the set of all streams in the process. Chemical reactions can be easily incorporated through artificial input and output streams. In matrix form equation (1) yields

$$A\, u = 0 \tag{2}$$

The vector u is an n+m dimensional vector which can be partitioned into two vectors; the n-dimensional vector x of measured parameters and the m-dimensional vector of unmeasured ones. Some of the unmeasured variables can be evaluated from the measurement of the others variables using the balance equations, and some not. Thus, the unmeasured parameters may be classified as "determinable" or "indeterminable". On the other hand, some of the elements of vector x of measured variables can be computed from the balances and the rest of the measured parameters. Such measured variables will be called "overdetermined". The rest of the elements of vector x will be called "just determined". Measurement of x is denoted by $\tilde{x}$, and the difference of any measured system parameter and its true value is called the "error" denoted by δ, i.e.

$$\delta = x - \tilde{x} \tag{3}$$

The estimated value of x is denoted by $\hat{x}$ and the difference between any estimate and its measured value $\tilde{x}$ is called the "correction" for that parameter and is represented by $\hat{\delta}$, i.e.

$$\hat{\delta} = \hat{x} - \tilde{x} \tag{4}$$

In general practical situations the mass and energy balances do not yield always linear expressions like equation (2). For example, one device serves for the measurement of mass flow rate and separate analyzers are used to determine the compositions. Consequently the balance may result in bilinear forms as shown in equation (5).

$$\sum_{j=1}^{J} a_{jk} M_j \xi_{ji} = 0 \qquad \begin{array}{l} i = 1,2,\ldots,I + j \\ k = 1,2,\ldots, K \end{array} \tag{5}$$

where, ξ_{ji} for $i = 1,2,\ldots I$ are the concentrations of the I components in stream j, and $\xi_{j(I+1)}$ is the enthalpy of stream j, with M_j its flowrate.

In the subsequent sections the following problems will be tackled:

- Classify the measured variables into overdetermined and just determined.
- Classify the unmeasured variables into determinable and indeterminable.
- Define the subset of balance equations which should be used for the rectification of the measurement errors.
- Rectify the measurement errors.
- Check the presence of gross biased errors in the measurements and identify their sources.

The above development can be extended to the dynamic systems. The dynamic equivalent of equations (2) and (5) are:

$$\dot{w} = Gw(t) + Br(t) \tag{6}$$

$$v = Cw(t) + Dr(t) \tag{7}$$

where, w is the vector of state variables, r the vector of input variables and v the vector of output variables. Discretize equations (6) and (7) in time and introducing the shift-operator z. Eliminate the state variables and develop an input-output relationship

$$v = [\ C(I - z^{-1}\Phi)^{-1}\ z^{-1}\ \Psi + D\ ]\ r \tag{8}$$

where, Φ is the transition matrix and $\Psi(\Delta t) = \left[\int_{t-\Delta t}^{t} \Phi(t-\tau)d\tau\right] B$

Expanding the inverse matrix [5] appearing in the above equation as a fraction of polynomial of z^{-i}, leads to the input-output relationship of the form

$$v = \frac{\sum_{i=0}^{n} R_i z^{-i}}{\sum_{i=0}^{n} h_i z^{-i}} r \qquad (9)$$

where, $R_i = C F_i \Psi Dh_i$ with $F_0, F_1, \ldots F_{n-1}$ n x n matrices with constant elements and coefficients F_i^{n-1} and h_i depend on elements of matrix Φ with the definition $F_n = 0$. We can see that in each equation both the input and the output are represented by their present values and as many previous values as the order of the system, i.e.

$$h_o v(k) + \sum_{i=1}^{n} z^{-i} h_i v = R_o r(k) + \sum_{i=1}^{n} z^{-i} R_i r \qquad (10)$$

Defining the vector $e = \sum_{i=1}^{n} z^{-i} h_i v - \sum_{i=1}^{n} z^{-i} R_i r$ and lumping into the same vector u we have the following balances equations

$$K u + e = 0 \qquad (11)$$

where, $u = \begin{bmatrix} r \\ v \end{bmatrix}$ $K = \begin{bmatrix} R_o ; \cdot h_o I \end{bmatrix}$ and $e = \sum_{i=1}^{n} z^{-i} K_i u.$

Equation (11) represents the time-discrete dynamic equivalent of the steady-state balance equations (2). The dynamic balance equations (11) present some characteristic properties of the sampled-data input and output relationship, that are not present in the corresponding steady-state equations: 1) There are as many equations as the number of outputs 2) Each equation contain only one output 3) Each equation contain, except for special cases, all the inputs variables.

Classification of the Operational Parameters

Starting with the balances as they are given by eq. (2) for static system (or equation (11) for dynamic systems), we can partition matrix A into A_1 and A_2 (or equivalent partition K into K_1 and K_2), in such a way that eq. (2) becomes

$$A_1 x + A_2 y = 0 \qquad (12)$$

Where, x and y are the measured and unmeasured parameters respectively, and A_1 and A_2 are compatibles matrices. The topology of the balance equations is represented by the structure of these matrices. In order to classify the parameters one must first establish what information each equation is to supply, that is, to obtain the output set assignment for the balance equations. With the output set assignment we assign to any unmeasured parameter one equation and it will produce a directed graph, as is shown in Figure 1 and 2.

Then for the unmeasured parameters we will have the following two cases:

a. If an unmeasured parameter is assigned as output to one equation, then it is determinable.
b. If the unmeasured parameter is not assigned, then it is not determinable.

The output set assignment is not unique but whether an unmeasured parameter is determinable or not does not depend on it. As Steward (13) has proved, all the unmeasured variables will be assigned to every possible output set assignment if they are determinables, i.e. if there is no structural singularity. This result is general and does not depend on the functional form of the balance equations. Consequently, except for isolated numerical singularities the determinability of an unmeasured parameter is specified from whether it can be assigned in an output set assignment or not.

After the classification of the unmeasured parameters is carried out, the following task is to classify the measured ones. The overmeasured parameters can be found by the following two step procedure.

First Step. From the output set assignment we obtain equations that are not assigned. This set E of not assigned equations is composed of three subsets of equations.

1. Subset E_1, of equations not assigned which contain only measured parameters.
2. Subset E_2 of not assigned equations which contain unmeasured but determinable parameters (i.e. they have been assigned as output to equations not included in set E).
3. Subset E_3 of equations not assigned which contain unmeasured and indeterminable parameters.

Then, it is easy to prove that the following is true:

i. All measured parameters contained in subset E_1 of unassigned equations, are overmeasured (overdetermined) and will be available for measurement correction.
ii. For steady balances, all the measured parameters (variables) contained in an equation belonging to subset E_2 are overmeasured if this equation does not contain an unmeasured composition (enthalpy). That means we have new independent equations which contain new overmeasured parameters along with the previous ones.
iii. For dynamic balances all the measured parameters contained in an equation belonging to E_2 are overmeasured. That results from the special configuration of the elements in matrix K, which eliminates the probability of numerical singularities.
iv. The subset E_3 does not introduce any new overmeasured parameters.

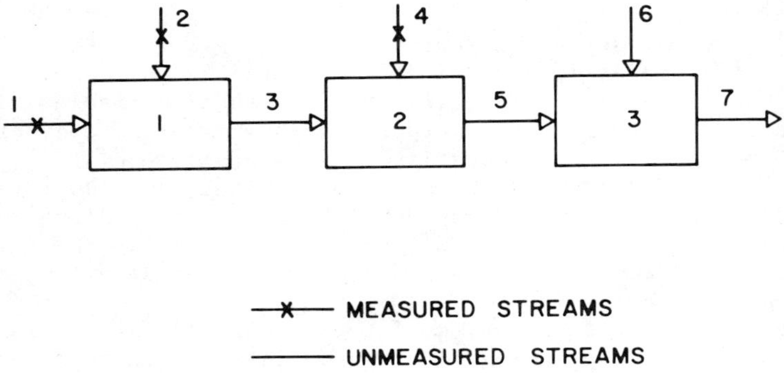

Figure 1. Flow diagram for a serial system

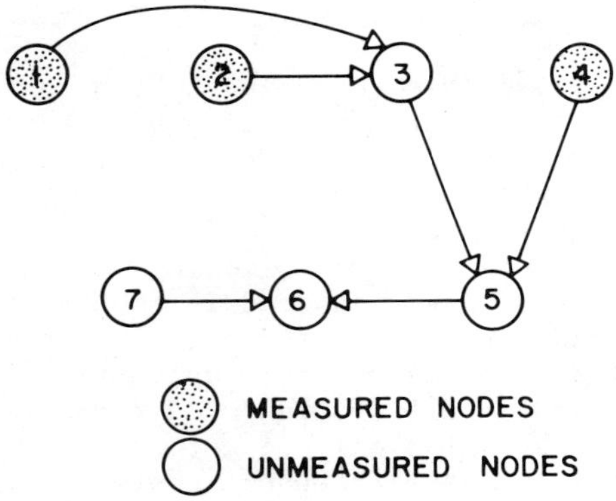

Figure 2. Signal graph representation for the system in Figure 1

Second Step. We identify additional overmeasured parameters by aggregating the balance equations to form a chain of sequentially solvables sets of equations. Let us consider the case shown in Fig. 3. Assume one component in each stream. The flowrates of streams 1,2,5 and 6 are overmeasured as it can be seen from an overall material balance (over the three units) which contains only measured variables. Such overmeasured variables cannot be detected in the first step of the classification procedure for the measured variables described above. Note that the three units in Fig. 3 constitute a disjoint system of a larger complex and they can be solved independently. The same hold for dynamic balances, but note in this case that for the general case of several inputs and outputs we always should have an additional equation belonging to set E_2. Consequently, in this second step we attempt to identify all the disjoint subsystems (such as the one in Fig. 3) and this can be easily done from the occurrence matrix once an output set assignment has been specified. Let us assume that the occurrence matrix is arranged into a block diagonal form

$$D = \begin{bmatrix} D_{11} & 0 \\ & \\ 0 & D_{pp} \end{bmatrix}$$

Where each D_{ii}, $i = 1,2...,p$ represents the topology of each disjoint subsystem. The agregation of the balance equations to form the chain of disjoint subsystems, eliminate the streams among the units of each disjoint system. This leaves only the streams that will connect a particular disjoint system with similar others. These streams together with the streams connected to one unit are called external and the parameters belonging to these streams are called external parameters.

Classification Algorithm. In this section we summarize the various aspects discussed earlier in a systematized algorithmic approach in order to classify the parameters for a given system.

1) Consider all the measurable parameters as measured and the set $\bar{C}$ of the required parameters.
2) Construct the occurrence matrix of the balance equations for the system, dividing the parameters in unmeasured and measured ones.
3) Apply an algorithm to assign the unmeasured parameters, as output of the balance equations. (See appendix A)
4) Classify the unmeasured parameters (set $\bar{U}$):
   a) The unmeasured parameters which are assigned as outputs in an output set assignment are defined as determinables and constitute the elements of the set $\bar{U}_1$
   b) The remaining unmeasured parameters which have not been assigned, are not determinables and constitute the elements

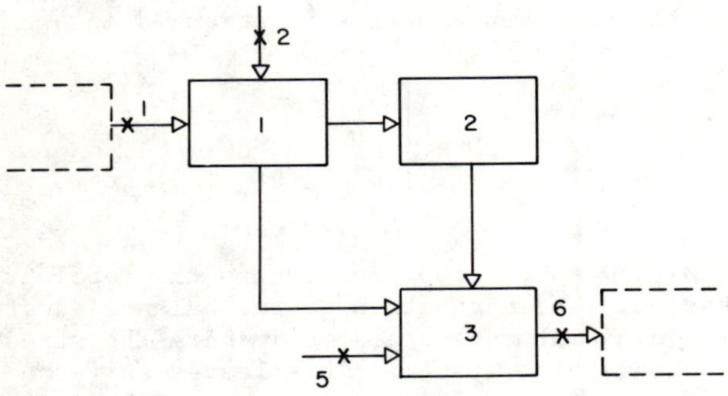

Figure 3. Disjoint subsystem

of set $\bar{U}_2$.

5) Classify the unassigned equations (set E):
   a) Unassigned equations which do not contain unmeasured parameters form the set E_1
   b) The subset E_2 is formed by unassigned equations which contain unmeasured but determinable parameters.
   c) For multicomponent mass and energy static balances, form the subset E_2' with the unassigned component or energy balance, which do not contain unmeasured compositions or enthalpies respectively.
   d) The unassigned equations which contain undeterminable parameters form the subset E_3.
6) Classify the measured parameters.

Let the set $\bar{M}$ represent all the measured parameters. We want to form the set $\bar{M}_1$ (overdetermined parameters) and $\bar{M}_2$ (just determined).

For steady balances

a) A measured parameter belongs to set $\bar{M}_1$ if it is contained in any equation of sets E_1 and E_2'.
b) Find all the possible disjoint subsystems D_{ii}.
   - If a given D_{ii} contain at least one external unmeasured parameter, then this disjoint subsystem does not add new parameters to $\bar{M}_1$. Repeat for all D_{ii}'s.
   - If a given disjoint subsystem has only measured external parameters then they are all overmeasured, and they are incorporated in the set $\bar{M}_1$.
c) Repeat parts a) and b) considering only flow rates, compositions and enthalpy separately.

For dynamic balances

a) A measured parameter belong to set $\bar{M}_1$ if it is contained in any equation of sets E_1 and E_2.
b) All the measured parameters in the considered D_{ii}, belong to set $\bar{M}_1$ if the corresponding D_{ii} contain at least one equation of set E_2.
c) The remaining measured parameters constitute the elements of set $\bar{M}_2$, i.e. the justdetermined parameters.

Selection of the Necessary Measurements for the Required Parameters to be Determinables. It follows from the above development that the parameters in set $\bar{M}_1$ are redundant and can be reconciled. Also the parameters in set $\bar{M}_2$ are not redundant and there is no other indirect way of obtaining their values. Hence this classification of parameters can serve to test a given design of measurement places. There is however another important task concerning this problem. We want ot present an algorithm to select the necessary measurement for a set of required parameters to be completely determinable. Let $\bar{C}$ be this set of parameters which for various reason should be known. $\bar{C}$ may be composed of measured and unmeasured parameters. The necessary measurements can be found using the following scheme:

1) Check if the disjunction $\overline{C} \cap \overline{U}_2$ is an empty set. If this condition is not satisfied the problem is not solvable and has to be reformulated.
2) Consider the set $\overline{N}_x$ of measured and required parameters, i.e. $\overline{N}_x = \overline{M} \cap \overline{C}$
3) Identify the "interval" of the required unmeasured parameters ($\overline{N}_y$) (see Identification algorithm).
4) Check if $\overline{N}_y \cap \overline{U}_1 \neq 0$, if it is incorporate these parameters to set $\overline{N}_m$ and eliminate them from $\overline{N}_y$.
5) If $\overline{N}_y \cap \overline{U}_1 = \phi$, identify the set of parameters to be measured by $\overline{M}_x = \overline{N}_y \cap \overline{N}_x$

The results are given in the form of three sets of parameters: the parameters which must be, need not be and should be measured

$\overline{D}_2 = \overline{M}_x \cap \overline{M}_1$ = do not need to be measured in whole extent

$\overline{D}_1 = \overline{M}_x - D_2$ = have to be measured

D_3 = are not measured (includes the remaining parameters)

Algorithm to identify the"interval" ($\overline{N}_y$). The set $\overline{C}$ of required parameters is known for a given system.

1) Identify all the required determinable parameters Y_j which are determined by the intersection of the sets $\overline{U}_1$ and $\overline{C}$, i.e. $Y_j \ \varepsilon \ (\overline{U}_1 \cap \overline{C}) = \overline{N}_m$
2) From the occurrence matrix for each Y_j find the now which was assigned to this Y_j and take all the parameters contained in this equation, except Y_j
3) The set $\overline{N}_y$ of all the parameters found in 2) constitute the "interval" of required unmeasured parameters.

Rectification of Random Measurement Errors

In the previous section we classified the measured parameters into two categories 1) overmeasured (overdetermined) 2) not overmeasured (just determined), so we can correct only the parameters belonging to the first category. In this way the accuracy of measured data can be improved. The next question is: how to formulate the appropiate subset of balances to be used in the rectification by solving the accompanying least square problem. Let us start with equation (12). Divide x into two vectors, $x' \varepsilon \overline{M}_1$, and $p \notin \overline{M}_1$. Consequently

$$A_1 \ x' \equiv A_{11} \ x' + A_{12} \ p = A_{11} \ x' + q \tag{13}$$

with $q = A_{12}$ p. Introducing the measurement errors. The balance equations (12) becomes

$$A_{11} \ (x' + \delta) + A_2 y + q = 0 \tag{14}$$

But from the classification strategy outlined earlier we can find those equations, which contain only measured (overmeasured) parameters i.e. the system of balance equation can be regrouped into the following form

$$A_o \, (\tilde{x}' + \delta) = 0 \qquad (15)$$

where; A_o: matrix whose rows (equations) contain only measured parameters and $x' \varepsilon \bar{M}_1$. If we allow bilinear balances as we described before, following the same procedure we arrive to a similar result

$$\sum_j a_{ij}^T \, x' + \sum_j \sum_\ell e_{ji\ell} \, x'_{ij} \, x'_{i\ell} = 0 \quad i = 1,2...K \qquad (16)$$

where; K: number of equations. We notice that the equations(16) in addition to the linear terms of the measured parameters, they also contain quadratic terms. In matrix natation and introducing the measurement noise, each row may be written as

$$g_i = a_{io}^T \, (\tilde{x}' + \delta) + (\tilde{x}' + \delta)^T \, E_{io} \, (\tilde{x}' + \delta) = 0 \qquad (17)$$

This can be extended to dynamic systems. The reduced set of balance equations can be expressed in this case as

$$L_o \, (\tilde{x}' + \delta) + e = L_o \delta + e' = 0 \qquad (18)$$

where; L_o: matrix whose rows (equations) contain only overmeasured parameters and $e' = L_o \tilde{x} + \sum_{i=1}^{n} K^i \, z^{-i} u$. Now when a new sample is taken, the present value of δ are computed while the present and previous values of $\tilde{x}$ are available from the measurements and the previous values of δ are taken from previous computations. The reconciliation problem is equivalent to that formulated for steady state case and has been solved by Gertler [5] using the weighted least squares.

It is clear from the above procedure that equations (15),(17) or (18) should be linearly independent. The selection of the linearly independent balances can be done rather easily, using the earlier method of parameters classification. From the classification problem we obtained a set of unassigned equations E which contain the subsets E_1, E_2 and E_2' of unassigned equations. For steady balances all the equations in subset E_1 are considered for the correction of measurements. The equations in the subset E_2' are modified by substituting the unmeasured parameters by the measured ones which belong to their "intervals". On the other hand these are not all the possible equations that should be considered. From the block-diagonal form D, we obtained the D_{ii}'s which are disjoint subsystems, and which are going to introduce more equations containing only overmeasured parameters. But we must be careful to select which equation to incorporate to avoid redundancies. If all the flow rates are measured we can incorporate a redundant overall mass balance equation and if we consider a system with n-components then we can only incorporate n-1 equations from the disjoint subsystems if all the compositions of the external streams are measured. We will have a similar situation if we consider the heat balances. For dynamic balances all the equations in subsets E_1 and E_2 are considered for the reconciliation problem with equations in E_2 modified by substituting the parameters in set $\overline{U_1}$ by their"intervals". The systematized algorithmic approach to construct the system matrix is as follow.

Algorithm.

1) Consider the set E_4 of the equation to be used for measurement reconciliation. E_4 is initially empty. For steady balances.
2) All the equations in set E_1 belong to set E_4
3) All the equations in set E_2' belong to set E_4 after all the parameter in set $\overline{U_1}$ have been replaced by their "intervals"
4) Identify all the D_{ii}'s whose external parameters are measured. Intoduce into the set E_4 the following balances resulting from each D_{ii}.
   a) An overall total mass balance if all the external flow rates are measured.
   b) A component mass (tatal energy) balance if all the external compositions (enthapies) are measured. For dynamic balances.
5) All the equations in set E_1 and E_2 belong to set E_4, after all the unmeasured determinables parameters have been replaced by their "intervals".

From the set E_4 of equations the construction of the system matrices follow directly. The approach outlined before, making use of the classification strategy allows the general reduction of the initial balances into a set of equations smaller in size than that suggested by Vaclavek. The reduced set of balance equations given by eq. (15), or (18) define now the following weighted least squares problem for the reconciliation of the measurement errors. In the linear case

$$\min_{\delta} \ \delta^T P^{-1} \delta$$

$$\text{s. t.} \quad A_o\delta + q = 0$$

where; P is the covariance matrix of measurement errors, and $q = A_o \tilde{x}'$. The Lagrangian solution of the above problem for the linear case yields

$$\hat{\delta} = \text{error estimate} = PA_o^T (A_oP A_o^T)^{-1} (A_o \tilde{x}') \qquad (19)$$

For the bilinear balances the objective remains the same but the equality constraints are now defined by equation (17). We may develop these equations in a Taylor series in a neighborhood of x_o, which denotes a first approximation of the true values x'

$$g_i(x') = g_i(x_o) + \sum_{j=1}^{n} \left(\frac{\partial g_i}{\partial x_j}\right)_{x_o} (x_j' - x_{jo}) \quad i=1,\ldots,n' \quad (20)$$

Introducing: $b_{k\ell} = \left(\frac{\partial g_i}{\partial x_\ell}\right)_{x_o}$, $B = \begin{bmatrix} b_{11} \cdots b_{1n} \\ b_{n'1} \cdots b_{n'n} \end{bmatrix}$

$$c_i = g_i(x_o) \qquad \alpha_i = (x_i' - x_{io})$$

Where; n': number of equations in the reduced system. We can rewrite equation (20) as follow

$$B\,\alpha + c = 0 \quad (21)$$

It is not difficult to find a first approximation x_o, since the $\tilde{x}_j'$ are measurements of the x_j'. We can therefore use them directly as a first approximation and set $x_o = \tilde{x}'$. The vector δ then becomes

$$\alpha = x' - x_o = x' - \tilde{x}' = \delta$$

Equation (21) therefore takes the form

$$B\,\delta + c = 0 \quad (22)$$

Which is equivalent to the linear case and the estimate $\hat{\delta}$ is given by equation (19). The result of each step is used to get a better approximation of x'. If the iteration is broken off after s steps we have

$$\hat{x}' = x'_s = x'_{s-1} + \hat{\delta}$$

This iteration process can be repeated until a satisfactory solution is obtained. In general is not easy to determine when a solution is satisfactory. The simplest method is to investigate the value of the covariance matrix of the error estimate and break off the iteration process if this covariance matrix falls below a given value or decreases less than a given fraction from one step to the next.

Rectification of Gross Errors

The major question related to the rectification of gross errors is the identification of the sources. Quite often this question will be set aside and the gross error will be distributed among several measurements. Although this approach may suffice in the short run, it is unacceptable in the long run. The detection of a gross error requires the construction of an appropiate statistical test. Two kind of approach can be follow 1) A single statistic for the entire process 2) A separate statistic for each equation. Consider in general the nonlinear balances.

$$g(x_1, x_2, \ldots\ldots x_K) = 0 \qquad (23)$$

The balance will be satisfied exactly in $x_1, x_2, \ldots x_K$ are the real values. But because the measurement errors, previously defined, the set of measured parameters is usually not internally consistent. So introducing the residuum in the balance equation, we can write

$$\varepsilon = g(x) - g(\tilde{x}) \qquad (24)$$

where; ε is the vector of residua in the balances. The problem is to construct the confidence region for the residuum value ε. Now it is assumed that the measurement noises are white and normally distributed. That is $x - \tilde{x}$ is sufficiently small, where $\tilde{x}$ is taken to be the arithmetic mean of the measurement indications. So that g can be expanded in a neighborhood of the point $\tilde{x}$ and only the linear terms need to be kept in the approximation. Thus

$$g(x) = g(\tilde{x}) + G(\tilde{x})\ (x - \tilde{x}) \qquad (25)$$

$$g(x) = g(\tilde{x}) + G(\tilde{x})\ \delta \qquad (26)$$

where; $G_i = \left[\sum_{j=1}^{k} \frac{\partial g_i}{\partial x_j} \right]_{x = \tilde{x}}$ is the i-th column of the matrix G. Combining (26) and (24) to obtain

$$\varepsilon = g(\tilde{x}) + G(\tilde{x})\delta - g(\tilde{x}) \qquad (27)$$

$$\varepsilon = G(\tilde{x})\delta \qquad (28)$$

Here the vector ε is known from the measurements. Now a postulate that the vector δ of measurement error is normally distributed is equivalent [10] to stating that the ε is also a normal vactor, because by (28) ε is approximated by a linear transfomation of a normal vector. Assume $E\,[\delta] = 0$, then ε will also have zero mean and covariance matrix

$$C_{\varepsilon\varepsilon} = E\ [\varepsilon\ \varepsilon^T] = G\ E\ [\delta\ \delta^T]\ \ G^T \qquad (29)$$

$$= G\ C_{\delta\delta} G^T$$

where; $C_{\delta\delta} = E\ [\delta\ \delta^T]$. Following the first approach we can formulate the following test function

$$h = \varepsilon^T C_{\varepsilon\varepsilon}^{-1} \varepsilon = \frac{\left| C_{\varepsilon\varepsilon} + \varepsilon\varepsilon^T \right|}{C_{\varepsilon\varepsilon}} - 1 \tag{30}$$

The latter of the quadratic form may be easier to evaluate, because the inverse of the covariance matrix. Since ε has n'elements (n'equations), h will have a chi-square distribution with n' degrees of freedom. Thus at specified significance level α

$$P\ (h \geq \chi^2_{1-\alpha}\ (n')) = \alpha$$

This means that the probability of the event that a particular h exceeds the critical χ^2 value belonging to $1-\alpha$ level for n' degrees of freedom is α. This provides us with a test for the inconsistency of a set of measurements in the presence of gross errors. We need to preassign an allowable error probability which gives us a critical value of h. The choice of the error probability depends on the process characteristics. For example, 0.10 is acceptable in many cases. If an extreme error is found by the test function, we need to identify which measurement has the gross error to guide instrument repair or similar corrective action. Nogita [14] proposed a serial elimination algorithm, that consist of discarding one element from the set of measurements at a time and check each time the value of the test function, choosing after that the consistent set of data with minimum value of h. Almasy and Sztano [6] proposed the following algorithm for searching the source of an extreme error. If r-th is the only measurement point for which

$$E(\delta^r) \neq 0 \quad \text{and} \quad E\ [\delta^j] = 0 \quad \text{for all } r \neq j$$

we define a vector z

$$z = D\ G^T\ (G\ C_{\delta\delta}\ G^T)^{-1}\ e \tag{31}$$

where;

$$D = \text{diag}\ (G_i^T\ (G\ C_{\delta\delta}\ G^T)^{-1}\ G_i)^{-\frac{1}{2}}$$

G_i is the i-th column of G. Then, vector z has the following properties

$$\text{abs}\ (E\ (z^r)) \geq \text{abs}\ (E\ (z^i))$$

From the above development, if only one measurement point has $E\ [\delta^r] \neq 0$ the measurement point with maximum absolute value of $E\ [z^i]$ is rejected. For practical cases we can use the equation for error detection with single ε_r vector values, if the ratio of the extreme error and the dispersion of regular error is not too small. Mah and his coworkers [12] studied extensively the problem of identifying the sources of gross errors. Their method, based on the second approach, utilizes the structural characteristics of the balance equations around the units and by developing intersections of appropriate sets of measured variables attempt to narrow the origin of the gross errors. The method is limited to one component mass balance systems. In the present work a more general algorithm is employed by using a separate statistic for each balance equation. Consider in general the i-th balance.

$$g_i\ (x_1,\ x_2\ \ldots\ldots x_K) = 0 \qquad i = 1,2,\ldots\ n \tag{32}$$

According to equation (28) this can be written as

$$\varepsilon_i = \sum_{j=1}^{K} G_{ij}\ \delta_j \tag{33}$$

Assuming the hypotesis that $\varepsilon = N\ (0,\ C_{\varepsilon\varepsilon})$ to be true, if the variable is measured from the mean, using the standard deviation as unit, we get a new variable which is called the standarized variable. Defining [14] the standarized variable N_i, we have

$$N_i = \frac{\varepsilon_i - E[\varepsilon_i]}{\sigma[\varepsilon_i]} = \frac{\varepsilon_i}{\sigma[\varepsilon_i]} \tag{34}$$

where; $\sigma[\varepsilon_i]$ is the variance of the i-th residuom. If we substitute the observed value ε_i, we find a value of N_i which enable us to answer the question whether the test hypothesis seems reasonable or not. The critical region of the α-level of significance of the two tailed test is determined by the inequality

$$\left|N_i\right| = \left|\frac{\varepsilon_i}{\sigma[\varepsilon_i]}\right| > (N_{i,\ 1-\alpha})\ \text{critic}$$

we detect a gross error in equation (33) with a probability of occurrence at least 90% if

$$|N_i| > 1.645$$

This simple stochastic criterion is now imbeded in a generalized structural search. The set of linearized balance equations is represented by a directed bipartite graph. We develop an output set assignment with ε_i the output from equation i and $\delta_1,\ \delta_2 \ldots \delta_K$ the variables of the set of equations. If the stochastic test indicates a gross error in the i-th equation, then it is clearly originating from one of the $\delta_1,\ \delta_2 \ldots,\delta_k$ errors. Let

$$\Delta i = \{\delta_j \mid j = 1,2,\ldots,K \text{ and } \delta_j \text{ is in the i-th balance}\}$$

Then if $\varepsilon_\ell = 0$, it is clear that the origin of the gross error for the i-th balance is not in the set

$$\Delta_i \cap \Delta_\ell$$

The precedure continues until the source of the gross error has been uniquely identified. The method fails to identify uniquely the source of the gross error when the source of gross error is structurally non-accessible or it is part of sets of balances with structural singularities. Both conditions result to the lack of structural observability for the source of gross error. The complete strategy and its implementation can be found in Romagnoli [16].

Numerical Examples

Example 1: Multicomponent Mass and Enthalpy Balances at steady-state.

In this example we will apply the method to a process considering multicomponent mass and heat balances. In Figure 4 the system under discussion is shown, which consists of 12 units with 28 streams and two components. The system equations are

$$\sum_{j=1}^{J} a_{jk} M_j = 0 \quad \text{Total mass balance for unit k}$$

$$\sum_{j=1}^{J} a_{jk} M_j x_{ji} = 0 \quad \text{Component mass balance for unit k}$$

$$\sum_{j=1}^{J} a_{jk} M_j H_j = 0 \quad \text{Total heat balance for unit k}$$

The structural matrix and the block diagonal form D are given in Figure 5. From the classification algorithm we obtain the subsets $\overline{M}_1$, $\overline{M}_2$, $\overline{U}_1$, $\overline{U}_2$ of parameters given in table 1.

	M_j	x_j	H_j
$\overline{M}_1$	1,3,4,7,15,25,30,31 32,33,34,35,36,37	1,3,4,7,15,25	1,3,4,7,15,25
$\overline{M}_2$	10,20,24,13	9,18,24,13	11,16,19,22,29,13,24
$\overline{U}_1$	2,6,8,9,11,14,16 18,19,22,23	2,8,6,10,11,14 16,19,20,22,23	2,8,6,9,10,14,18,20 23
$\overline{U}_2$			

Table 1. The classification of measured and unmeasured parameters for example 1.

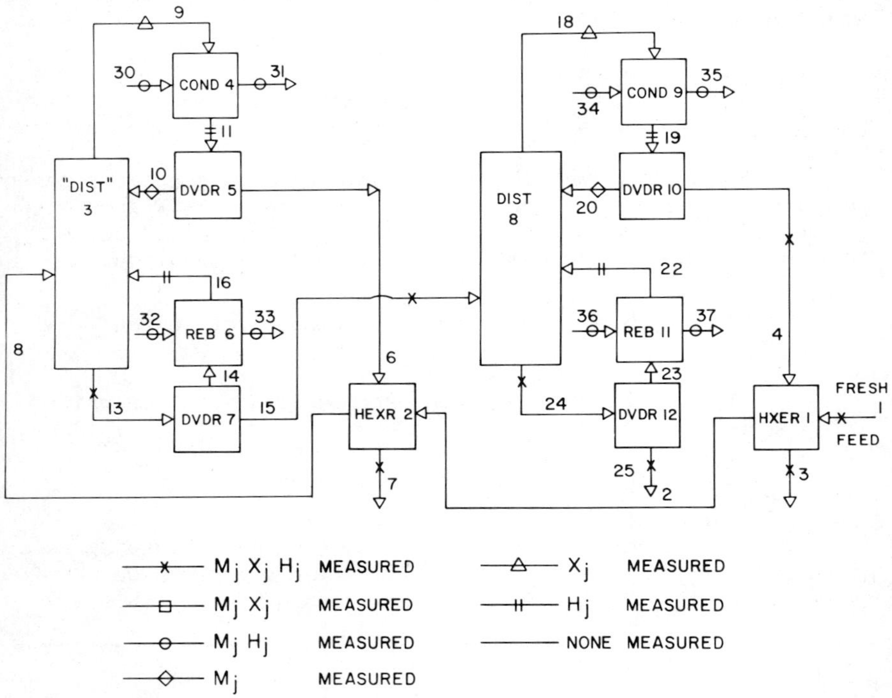

Figure 4. Flow diagram for Example 1

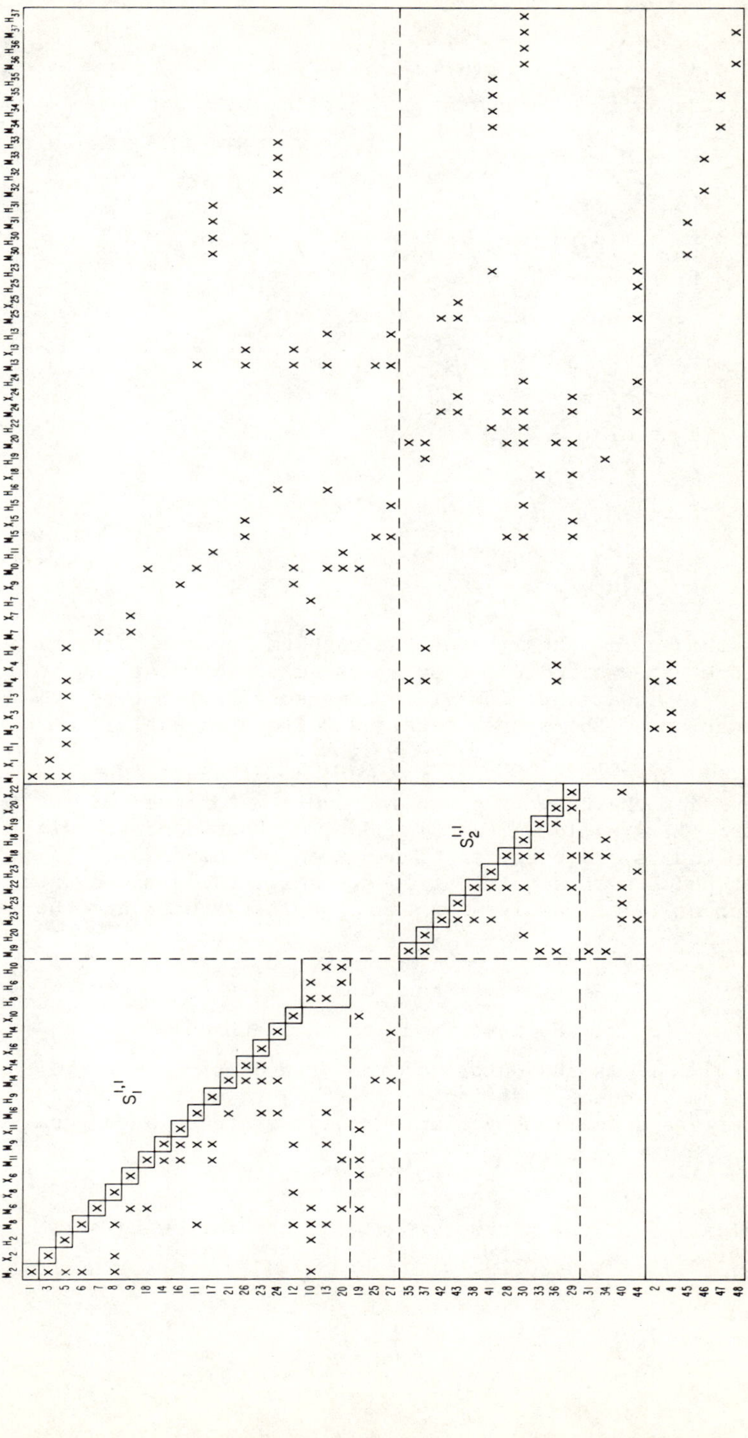

Figure 5. Modified occurrence matrix (block diagonal form) for the system in Example 1

The subsets E_1, E_2, E_3 of not assigned equations are

E_1 = 2,5,45 46,47,48 E_2 = 19,25,27,31 34,40,44 $E_2' = \phi$ $E_3 = \phi$

The equations with only overmeasured parameters are:

$$
\begin{aligned}
f_1 &= M_1 - M_7 - M_{15} = 0\\
f_2 &= M_1X_1 - M_7X_7 - M_{15}X_{15} = 0\\
f_3 &= M_1H_1 - M_3H_3 + M_4H_4 - M_7H_7 - M_{15}H_{15} = 0\\
f_4 &= -M_4 + M_{15} - M_{25} = 0\\
f_5 &= -M_4X_4 + M_{15}X_{15} - M_{25}X_{25} = 0\\
f_6 &= -M_4H_4 + M_{15}H_{15} - M_{25}H_{25} = 0\\
f_7 &= M_4 - M_3 = 0\\
f_8 &= M_4X_4 - M_3X_3 = 0\\
f_9 &= M_{30} - M_{13} = 0\\
f_{10} &= M_{32} - M_{33} = 0\\
f_{11} &= M_{34} - M_{35} = 0\\
f_{12} &= M_{36} - M_{37} = 0
\end{aligned}
$$

At this point the size of the problem for the correction of the parameters was reduced from a system of 48 equations to a system of 12 equations with all overmeasured parameters. The determination of the system matrices follows directly from here.

Example 2: Studies on a System Composed of Dynamic Units

Consider the network given in Figure 6 composed of two types of units: 1) dynamic units 2) nodes, without dynamics with nultiple inputs and outputs. The system is composed of 11 dynamic units, 4 nodes with a total of 28 streams. The sampled data input-output relationships for the dynamic systems have the form.

$$y^{(j)} = \sum_{i=1}^{m} \frac{a_i^{(j)}}{1 - z^{-1}\, b_i^{(j)}}\, u_i^{(j)}$$

where: $u_i^{(j)}$: is the i-th input to the j-th unit

$y^{(j)}$: is the output of the j-th unit

m : number of inputs to the j-th unit

The nodes are described by simple steady-state balance equations

$$\sum_i u_i^{(j)} = \sum_i y_i^{(j)}$$

Classification of the parameters yields the following sets

$\overline{M}_1$ = {1,3,4,15,24,22,20,19,10} $\overline{M}_2$ = {6,8,12,17,26}

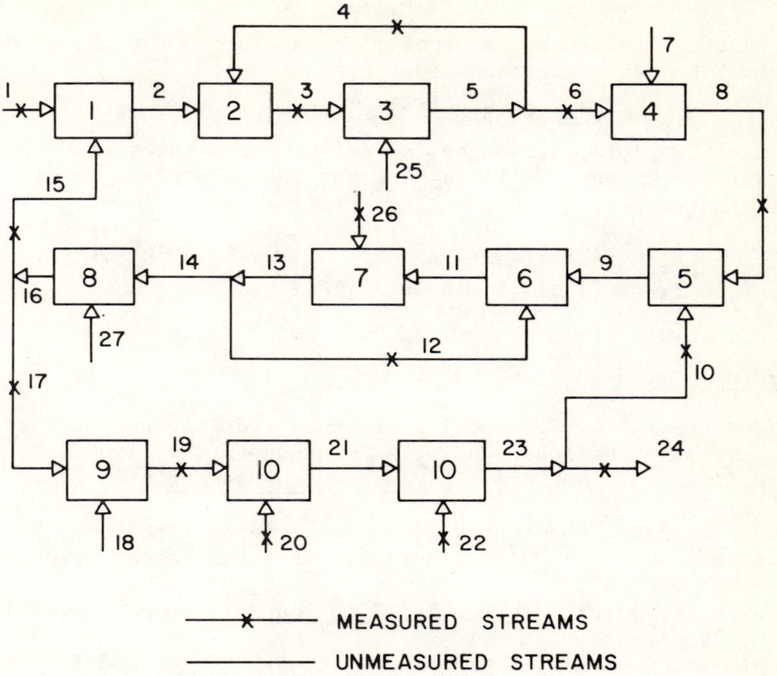

Figure 6. Flow diagram for Example 2

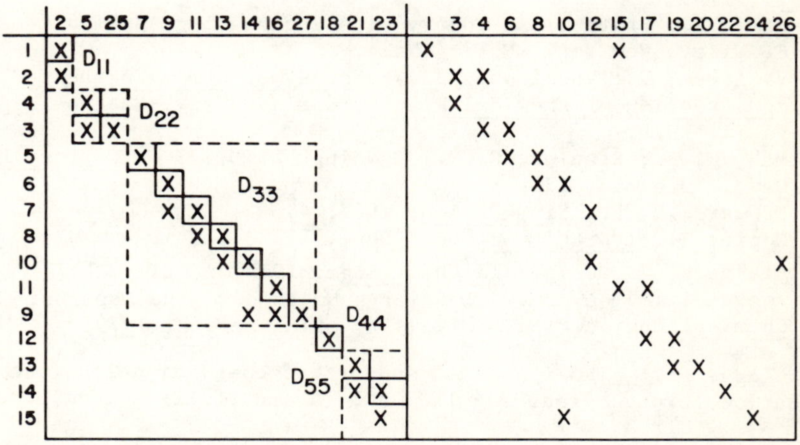

Figure 7. Modified (block diagonal) occurrence matrix for the system in Example 2

$\overline{U}_1$ = {2,25,5,7,9,11,13,14,27,18,21,23} $\overline{U}_2$ = ϕ

Figure 7 presents the block diagonal form D for the given system. The unassigned equations are

$E_1 = \phi$; $E_2 = 2,15$; $E_3 = \phi$

Let $\overline{C}$ = {1,4,7,12,23,24} be the set of the parameters that are needed to be known. Following the outlined algorithm we find easily that:

$\overline{D}_1$ = {1,4,24,22,20,19} $\overline{D}_2$ = {12,6,8}

$\overline{D}_3$ = {3,5,7,9,10,11,13,14,15,16,17,18,21,23,2}

Literature Cited

1. Swenker, A. G.; Proceeding of the Third International Measurement Conference, ACTO,IMEKO, Budapest, 1, 1964.
2. Hoffman, R.; Müller, R., Messen-Steuern-Regeln 9, 233 (1966).
3. Vaclavek, V. Chemical Engineering Science, 1969, 24, 947.
4. Almasy, G., 3rd CHISA Congress, Marianske Lazne, Czechoslovakia 1969.
5. Gertler, J.; Almasy, G. A., IFAC Journal Automatica, 1973, 9, 79
6. Almasy, G. A.; T. Sztanó, Problem of Control and Inf. Theory, 1974, 4, 1
7. Umeda, T.; Nishio, M.; Kamatsu, S., Industrial Engineering Chemical Process Design Development, 1971, 10, 2
8. Box, G. E.; Praper, N. R., Biometrica, 1965, 52, 355
9. Vaclavek, V.; Loucka, M., Chemical Engineering Science, 1976, 31, 1199
10. Hald, A., Statistical Theory with Engineering Applications, J. Wiley, New York, N.Y. (1952).
11. Vaclavek, V.; Vosolsobe, J., Proc. Symp. Comp. in the Design and Erection of Chemical Plants, Karlovy Vary, Czechoslovakia, 1975.
12. Mah, R. S.; Stanley, G. M.; Downing, D. M., I & EC Proc. Des. Dev., 1976, 15, 175
13. Steward, D. V., SIAM Review, 1962, 4, 321
14. Nogita S., I & EC Proc. Des. Dev., 1972, 11, 197
15. Hlavacek, V., Comp. and Chem. Eng., 1977, 1, 75
16. Romagnoli, J. A. Ph.D. Dossier submitted to the Dept. of Chemical Engineering, University of Minnesota, 1979.

Acknowledgement: The financial support of the National Science Foundation through grant 75-11165-A01 is gratefully acknowledged.

RECEIVED December 10, 1979.

8

Prolegomena to the Simulation of Choked Flow in Gas Pipe Networks

O. A. ASBJØRNSEN

Norsk Hydro a.s, Porsgrunn Fabrikker, P.O. Box 110, N 3901, Porsgrunn, Norway

The distribution system for pure gases and liquids is a classical example of a network system described by large and weakly non-linear systems of algebraic equations. The network comprises pipes, fittings and valves, pumps and compressors, turbines etc., and is subject to boundary conditions at the terminal inlet and outlet points. At these points, the pressure or flow or both are specified.

In the steady state simulation and design, the state variables are the flowrates and the pressure drops or terminal pressures for each branch of the net. Each of the branches between two nodepoints are described by a mathematical model of the hydrodynamics relating the pressure drop to the fluid flow between the nodes. The material balance sums up the flow into and out of a node no i:

$$\Sigma\delta_{i,j}W_j = 0 \;;\; W = \text{mass flow} \tag{1}$$

and the momentum balance in a loop no k composed of a number of nodes and line segments gives:

$$\sum_i \Delta P_{k,i} = \sum_i \Delta P_{k,i}\left(\sum_j \delta_{i,j}W_j\right) = 0 \tag{2}$$

Several models are used for the pressure drop and fluid flow correlation, but the most commonly used for pipes and fittings is simply:

$$\Delta P = \frac{L}{D\rho}\, f(\mathrm{Re})\,|w|w \;;\; \mathrm{Re} = 4|w|/\pi D\mu,\; w = \frac{W}{A} = \text{mass flux} \tag{3}$$

This model is useful strictly for incompressible fluids only, but for small pressure changes it may be used for gas networks as well. The equations for general network simulation are based on eqn (1) and eqn (2) and a variety of models analogous to eqn (3). It goes beyond the scope of this paper to elaborate on networks with loops, this is dealt with in details by for example Mach (1), Gay and Preece (2,3) and Carnahan and Wilkes (4). A variety of

0-8412-0549-3/80/47-124-175$05.00/0

available computer programs for the simulation of fluid networks are summarized by Petersen, Cheu and Evans (5).

For compressible fluids at isothermal conditions, more accurate computations may be obtained by assuming proportionality between pressure and density, whereby eqn (3) is slightly modified:

$$\Delta(P^2) = \frac{2LP_o}{\rho_o D} f(Re)|w|w \; ; \; \rho = \rho_o P/P_o \qquad (4)$$

This is the most commonly used model for natural gas nets, and most algorithms for incompressible networks may be used for gas networks as well, simply by replacing eqn (3) by eqn (4) in the library of pressure drop correlations. There are several commercially available computer programs for gas networks, among which the ones from the British Gas Corporation (6) and Intercomp (7) are found.

In many gas networks, the net has a special tree structure, with one source (a supplier) and several sinks (consumers), as illustrated in figure 1. This simplifies the simulation of the gas net to a very large extent. A gas net of such a tree structure is common for the nitrogen purge systems in petrochemical plants, which is the practical example for this analysis.

The normal load on the nitrogen net meets the specifications of the net with reasonable safety margins, but under special occations very high spot loads may occur. In such a case the pressure at some load points may drop below safe limits and cause the plants to trip out. Hence, the maximum spot load capacity as well as the pressure distribution under such a load are important for the safety of the plant as a system of interconnected units. This kind of information may be easily obtained by simulation, as outlined in this paper.

The simulation of a gas distribution system having a tree structure with one supplier may be reduced to a system of equations which has as many dependent variables as there are sinks or consumers. At each sink, the pressure (pressure control loop termination) or the mass flowrate (mass flowrate control loop termination) must be specified. The purpose of this short paper is to demonstrate some of the most commonly used mathematical models for the simulation of a gas net with a tree structure and point out some of the problems which arise if the net has an overload, a severe leakage or a bleeding at one point or another leading to choked flow.

Mathematical models for steady state subsonic gas flow in a straight pipe.

If the pressure drop in a pipeline is reasonable and the linear gas velosity is well below the speed of sound, the acceleration term in the equation of motion is usually neglected, and the pressure to mass flow correlation is described by eqn (4). For a given pressure drop the mass flowrate is easily computed

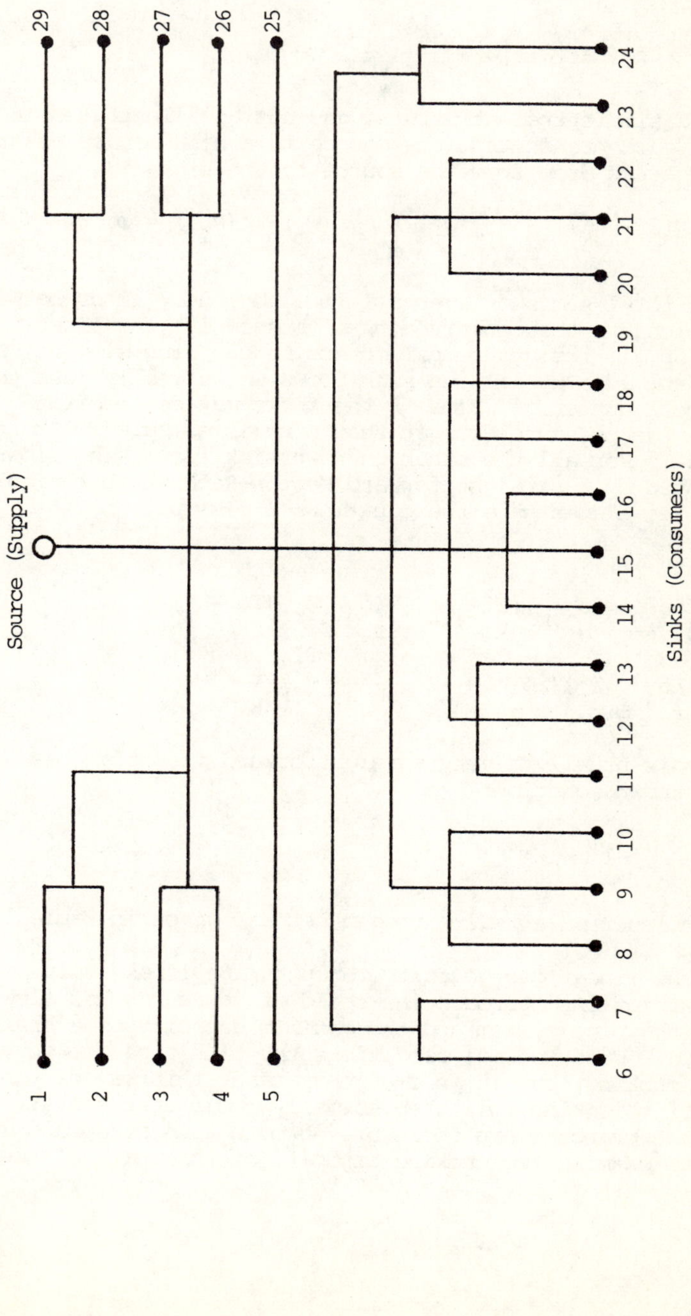

Figure 1. Gas-distribution network with a tree structure

by a standard Newton iteration in the flowrate, where the derivative is simply:

$$\varphi = f^*(Re)\,|w|w - \Delta(P^2)\ ; \qquad f^* = \frac{2LP_o}{D\rho_o} \cdot \frac{1}{A^2} \tag{5}$$

$$\frac{d\varphi}{dw} = 2f^*(Re)\,|w| + |w|w\,\frac{df^*}{dRe}\,\frac{4}{\pi D\mu}$$

For a simple tree structured gas net as illustrated in figure 1, each branch from the source to a sink may be described by the pressure drop from the source to the sink, i:

$$\sum_j \delta_{i,j}[\sum_k f^*_{j,k}]\,|\sum_m \delta_{j,m} W_m|\,(\sum_m \delta_{j,m} W_m) - \Delta(P_i^{\,2}) = \varphi_i \tag{6}$$

where the total sum sums over all line segments (lines between nodes) which are the constituents of branch i. The first inner sum adds up all resistance parameters in one line segment, and the last sum adds up all the sink flows which really pass through line segment number j. Some of the flows may be specified, in that case the pressure drop at that sink is computed without any iterations. For all the other unknown sink flows, the flowrates are computed by a straight forward Newton-Raphson iteration where the Jacobian elements to the equations are simply:

$$\frac{\partial \varphi_i}{\partial W_m} = \sum_j \delta_{i,j}\,\{2\delta_{j,m}\,[\sum_k f^*_{j,k}]\,|\sum_r \delta_{j,r} W_r|\,(\sum_r \delta_{j,r} W_r) + |\sum_r \delta_{j,r} W_r|\,(\sum_r \delta_{j,r} W_r) \sum_k \frac{4}{\pi D_k \mu}\,\frac{\partial f^*_{j,k}}{\partial Re_k}\} \tag{7}$$

This is usually a very straightforward iteration based on the well known equation:

$$\Delta W_m = J^{-1}_{m-1}\varphi_{m-1} \tag{8}$$

Some examples from a real nitrogen net are given in table 1 below.

If the linear gas velocity through pipe lines, valves or other restrictions approach the speed of sound, choked flow may occur in the pipe system and the maximum capacity of a branch is reached. Under normal conditions the pipelines are very far from choking, but in an emergency case, this may happen. Choking of this kind will then normally occur in the last part of the branch where there are no more take offs, and it suffice to examine the choking capacity of that part of the

Table 1. A simple industrial example showing flowrates in SCM/hr and pressures in bar for a nitrogen purge system.

Branch nr	1*	2*	3*	4*	5*
Flowrate	2300	60	75	75	300
Pressure	9.49	9.56	9.55	9.55	9.55
Branch nr	6 *	7*	8*	9*	10*
Flowrate	500	160	250	50	551
Pressure	9.56	9.62	9.63	9.63	9.62
Branch nr	11	12	13	14	15
Flowrate	404.7	402.0	166.1	99.1	383.9
Pressure	1.0	1.0	1.0	1.0	1.0

* Branches with specified flowrate (flowrate control).
Source pressure 9.7 bar.

branch. Furthermore, the simulation of a high spot load (choking conditions) is carried out for the other part of the net as before, but now with the choking capacity as a specified load at the particular sink in question. The examination of the choking conditions of a line segment or a single branch is hence essentially reduced to a series of resistances, pipes and restrictions.

Choking in a straight pipeline.

If the linear gas velocity approaches the speed of sound, the simple mathematical model used in equation (4) breaks down. The acceleration term must be taken into account, and the steady-state equation of motion for a straight pipeline with constant diameter may be written (8):

$$\rho v \frac{\partial v}{\partial \xi} + \frac{L}{D} f(Re) \rho v^2 + \frac{\partial P}{\partial \xi} = 0 \ ;\xi = l/L \qquad (10)$$

Furthermore, the equation of continuity and the equation of state eliminate the linear velocity and the gas density:

$$w = \rho v = \text{constant}; \ \rho = \rho(P,T) \qquad (11)$$

At some critical pressure drop, the mass-flux, w, reaches a maximum value. This takes place at the exit of the straight pipeline and is described by the general set of equations:

$$w^2(\ln(\rho_o/\rho_1) + \frac{L}{D}f(Re)) - \int_{P_1}^{Po} \rho dP = 0 \qquad (12)$$

$$w^2 \partial\rho/\partial P|_1 - \rho_1^{\ 2} = 0 \ ; \ \partial P/\partial\rho|_1 = c_1^{\ 2} \qquad (13)$$

which applies to the choked flow of any non ideal compressible fluid provided the density function of pressure and temperature is known.

At this stage it should be emphasized that eqn (10) strictly speaking implies a uniform velocity profile. This does not apply to real turbulent flow, but eqn (10) is a reasonable approximation for the average velocity under subsonic conditions. Under sonic flow however, the approximation becomes more doubtful if sonic shocks develop, because the boundary layers play an important role in such shocks. Nevertheless, eqn (10) is used throughout this paper as the first order model approximation to get a reasonably simple model for industrial computations and for the illustration of some of the main effects of choked flow.

The derivative of the pressure with respect to the density is usually referred to as the square of the speed of sound, which however is very misleading for two-phase problems and for single phase fluids in an elastic pipe. The velocity c, should rather be taken as the speed of propagation of large scale pressure transients in the pipeline.

A few special cases may illustrate choking in the straight pipeline.

a. Isothermal conditions, constant friction factor. In this case the pressure and the mass flux may be written in dimensionless forms:

$$x = P_1/P_o \quad ; \quad y = w\sqrt{RT}/P_o \quad \text{and } z = x^2 \tag{14}$$

At choking flow, y = x always, and the choking pressure drop is determined by:

$$z(2f'-\ln(z)) - (1-z) = 0 \; ; \quad f' = \frac{L}{D} \cdot f \tag{15}$$

If the friction factor, f, approaches zero, then z aproaches unity as does y. Hence, y is equal to the Mach number at the inlet of the pipeline. In figure 2 is shown how the inlet Mach number and the choking pressure drop change with the pipe length for isothermal conditions.

b. Isothermal conditions, variable friction factor. The friction factor f is not constant, but may be expressed as a function of the Reynolds number and the surface roughness of the pipe, by a Moody (9) or Fanning (10) chart:

$$f'(Re) = \frac{L}{D}\,\psi(Re) \quad ; \quad Re = (Re)_s \sqrt{z} \tag{16}$$

where the sonic Reynolds number at the inlet of the pipeline is:

$$(Re)_s = c\rho_o\, D/\mu \tag{17}$$

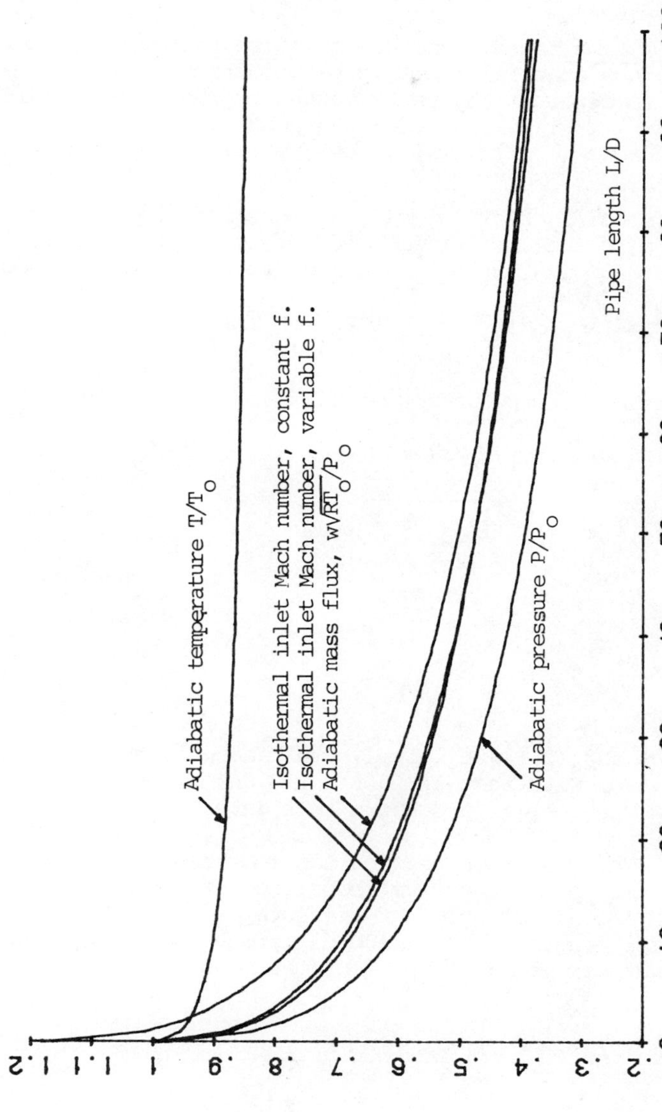

Figure 2. Different models for choked flow in a straight pipe

Introducing the general correlation for the friction factor suggested by Churchill (11), figure 2 shows how the choking conditions now depend on the pipe length for isothermal conditions and variable friction factor. As it is seen, the effect of the variable friction factor is negligible.

c. Adiabatic conditions, variable friction factor. Under adiabatic conditions, the relationship between the linear gas velocity and the temperature may be expressed by the enthalpy (12):

$$v^2 = v_o^2 + 2c_p(T_o - T) \ ; \quad v_o = cy \ ; \ y = w\sqrt{RT_o}/P_o \qquad (18)$$

provided the specific heat capacity at constant pressure is constant and c is the isothermal speed of sound at the inlet of the pipe. Introducing the dimensionless temperature u, gives:

$$u^2 + 2\alpha \frac{x^2}{y^2} u - (x^2 + 2\alpha \frac{x^2}{y^2} = 0 \ ; \quad u = T/T_o \qquad (19)$$

and the conditions for maximum mass flux is:

$$y^2(f' - \ln(x/u)) - \int_x^1 \zeta d\zeta/\eta(\zeta) = 0 \ ; \qquad \zeta = P/P_o \qquad (20)$$

$$\eta(\xi) = [1 + \sqrt{1 + y^2(2\alpha + y^2)/\xi^2}]\alpha/(2\alpha + y^2)$$

$$\alpha = c_p/R$$

$$y^2 u - x^2 - x\ \partial u/\partial x = 0 \qquad (21)$$

Eqns (20) and (21) are solved simultaneously by the Newton-Raphson iteration with respect to the two unknown variables y and x. The results are shown in figure 2 where the choking pressure drop and the temperature are shown as functions of the pipe length. The deviation between the isothermal and adiabatic mass flux at zero pipe length reflects the ratio between the adiabatic and isothermal speed of pressure propagation. Furthermore, at a pipe length of about 100 diameters, the discrepancy between the various models to predict the mass flux is not very significant.

d. Isothermal flows with a series of restrictions. A series of restrictions may be a line segment (between two nodes) with valves, orifices, pipes with different lengths and diameters etc. If the pressure drop across this line segment is below critical, the flow in this line segment will be everywhere subsonic. According to eqn (20) the mass flux through element number i under isothermal conditions is:

$$y_i^2 = \int_{x_i}^{x_{i-1}} \zeta d\zeta / (f_i + \ln(x_{i-1}/x_i)) \quad (22)$$

Then the intermediate pressures are given by the set of equations expressing constant mass flow at steady state:

$$A_i^2 y_i^2 = A_{i+1}^2 y_{i+1}^2 \; ; \quad f_i' = \frac{L_i}{D_i} f(Re_i) \quad ; \; Re_i = \frac{WD_i}{A_i \mu} \quad (23)$$

leading to:

$$\varphi_i = \frac{x_{i-1}^2 - x_i^2}{f_i' + \ln(x_{i-1}/x_i)} - \beta_i \frac{x_i^2 - x_{i+1}^2}{f_{i+1}' + \ln(x_i/x_{i+1})} = 0 \quad (24)$$

$$x_i = P_i/P_o; \quad x_o = 1; \quad x_n = P_n/P_o; \quad \beta_i = A_{i+1}^2/A_i^2$$

Eqn (23) is a second order nonlinear difference equation the Jacobian of which is easily established as a regular tridiagonal matrix with a dominating diagonal, similar to system matrices found in the simulation of distillation columns. The analytical derivation of the Jacobian and the Newton-Raphson iteration is trivial. In figure 3 is shown an example where the intermediate pressures are plotted as functions of the total pressure drop across the line segment. The example is artificially chosen such that all ε-parameters are the same, i.e.:

$$\varepsilon_i = \text{constant} = 1.0 \; ; \quad \text{i.e.} \quad [LA^2/D]_i = \text{constant} \quad (25)$$

and all the γ-parameters are the same, i.e. the pipe cross section increase as geometric progression:

$$\gamma_i = \text{constant} = 1.2 \; ; \quad \text{i.e.} \; A_i = 1.2 \, A_{i-1} \quad (26)$$

Notice the approach to maximum mass flux.

If the total pressure drop increases, choked flow will occur in one of the restrictions along the line, usually the last. The normalized mass flowrate is then:

$$y^2 = (1-x_n^2)/\sum_{i=1}^{n}(\varepsilon_i + \gamma_i \ln(x_{i-1}/x_i)) \; ; \; y = W\sqrt{RT_o}/(P_o A_o) \quad (27)$$

$$\varepsilon_i = f'_i \beta_i \; ; \; \gamma_i = A^2_{i+1}/A^2$$

where choking conditions satisfies the criterion:

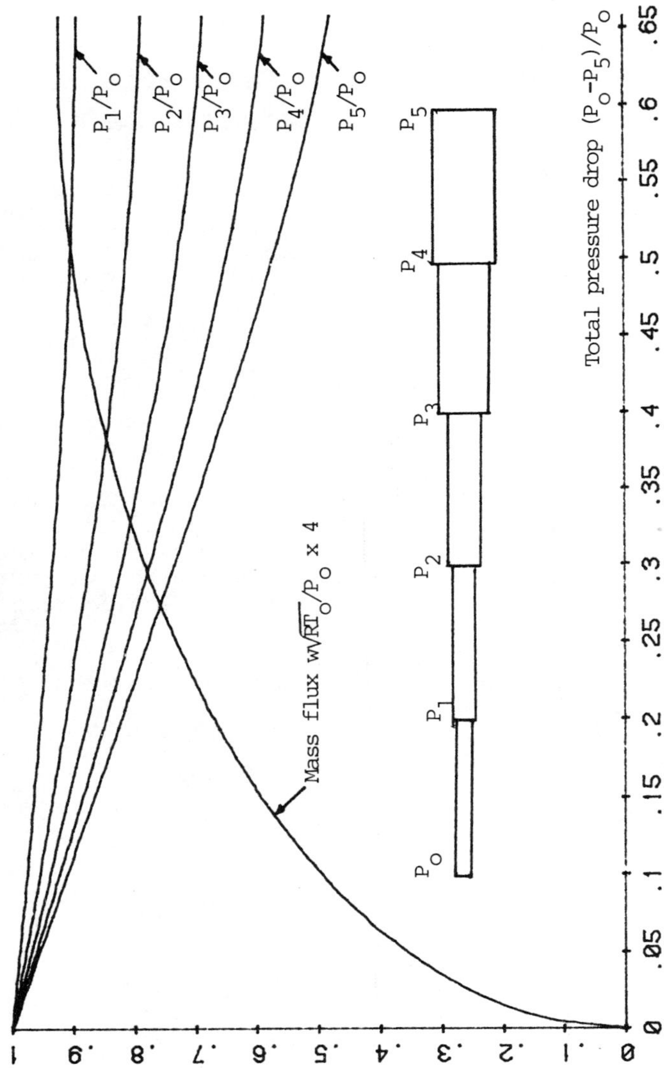

Figure 3. Subsonic flow in a series of expanding pipes

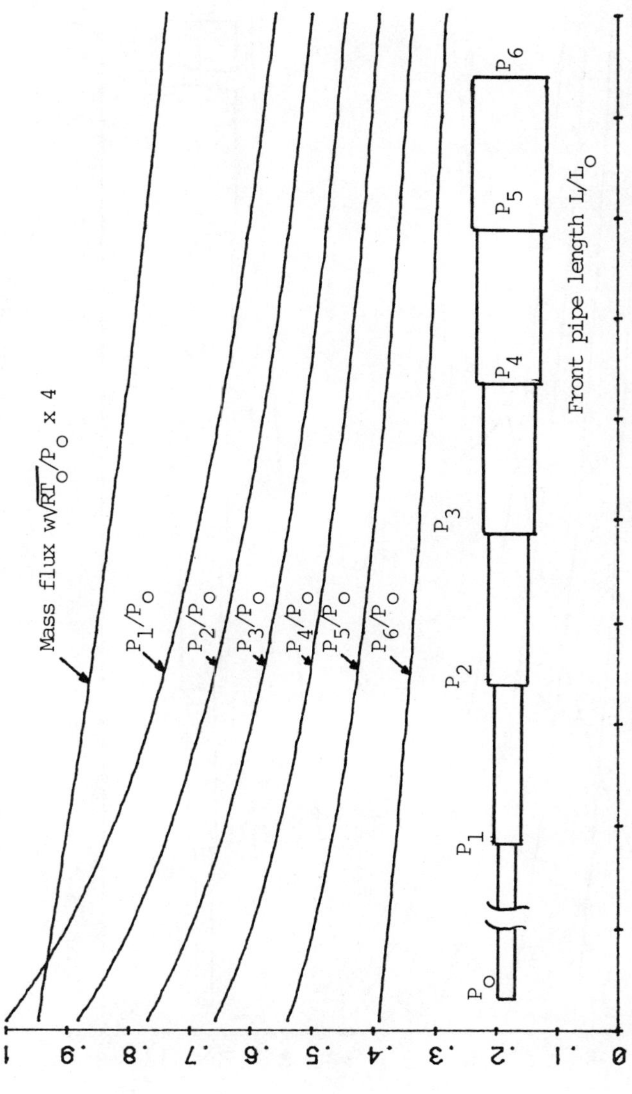

Figure 4. Sonic flow in a series of expanding pipes, front pipe extension

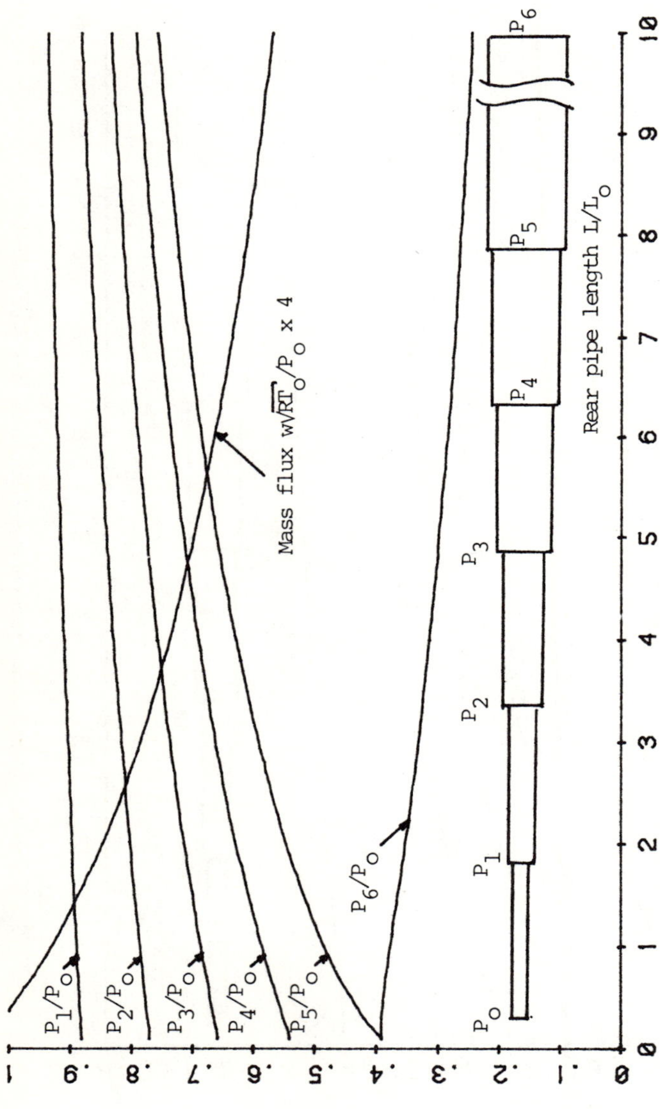

Figure 5. Sonic flow in a series of expanding pipes, rear pipe extension

$$\frac{\partial y}{\partial x_n} = 0 \text{ or equivalently: } \quad \frac{\partial (y^2)}{\partial (x_n^2)} = 0 \tag{28}$$

At this stage one may continue with an analytical derivation of the mass flux with respect to the pressure drop:

$$\sum_{i=1}^{n} (\varepsilon_i+\gamma_i \ln(x_{i-1}/x_i)) - (1-x_n^2)\left(\sum_{i=1}^{n-1} (\gamma_{i+1}-\gamma_i)\frac{1}{x_i}\,\frac{\partial x_i}{\partial x_n} - \frac{\gamma_n}{\partial x_n}\right) \tag{29}$$

where the n-1 derivatives in the last sum are computed from the n-1 equations specified by eqn (23):

$$\frac{\partial \underline{x}}{\partial x_n} = J^{-1} \frac{\partial \underline{\varphi}}{\partial x_n} \tag{30}$$

Eqn (29) may be included into the Newton-Raphson iteration as the n-th equation to determine all the intermediate as well as the final pressure. This however, requires subsequent derivation of the extra row in the Jacobian by the differentiation of eqn (29) with respect to the vector $\underline{x}$ of all pressures. This leads to fairly involved algebraic expression, so the quickest and safest method of calculating the choked flow conditions in the line segment is by a simple single variable optimization of y from eqn (27) with respect to the final pressure. The vector $\underline{x}$ is computed from eqn (24) by the straight forward Newton-Raphson iteration for each step in the single variable hillclimbing.

In figure 4 and 5 is shown the effect of inserting a longer pipeline in a line segment subject to choked flow. The configuration of the expanding pipes is the same as given by eqn (25) and (26), but the length of the front pipe is changed from zero to ten times its original value in figure 4. In figure 5 the length of the rear pipe is changed over the same range. As it is seen, the effects of the extra pipeline on the total mass flow is not the same, and this may be of vital importance for safety valve and flare installations as well as for the limiting choking capacity of pipelines in general.

Conclusion

By the use of extremely simple models for the gas flow in pipelines, it has been demonstrated how important the acceleration term becomes when sonic conditions are approached in a gas network. This term is usually neglected in most design computations, but the simple examples in this paper show that this may not be justified.

The numerical approach and the computational algorithms may be further developed along the lines suggested in this paper, and the problem of choking is found to be a very significant one in flare gas systems, safety relief systems and purge systems for emergency. Further experimental results may be reported at a later date.

Acknowledgement.

The author wishes to acknowledge the valuable comments made by T. Fannelöp and Ö. Jakobsen at the Institute of Aero and Gas Dynamics at the University of Trondheim.

List of symbols

A cross-section of flow
c speed of propagation of pressure transients
c_p specific heat capacity at constant pressure
c_v " " " " " volume
D pipe diameter
f friction factor
i,j,k,m,o indices
n number of elements in a line segment
P pressure
R gas constant
Re Reynolds number
T temperature
u dimensionless temperature
v linear gas velocity
w mass flux
W mass flowrate
x dimensionless pressure
y dimensionsless mass flux, isothermal Mach number
β,γ,ε line segment parameters as explained in text
δ Kronecker variable, 0 or 1.
μ gas viscosity
ρ density
φ residual function
ξ dimensionless pipe length coordinate
ζ dimensionless pressure

Literature Cited

1. Mach, R.S.H.: "Pipeline Network Calculations using Sparse Computation Technique." Chem. Engg Sci. (1974), 29, pp 1629-1638.
2. Gay, B. and P.E. Preece: "Matrix Methods for the Solution of Fluid Network Problems: Part I - Mesh Methods." Trans. Instn. Chem. Engrs, (1975), 53, pp 12-15.
3. Gay, B and P.E. Preece: "Matrix Methods for the Solution of Fluid Network Problems: Part II - Diakoptic Methods." Ibidum (1977) 55, pp 38-45.

4. Carnahan, B. and J.O. Wilkes: "Simulation of a General Piping and Pumping Network." CACHE Problem VI-4, pp 71-89.
5. Peterson, J.N., C.-C. Cheu and L.B. Evans: "Computer Programs for Chemical Engineers: 1978 - Part 4." Chem. Engg. (1978), Aug. 28, pp 107-115.
6. Abramson, B.: "A Guide to Network Analysis Programs." British Gas Corporation, LRS T 177, (1974)
7. INTERCOMP Resource Development and Engineering, Inc.: "SSTRAN, Steady-State Gas Transmission/Gathering System Model." (1974).
8. Bird, R.B., W.E. Stewart and E.N. Lightfoot: "Transport Phenomena" John Wiley, New York, 1960.
9. Moody, Trans. ASME, (1944), 66, pp 671-684.
10. Perry, R.H. (ed): "Chemical Engineers' Handbook." Fourth Ed. McGraw-Hill, New York, (1963).
11. Churchill, S.W.: "Friction-factor Equation Spans All Fluid-flow Regimes." Chem. Engg, (1977), Nov. 7, pp 91-92.
12. Schmidt, E.: "Einführung in die Technische Termodynamik." Springer-Verlag, Berlin (1953).

RECEIVED November 5, 1979.

COMPUTATIONAL METHODS

9

Reaction Path Synthesis in Industrial Chemistry

R. B. AGNIHOTRI
AAA Technology, Houston, TX 77036

R. L. MOTARD
Department of Chemical Engineering, Washington University, St. Louis, MO 63130

Many industrial researchers would like to have the capability of outlining the various alternative routes a process could take. Some would like to predict all the products or byproducts formed in a process while others might be interested in exploring the different ways to synthesize a particular compound. In recent years the field of computer-aided organic synthesis has become increasingly sophisticated. Chemists, in conjunction with computer scientists, are trying to construct synthetic steps leading to the formation of complex molecules from simple starting reactants. These attempts fall more or less into two broad categories (1, 2) i) the direct-associative and ii) the logic centered approach. In the direct-associative approach the chemist has available a number of subunits which he can bring together using standard laboratory reactions with which he is already familiar. This empirical approach is obviously limited to known reactions and subunits. The logic-centered approach on the other hand consists of the generation of sets of intermediates which form a synthetic tree which is used to lead to the target molecule. The different branches of this tree are the alternative routes one would choose or reject. In practice, most chemists use an approach which is a mixture of both.

A number of different systems have been developed to conduct computer-aided organic synthesis. Of these, one of the most extensive projects is LHASA (3 - 11). The method starts from a target molecule and derives a set of precursor molecules which can be expected to be converted to the target by one synthetic reaction or a simple sequence of reactions. Each precursor molecule so generated serves as the next target and the procedure is repeated, thus generating a tree of synthetic intermediates. Each precursor is somewhat simpler than its parent target molecule. The analysis terminates when those precursors are found which are considered to be relatively simple or readily available.

This analysis is termed "retrosynthetic" or "antithetic" because it runs counter to the laboratory route which is termed

0-8412-0549-3/80/47-124-193$05.00/0

"synthetic." A process in the antithetic direction is called a "transform" while a process in the forward synthetic direction is a "reaction."

The chemical transforms are the heart and soul (11) of LHASA which has a database of over 600 common chemical reactions. This feature, ironically, shows its limitations. The empirical information about known chemical reactions limits the generation of synthetic pathways. The value of LHASA, however, is in its ability to investigate all possible known synthetic routes, a task which a chemist would find exceedingly tedious.

Gelernter et al (12, 13) developed a program SYNCHEM to produce reasonable problem-solving trees containing a high density of complete and satisfactory synthesis. Like LHASA, this also depends a lot on empirical chemical transforms.

Hendrickson's (14, 15, 16, 17, 18) approach is much more systematic. An organic molecular structure is viewed as a collection of functional sites arranged on a skeletal lattice. The most important step in synthesis is skeletal construction. Creating functionality by attaching or replacing functional groups, i.e. functionalization steps, are primarily necessary to prepare the next construction step. Ideal synthesis would eliminate discrete refunctionalization steps altogether since each construction step would leave just the right functional groups on the skeleton for the next construction step. Hendrickson has invented a complete library of generalized reaction paths, independent of empirical named organic chemistry, to be executed without computer assistance. With his notation and strategy he can teach organic synthesis methods.

SECS(19, 20) was developed by Wipke and coworkers to focus on stereochemical features originally missing in LHASA, however stereochemical features were introduced into both programs at the same time. REACT (21) uses LHASA chemical transformations to generate complete reaction paths which are constrained to use industrially available intermediates whose prices appear in the Chemical Marketing Report.

A basic difficulty that we find with the LHASA - SECS - REACT strategies is that the reaction paths are more typical of a discrete step-wise laboratory procedure. The use of named organic chemical transforms leads to a bench scale procedure. Reactions of industrial interest on the other hand are generally carried on in one to three continuous steps, with a minimal use of reagents and purification steps. Solid catalyzed vapor or liquid phase reactions are favored. Intermediates and raw materials are far more limited in variety on the scale of industrial production. We are also dealing with simple molecules. Frequently, the issue is simply how can one produce a desired set of products from limited available set of raw materials. Reaction paths should be short and highly selective otherwise the purification steps destroy the economics of the process. Separations, principally distillation, already consume 50% of the energy used in the petrochemical industry.

Our goal is to identify, on the basis of thermodynamic feasibility, the most effective paths between specific reactants and specific products. Side reactions which are equally or more feasible than the desired reaction steps are candidates for inhibition either through an appropriate choice of operating conditions or the design of a catalyst or other reaction technology. Since we do not wish to be hampered in the search for reaction paths by an allusion to laboratory-based organic chemistry we are attracted to the work of Ugi and co-workers (22 - 28) as the most powerful general representation for reaction synthesis. Since the CICLOPS, EROS and MATCHEM programs were not available we developed our own building block package based on Ugi's work, called CHIRP, as the inner core routine for the ultimate total system that we envision.

This paper is a report on the first phase of the project.

Connectivity Matrix Approach to Organic Synthesis

Ugi et al (22, 23, 24) perceived an underlying mathematical model of constitutional chemistry (24) in that there was some order in the relations between chemical systems. They extended the concept of isomerism (of molecules) to isomerism of ensembles of molecules (EM). The left and right sides of chemical equations refer to individual EM belonging to the same family of isomeric ensembles of molecules (FIEM) i.e. all chemical reactions occur within the same pertinent FIEM. The atoms in a FIEM remain the same, the relationships between them may change to give different isomeric EM. A multistep synthesis can be looked upon as a sequence of isomeric ensembles of molecules which begins with the set of starting materials EM_A and ends with the target molecule and all byproducts which are formed, EM_Z.

The chemical constitution of an EM is represented by a "be" matrix ["bond-electron" matrix]. A be-matrix B representing an EM(B) consisting of n atoms is an nXn matrix with integral entries where the off-diagonal entries b_{ij} represent covalent bonds between the atoms A_i and A_j, and the diagonal entries b_{ii} correspond to the numbers of free unshared valence electrons on the atom A_i. It is easy to see that B is symmetric, that the row/column sums, $b_i = \sum_j b_{ij} = \sum_j b_{ji}$ are the numbers of valence electrons belonging to the atom A_i and the cross sums $c_i = 2b_i - b_{ii}$ are the overall numbers of electrons in the valence shell of the atom A_i. Two be-matrices represent the same EM if they are interconvertible by permuting the atom indices i.e. permuting the rows/columns.

If B is the be-matrix for the starting materials, and E the be-matrix for the end products in a chemical reaction, the chemical reaction $B \rightarrow E$ is represented by a reaction matrix $R = E - B$. Since R is the difference of two symmetric matrices, it is also symmetric. The entries $r_{ij} = r_{ji}$ of an R-matrix indicate which bonds are made and broken as well as the changes in the distribution of free electrons. The off-diagonal entries r_{ij} $(i \neq j)$ indicate how many bonds between A_i and A_j are made (+) or broken (-). The diagonal entry r_{ii} indicates how many free electrons A_i gains (+) or loses (-). Each reaction matrix represents a class of chemical reactions with the same electron redistribution pattern i.e. an R matrix is an "electron-pushing" device. Each reaction matrix represents s type of chemical reaction which is independent of the individual reaction systems.

The application of the be-matrices and R-matrix is illustrated by the following example (25) of the conversion of α-hydroxy acetonitrile (1) into its components (2) and (3):

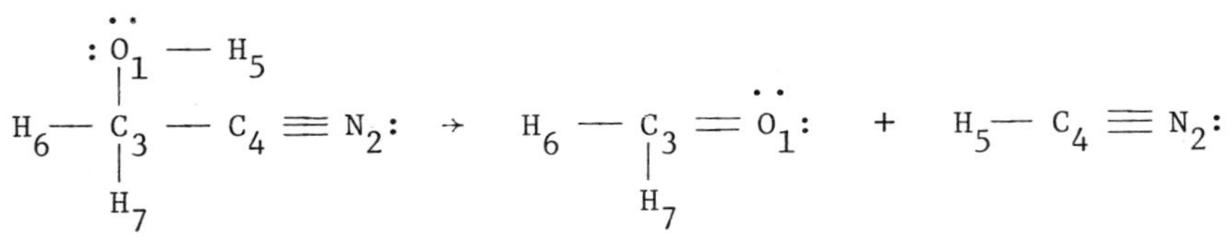

α-hydroxy acetonitrile(1) Formaldehyde(2) Hydrogen Cyanide(3)

$B = E_{(1)}$

	1	2	3	4	5	6	7
1	4		1		1		
2		2		3			
3	1			1		1	1
4		3	1				
5	1						
6			1				
7			1				

\+ R

	1	2	3	4	5	6	7
1			1		-1		
2							
3	1			-1			
4			-1		1		
5	-1			1			
6							
7							

→ $E = E_{(2)+(3)}$

	1	2	3	4	5	6	7
1	4		2				
2		2		3			
3	2					1	1
4	3				1		
5				1			
6			1				
7			1				

[The zeroes have been omitted from these matrices.]

Since an EM cannot have negative integer entries it is obvious that only that R matrix can act on the matrix B in which r_{ij} has negative integers such that $|r_{ij}| \leq b_{ij}$. This fitting requirement for an R matrix can be used in the manufacture of R-matrices.

Description of the CHIRP Program

As a preliminary step towards computer-aided investigation of reaction paths where reaction conditions could be prescribed to satisfy the needs of a particular process, it was felt necessary to develop a procedure to generate alternate reaction paths many of which may not be obvious in non-aided design. CHIRP (Chemical Engineering Investigation of Reaction Paths) is the initial outcome of such efforts and serves to find alternate reaction paths, products and byproducts for a given process.

Our objective is to look at all the possible products or byproducts emanating from a given reactant set and undergoing a given process. Once we have the various paths leading to the different products we can choose to promote some paths and to inhibit others by influencing the reaction conditions. This is essentially a forward search strategy. We did not want our choices to be limited by the data input and bias of the user, because that would detract from the value of the free search. Ugi's (24) R-matrices, therefore, offered us a valuable tool to investigate the tremendous number of reaction paths, each a result of a simple algebraic manipulation of the atoms and bonds contained in a reactant set.

In searching for a notation for the representation of chemical compounds we found that both the Wiswesser Line Notation and the IUPAC (i.e. Dyson) notation were inadequate to the problem of recognizing unknown molecules generated by structural rearrangements. The representation of molecules is thus based on the concept of a connectivity matrix of atoms as described by Ugi (22). This representation allows us to code a given molecule in a unique linear form suitable to computer storage, to apply transformation (i.e. reactions) on the molecule and then to recognize the product molecules by comparing them against a catalog of compounds.

In our system, chemical reactions correspond to the selective manipulation and rearrangement of the bonds between atoms in the reactant set, i.e. reactant matrix. This rearrangement is directed by some preselected R-matrices. The resultant product matrix is first sorted to yield separate and distinct product molecules that are subsequently checked against the compound catalog. The system is flexible enough to accommodate new molecules that may be formed during a reaction but which do not currently exist in the compound catalog.

Concurrent with the process of reaction and product formation are the evaluation and restriction procedures built into CHIRP. These procedures are designed to:

i) eliminate those reactions that generate compounds deemed unstable on chemical valence criteria,

ii) prevent formation of redundant product sets. Specifically, a) no product sets which contain molecules already present in the reactant set are saved and b) product sets that are identical to previously generated sets are eliminated.

iii) eliminate those reactions which involve reactant sets (or their equivalent atoms) that have already been reacted.

iv) eliminate reactions which involve
   a) replacement of an atom set by an equivalent atom set
   b) replacement of a monatomic element by one of the same species

As is evident, these restrictions serve to limit the profusion of unnecessary and redundant computer output. The modular structure of the program permits us to add to or subtract from the current built-in limitations depending on new available information.

Despite the limited number of R-matrices we have made available for reaction and the restriction procedures outlined above, the number of products generated is tremendous. In order to restrict these reactions along fairly identifiable patterns it was felt necessary to channel reactions along pre-ordered lines.

The R-matrices that can act on a reactant matrix B must contain negative integers r_{ij} such that $|r_{ij}| \leq b_{ij}$. The negative entries in R represent the bonds broken in the reactant set represented by B. We can restrict reactions by specifying which bonds must be broken in any reaction. We can also augment the reactant set by a specified species such as hydrogen or chlorine, etc. These two above requirements are specified by a 'reaction block.' There are 9 reaction blocks currently in the system. These are hydrogenation, cracking, oxygenation, alkylation, hydration, halogenation, dehydrogenation, dehydration and dehydrohalogenation. Every reaction must involve the breakage of the bonds that the applicable reaction block has flagged as breakable. The reaction block restriction is very flexible and new reaction blocks can be added to the system.

The system has the capability to carry on the reaction to the desired number of stages. The first stage is the application of the R-matrices to the starting reactant set. The specified reaction block will cause a certain number of bonds to be flagged as breakable and may add some additional desired reactants. After the reaction is over, we will obtain many product sets, each such set having one or more product molecules. The question now is - what constitutes the reactant set for the second stage? In a closed environment many reactions are possible, reactions involving each reactant only, all reactants, reactant-product, product-product, etc. A suitable choice is to isolate those product sets that are thermodynamically the most feasible in terms of least Gibbs free energy of reaction. The user can specify the number of such sets. In each of these sets we select a number of molecules that are the largest in the product set. The assumption here is that size is an indicator of the complexitv of the molecule and that the user is interested in the more complicated molecules to reduce the length of the reaction path. These selected molecules are then stacked in the reactant pile for the next stage.

In the next stage, each of these molecules undergoes the same reaction that the parent reactant set underwent, e.g. if the parent reactant set was one molecule of 1,3 butadiene and the reaction block desired was hydrogenation with the addition of one (user specified) molecule of H_2, one typical product would be 1-butene. In the next stage, 1-butene would be combined with one molecule of hydrogen to give one molecule of n-butane. This procedure is repeated at every level. The program terminates when either i) the specified number of stages are completed or ii) there are no more available compounds to be placed in the reactant pile for the next succeeding level (stage). The second alternative stems from a situation where either the set of breakable bonds is empty of when no products were formed at the previous level.

Two items are central to the CHIRP system. These are i) the compound recognition algorithm and ii) the R-matrices that are applied on reactant sets. These are discussed in turn.

Compound Recognition

When we 'react' the starting molecules and obtain a product set it is imperative that the product molecules be isolated and numbered in a unique manner so that they can be recognized. Previous researchers (25, 27) have presented algorithms for linear (i.e. acyclic) molecules and the latter paper (27) gives an example for a simple ring molecule. We have developed an algorithm for a unique representation of molecules. It is divided into two parts, one for acyclic molecules and the other for molecules containing rings. The acyclic part is inspired partly by the Wiswesser Line Notation in that it isolates the longest chain and partly by the Barkley and Motard (29) technique for finding tear streams in a complex flowsheet in that successive precursors are found for each atom, the last precursor being the center of the longest chain. Details of this method are given in Agnihotri (30). The algorithm for compounds containing rings is extremely versatile. It can uniquely number atoms in any ring system. There are no restrictions on the number or order (i.e. number of edges) of the rings. The rings can be simple, fused, spiro, bridged, etc. or they may have linear segments in them. There is also the advantage that we can easily represent, recognize and number resonant aromatic structures without any loss of generality in positioning the double bonds contained therein. Again, details are given in the above reference (30).

Classification of R-Matrices

Since an R-matrix represents an electron or bond redistribution pattern, it can represent not only one specific reaction but a general class of reactions. Because of the properties of R-matrices, they may be grouped in several classes. Brandt, et al (28) have defined an R-category as an equivalence class of chemical reactions which have in common the same electron relocation pattern and certain features of the participating bond systems. An irreducible R-matrix is one that does not have a row/column containing all zeroes. Any two chemical reactions which are represented by the same irreducible R-matrix belong to the same R-category. The Brandt categories are characterized by R-matrices which may have up to three off-diagonal pairs of positive and negative entries.

We have concentrated on a restricted chemistry (24) and the constraints imposed on the R-matrices involved therein are:

1) The R-matrix is symmetric
2) The diagonal elements $r_{ii} = 0$ (no unshared electron redistribution
3) The sum of every row/column is zero

Since the off-diagonal entries r_{ij} $(i \neq j)$ represent the change in bonds between atoms A_i and A_j, the absolute sum of all negative entries equals the absolute sum of all positive entries, i.e. the number of bonds broken equals the number of bonds made (= M).

In restricted chemistry, no atom carries a net charge. Hence any diagonal entry for an atom in a BE matrix will be the free pair of electrons the atom possesses. Thus, the diagonal entry for Nitrogen could be 2 whereas for Hydrogen, Chlorine and Carbon, etc. it is zero. As a consequence, a reaction matrix in restricted chemistry will have all diagonal entries zero and each row sum equal to zero. Dugundji and Ugi (24) came to the following conclusions about R-matrices in restricted chemistry. If LR(n) be the set of all n x n symmetric integral matrices with all diagonal entries zero and each row sum equal to zero, then LR(n) is a free Abelian group of rank $\frac{n(n-3)}{2}$.

This means that there are $\frac{n(n-3)}{2}$ distinct basic reactions in the restricted chemistry of an n-atom FIEM, all others follow by linear combination. This criterion leads to another obvious theorem that there is no nontrivial restricted reaction matrix for $n \leq 3$. Hence, the lowest order possible is for n = 4.

Using these conclusions, we have classified irreducible R-matrices in restricted chemistry into R-categories. The terminology is so chosen as to conform with that of Brandt, et. al. (28).

The nomenclature for describing an R-category is given by RMMNi, where

M: Number of bonds broken/made.

N: Number of atoms involved in the reaction.

i: An integer denoting a sub-category if more than one exist.

In the following discussion, (we will develop R-categories by a process of logical deduction). Henceforth, we will refer to an irreducible R-matrix simply as an R-matrix. All R-matrices that can be obtained by a simple row/column transposition fall into the same R-category. There are some general conditions governing the composition of an R-category. The designated bond associated with each participating atom has to be broken. If not, that bond is redundant because it will not contribute to the R-matrix. Also, since any atom has to have bonds broken it will have to have the same number of new bonds made, or else the sum for the row/column corresponding to that atom will not equal zero. All designated atoms have to participate in the R-matrix otherwise the R-matrix is reducible by eliminating null columns and null rows.

There are certain existence requirements for an atom set of an R-category. We will refer to valency of an atom as the number of its designated bonds participating in the R-matrix. Other bonds associated with the atom but not taking part in the reaction are of no concern to us.

In an R-category RMMN,

let S_1 = number of monovalent atoms

S_2 = number of divalent atoms

S_r = number of r-valent atoms

$$\therefore \text{ Number of bonds} = \frac{S_1 + 2S_2 + 3S_3 + \ldots + rS_r}{2} = M \quad \text{for } r < M$$

$$2M = \sum_{i=1}^{r} iS_i \tag{1}$$

Also, $S_1 + S_2 + \ldots + S_r = N$ for $r < M$

$$\text{or,} \qquad N = \sum_{i=1}^{r} S_i \tag{2}$$

Subtracting (2) from (1) yields

$$2M - N = \sum_{i=1}^{r} (i-1)\, S_i \qquad \text{for } R < M \tag{3}$$

Taking into consideration the number of atoms participating in the vast majority of reactions and the fact that higher R-matrices can be represented as combinations of simpler R-matrices (25), we have limited N to 6 and M to 4.

Given that the minimum value of N can be 4, the minimum value of M has to be 2 since for N = 4 four atoms are involved and so the minimum number of bonds that can be broken is 2.

We will now investigate the R-category, R224.

R224

M = 2

N = 4

Using eqn (3),

$$2M - N = (2)(2) - 4 = 0 = S_2 + 2S_3 + \ldots (r-1) S_r \quad r < M$$

$$\therefore \quad S_2 = S_3 = \ldots = S_r = 0$$

$\therefore$ Only monovalent atoms are involved

$$\therefore \quad S_1 = N = 4$$

Let us say we have four atoms, A, B, C and D.

Assume that these four atoms are arranged in a configuration that we will refer to as the reactant set. Then a product set is simply another configuration of these four atoms. The new configuration does not contain any of the relationships (bonds) of the old one but has an entirely different bond structure. Thus, the transformation from one configuration to another is actually the R-matrix which has broken 2 bonds and formed 2 new ones in an atom set involving 4 monovalent atoms. This R-matrix obviously belongs to the category R224. If we have a reactant set that cannot be arranged in another configuration, no R-matrix is possible.

An interchange of the rows/columns of an atom set creates a situation where new configurations are possible. We also have the possibility of the same atom set arranged in different structures. Given an atom set for a given RMMN we can easily find the various possible structures by using a simple enumeration technique. Actually these structures cause the various subcategories within a given R-category.

Structure Generation for the R224 category is illustrated below:

R224: We have 4 atoms, A, B, C and D, all monovalent. We will list their bonds in linear order such that 1 represents a single bond between the 2 neighboring atoms, 2 represents a double bond, 0 for no bond and so on. The valency of an atom is thus the simple addition of the bonds adjacent to it

For R224, since we have 4 atoms,

i) No two adjacent entries can have a sum greater than 1 since all atoms are monovalent

ii) The number of such bonds can at best be 3 i.e. N-1

iii) The extreme positions have to be nonzero

iv) No two adjacent entries can be zero.

For a given ordering of the atoms, the possibilities for bond positions P_i are:

A		B		C		D
	P1		P2		P3	
	1		0		1	

This is the only structure possibility that meets the conditions set above. This translates to

A ———— B   C ———— D

This is the only structure possible.

Can we obtain other configurations by reordering the various atoms involved? The problem is one of selecting distinct 2-tuples from a set of 4 elements which lead to 6 ways of selecting 2 distinct elements from 4.

Since by selecting 2 elements we automatically select the remaining 2 elements, the number of distinct sets is 3. The 3 distinct configurations are

(1) A - B   C - D
(2) A - C   B - D
(3) C - B   A - D

The bonds in each of these sets are distinct from the other two.

We can now safely say that R224 exists. In the transformation

$$A - B + C - D \rightarrow A - D + B - C$$

the R-matrix is given by:

	A	B	C	D
A		-1		+1
B	-1		+1	
C		+1		-1
D	+1		-1	

If we enterchange columns and rows C with D but not the labels we will get another product set and hence another R-matrix. Both of these R-matrices, however, belong to the same R-category R224.

There can be no cases where N can be greater than 4 in R22N because if we had 5 atoms, one would be redundant and would be deleted to yield a matrix of order 4.

Similar considerations lead to the generation of the following additional R-categories: R335, R336, R444, R4451, R4452, R4461, R4462, and R4463. Typical reaction patterns are:

R335: $I - J + X - Y - Z \rightarrow I - Y - J + X - Z$
R336: $I - J + K - X + Y - Z \rightarrow I - K + J - Y + X - Z$
R444: $I = J + X = Y \rightarrow I = Y + J = X$
R4451: $I = J + X - Y - Z \rightarrow Y = J + X - I - Z$
R4452: $I = J + X - Y - Z \rightarrow X - I - Y - J - Z$

R4461: I - J - K + X - Y - Z → I - Y - K + X - J - Z
R4462: I = J + K - X + Y - Z → K - I - X + Y - J - Z
R4463: I - J - K + X - Y - Z → X - J - Y - I + K - Z

Example

A typical result of the current program involves the oxygenation of propylene at 400K. R224, R335 and R336 are not applicable. Using R444 the product set is formaldehyde and acetaldehyde with a reaction free energy of -74 Kcal. Using R4451, there are 7 product sets of which two have the most negative free energy: vinyl alcohol and formaldehyde (-199 Kcal); acrolein and water (-85 Kcal). This completes the first stage reaction steps.

At the second stage, the oxidation of vinyl alcohol yields formaldehyde and formic acid (48. Kcal) using R444 and five product sets using R4451 of which only ketene and hydrogen peroxide (118 Kcal) maintains an overall reaction free energy that is negative. The oxidation of acrolein yields 9 product sets, only two of which are of interest: vinyl alcohol and carbon dioxide (-237 Kcal); ketene and formic acid (-83 Kcal).

The thermodynamically favored reaction paths for one- or two-stage oxidation of propene are:

1. propene $\xrightarrow{O_2}$ vinyl alcohol + formaldehyde (-199Kcal)

2. propene $\xrightarrow{O_2}$ acrolein + water $\xrightarrow{O_2}$ vinyl alcohol + carbon dioxide (-322 Kcal)

3. propene $\xrightarrow{O_2}$ acrolein + water $\xrightarrow{O_2}$ ketene + formic acid (-168 Kcal)

Conclusion

In CHIRP, we have essentially used a forward breadth-first search. This is necessary if our goal is to look at the various product alternatives, any of which could be promoted (or inhibited) by a proper choice of reaction conditions. Today we can vary only the temperature to guage the effect on the free energy of reaction. We hope, however, that CHIRP when completed should be able to suggest other reaction conditions, including catalysts, for a process that converts some known starting materials to some desired products. Care must be taken to ensure that a proper balance is maintained between the empirical constraints inserted to make the system more accurate and the ability to search for a wide range of alternatives.

Acknowledgement

This work was supported in part by NSF Grant ENG 75-21544. The authors are grateful for the collaboration of Professor James F. Richardson at the University of Houston, Department of Chemical Engineering.

Literature Cited

1. Thakkar, A.J., Topics Curr. Chem., 1973, 39, 3.
2. Corey, E.J., Wipke, W.T., Science, 1969, 166, 178.
3. Corey, E.J., Quart Rev. (London), 1971, 25, 455.
4. Corey, E.J., Wipke, W.T. Cramer, R.D., Howe, W.J., J. Amer. Chem. Soc., 1972, 94, 421.
5. Corey, E.J., Wipke, W.T. et al, Ibid, 1972, 94, 431.
6. Corey, E.J., Cramer, R.D., Howe, W.J., Ibid, 1972, 94, 440.
7. Corey, E.J., Petersson, G.A., Ibid, 1972, 94, 460.
8. Corey, E.J., Howe, W.J., Pensak, D.A., Ibid, 1974, 96, 7724.
9. Corey, E.J., Howe, W.J., et al, Ibid, 1975, 97, 6116.
10. Corey, E.J., Jorgensen, W.L., Ibid, 1976, 98, 189.
11. Pensak, D.A., Corey, E.J., Compter-Assisted Organic Synthesis, ACS Symposium Series, 1977, 61, 1.
12. Galernter, H., Sridharan, N.S., Hart, A.J., Fowler, F.W., Shue, H.J., "A application of artificial intelligence to the problem of organic synthesis discovery", Technical Report, Depts. of Computer Science and Chemistry, SUNY at Stony Brook, August, 1971.
13. Galernter, H. et al, Topics Curr. Chem., 1973, 41, 113.
14. Hendrickson, J.B., J. Amer. Chem. Soc., 1971, 93, 6847.
15. Hendrickson, J.B., Ibid, 1971, 93, 6854.
16. Hendrickson, J.B., Ibid, 1975, 97. 5763.
17. Hendrickson, J.B., Ibid, 1975, 97, 5784.
18. Hendrickson, J.B., Topics Curr. Chem., 1976, 62, 49.
19. Wipke, W.T., Computer Representation and Manipulation of Chemical Information, ed. W. T. Wipke, et al, Wiley, 1974, p. 147-174.
20. Wipke, W.T., Braun, H., Smith, G., Choplin, F., Sieber, W., Computer Assisted Organic Synthesis, ACS Symposium Series, 1977, 61, p. 97.
21. Govind, R., Powers, G., Ibid, 1977, p. 81.
22. Ugi, I., Gillespie, P., Angew. Chem. Intern. Ed., 1971, 10, 914, 915.
23. Ugi, I., Gillespie, P., Gillespie, C., Trans. N.Y. Acad. Sci., 1972, 34, 416.
24. Dugundji, J., Ugi, I., Topics Curr. Chem., 1973, 39, 19.
25. Blair, J. Gasteiger, J., Gillespie, C., Gillespie, P., Ugi, I., Tetrahedron, 1974, 30, 1845.
26. - , Computer Representation and Manipulation of Chemical Information, Ed., W.T. Wipke et al, Wiley, NY, 1974.
27. Gasteiger, J., Gillespie, P., Marquarding, D., Ugi, I., Topics Curr. Chem., 1974, 48, 1.

28. Brandt, J., Friedrich, J., Gasteiger, J., Jochum., Schubert, W., Ugi, I., Computer Assisted Organic Synthesis, Ed. W.T. Wipke and W.J. Howe, ACS Symposium Series, 1977, 61, 33.
29. Barkley, E.W., Motard, R.L., The Chem. Eng. Jour., 1972, 3, 265.
30. Agnihotri, R.B., PhD Dissertation, 'Computer-aided Investigation of Reaction Path Synthesis', University of Houston, 1978.

RECEIVED November 5, 1979.

10

Design of Optimizing Control Structures for Chemical Plants

GEORGE STEPHANOPOULOS—Department of Chemical Engineering and Materials Science, University of Minnesota, Minneapolis, MN 55455

YAMAN ARKUN—Department of Chemical and Environmental Engineering, Rennselaer Polytechnic Institute, Troy, NY 12181

MANFRED MORARI—Department of Chemical Engineering, University of Wisconsin, Madison, WI 53706

Control objectives for a chemical process originate from certain regulation tasks (i.e. product quality control, material balance control, safety, environmental regulations, etc.) and economic objectives (i.e. optimizing the economic performance). Such a classification of control objectives automatically formulates the different design activities for the regulatory and optimizing control structures.

In the previous process control structure synthesis methods (1,2,3,4), the distinction between the different classes of control objectives and its impact on the design of the plant control structure have not been addressed. Articles representing the industrial views have recently indicated that the steady-state optimizing control constitutes the most fruitful control practice in the chemical process industry (5,6,7,8), whereas regulation of the chemical process units is accomplished by the practicing engineers with satisfactory degree of acceptance. Nevertheless, there has been no systematic approach towards the design and implementation of optimizing control structures. In the absence of any theoretical foundations and practical ramifications, the designer relies on his operating experience and intuition to select an optimizing control policy without exploring all the viable alternatives. In the present paper, we will lay down the theoretical foundations for the synthesis and design of steady-state optimizing control structures for chemical processes. The implementational problems will be also addressed to develop practical control strategies for the on-line application of the optimizing controllers.

Optimum design of chemical processes dictates that the optimal operating point of a well-designed plant lies at the intersection of operating constraints (8-15). Furthermore, current industrial practice indicates that the optimal operating point switches from the intersection of one set of active constraints to another as process disturbances change with time (8-18). Such a dynamic evolution of the process operation

0-8412-0549-3/80/47-124-207$05.00/0

constitutes the major thrust for the successful implementation of an optimizing control strategy.

Formulating the Optimizing Control Problem for a Single Unit

The optimization problem for the single unit considered as a self-standing plant is to determine the optimal operating point when the values of a set of external disturbances change.

The Steady State Optimization Problem. For a set of slowly varying external disturbances, we assume that the process is at pseudo-steady state. Then, the following static optimization problem can be formulated:

$$\min_{m} \ \Phi(x,m,d) \qquad \text{Objective function}$$

subject to

$$(P_1) \qquad \begin{array}{rll} h(x,m,d) &= 0 & \text{System's State equations} \\ g(x,m,d) &\leq b & \text{Process Design Constraints} \\ r(x,m,d) &= r_d & \text{Regulatory Control Tasks} \\ -d + d^* &= 0 & \text{Disturbance Specifications} \end{array}$$

where x is the vector of states, m is the vector of manipulated variables, and d is the vector of "slow" disturbances with major economic impact on the optimal process operation.

Selection of Process Controlled and Manipulated Variables. At the calculated optimum $X^* = (x^*, m^*, d^*)$ of the above problem (P_1), some of the inequality constraints will be active. The regulatory control objectives and the active design constraints (i.e. g_A) at the current optimum will constitute the class of primary controlled variables denoted by c^P, i.e.

$$c^P(x,m,d) = \begin{bmatrix} r\ (x,m,d) \\ g_A(x,m,d) \end{bmatrix} = \begin{bmatrix} c^P_{reg} \\ c^P_{opt} \end{bmatrix}$$

Some manipulated variables, m_D, will be selected by the algorithm in (19) to control c^P, which will partition the vector of the manipulated variables: $m^T = [m_D^T \vdots \tilde{m}^T]$; with $\dim(m_D) = \dim c^P$.

The available extra degrees of freedom at the optimum are:

$$\dim(\tilde{m}) = \dim(m) - \dim(c^P)$$

Let us now define;

$$\bar{x} = \begin{bmatrix} x \\ d \\ m_D \end{bmatrix} \quad \text{and} \quad f = \begin{bmatrix} h(\bar{x},\tilde{m}) \\ c^P(\bar{x},\tilde{m}) - c^{sp} \\ - d + d^* \end{bmatrix}$$

<u>The Lagrangian Formulation and the Kuhn-Tucker Conditions</u>. Formulate the Lagrangian function for the problem (P_1),

$$L(\bar{x},\tilde{m},\lambda) = \phi(\bar{x},\tilde{m}) - \lambda^T f(\bar{x},\tilde{m}) - \lambda_{IA}^T g_{IA}(\bar{x},\tilde{m})$$

where; λ is the vector of Lagrange multipliers for the equality constraints $f(\bar{x},\tilde{m}) = 0$,

λ_{IA} is the vector of Kuhn-Tucker multipliers for the inactive design constraints $g_{IA}(\bar{x},\tilde{m}) < 0$.

The Kuhn-Tucker conditions for the minimum are:

$$\nabla_{\tilde{m}} L = \frac{\partial \Phi}{\partial \tilde{m}} - \frac{\partial f^T}{\partial \tilde{m}} \lambda - \frac{\partial g_{IA}^T}{\partial \tilde{m}} \lambda_{IA} = 0 \qquad \nabla_{\bar{x}} L = \frac{\partial \Phi}{\partial \bar{x}} - \frac{\partial f^T}{\partial \bar{x}} \lambda - \frac{\partial g_{IA}^T}{\partial \bar{x}} \lambda_{IA} = 0$$

$$\begin{matrix} \lambda_A^T(g_A - b_A) = 0 \\ \lambda_{IA}^T(g_{IA} - b_{IA}) = 0 \end{matrix} \; ; \; \begin{bmatrix} \lambda_A \\ \lambda_{IA} \end{bmatrix} \leq 0 \; ; \; \begin{matrix} g(\bar{x},\tilde{m}) \leq b \\ r(\bar{x},\tilde{m}) - r_d = 0 \\ -d + d^* = 0 \\ h(\bar{x},\tilde{m}) = 0 \end{matrix}$$

Hence at the design optimum X^*: $g_A = b_A$, $g_{IA} < b_{IA}$, $\lambda_A < 0$, $\lambda_{IA} = 0$.

The vector of Kuhn-Tucker multipliers for the active inequality constraints is negative; Kuhn-Tucker multipliers for the inactive constraints are zero.

<u>Finding a Feasible Operating Point X^o</u>. Consider that $d = d' = d^* + \delta d$, i.e. slow disturbances with economic impacts deviate from their design values. We need to find a feasible operating point that satisfies the constraints of the original optimization problem (P_1).

<u>Modification of Constraint Sets at X^o</u>. At the calculated feasible point X^o, some of the inactive constraints (i.e. $\bar{g}_{IA}$) will become active. In order to preserve the determinancy features of the system of equations, some of the previously active constraints (i.e. $\tilde{c}^p_{opt}$) will become inactive. Starting at the current optimum X^* where c^p_{opt} is the set of active inequality constraints,

i) Release $\tilde{c}^p_{opt} \in c^p_{opt}$; ii) Keep $\bar{c}^p_{opt} \in c^p_{opt}$ tight

where $\bar{c}^p_{opt}$ is the set of constraints that remain active at X^o after d changes.

Partitioning of Initial Constraint Control Loops.

1) The controlled variables $\tilde{c}^p_{opt}$ that are released will identify the initial servo loops.

2) The controlled variables $\bar{c}^p_{opt}$ that are kept tight on their bounds will define the regulatory loops.

The feasible point will be attained by changing set-points of the servo loops while keeping $\bar{c}^p_{opt}$ and c^p_{reg} tight by the regulatory loops.

Control Structure Modification at X^o. We are no longer interested in controlling $\tilde{c}^p_{opt}$ which have moved into their feasible region at X^o. Instead, the new active constraints, $\bar{g}_{IA}$, become our critical variables to be controlled on their bounds. Restructuring c^p, alternative feasible control structures can be generated at X^o through the algorithm given in (19).

Selection of New Search Directions (NSD) at X^o. Evaluate $\lambda_{A,i}$ "Kuhn-Tucker Multipliers" $\lambda_{A,i} = \dfrac{\delta\Phi}{\delta c^{sp}_{i,opt}}$, $i = 1,2,\text{--}\dim(c^p_{opt})$

i) If $\lambda_{A,i} < 0$, keep $c^p_{i,opt}$ tight, ii) if $\lambda_{A,i} > 0$, release $c^p_{i,opt}$.

Repartitioning of Control Loops. $c^p_{opt} = \left[c^p_T : c^p_R\right]$ where;

c^p_T: Tight constraints which will identify new regulatory control loops

c^p_R: Released constraints which will identify the new servo loops

The Search Procedure Towards the New Optimum (20). After the new search directions are found, the initial optimization problem (P_1) for the new values of the disturbance (i.e., $d \neq d^*$) can now be casted into

$$\text{Min} \quad \Phi(c^{sp}_R,\ c^{sp}_T,\ c^{sp}_{reg},\ \tilde{m})$$

$$(P_2) \qquad \text{s.t.} \quad g_{IA}(c^{sp}_R,\ c^{sp}_T,\ c^{sp}_{reg},\ \tilde{m}) - b_{IA} \leq 0$$

The set points of the released constraints c^{sp}_R and $\tilde{m}$ (if available) are the only free variables to adjust at each optimization step. Whenever new constraints become active at a search step, constraint sets are modified to define the regulatory and the servo loops. If there remains no search direction at a search point due to encountering new constraints, Kuhn-Tucker multipliers are reevaluated at that point. If they are all negative, an apparent minimal is found and the search is terminated; otherwise, it is continued with the new search direction selected.

Formulating the Optimizing Control Problem for an Integrated Chemical Plant

The chemical plant will be first decomposed to its subsystems of unit operations with functional uniformity and common objectives in terms of economics, operation and control. Within each subsystem resulting from the decomposition of the overall plant, we employ the same general approach described in the previous section to; (a) classify the plant constraints at the current optimum, into active and inactive ones; (b) select the controlled variables for the decentralized subsystem controllers; (c) select the manipulated variables for each subsystem; (d) establish the initial search direction (ISD) to achieve feasibility and (e) to define the resulting new search directions (NSD) that will ultimately lead to the new optimum operation.

Since the plant is decomposed to its subsystems, the steady-state optimization problem is characterized by a multiechelon structure where the subsystem optimizing controllers communicate with a coordinator. For further details the reader is referred to the work by Arkun (20).

Implementational Strategies

The algorithms for the solution of the Kuhn-Tucker conditions for a single unit or an integrated system of processing units, point the direction for the development of alternative practical control strategies for the on-line implementation of the optimizing controllers.

The Sequencing Problem. Optimizing control theory and the search procedures as developed in previous sections, have delineated the sequencing strategies (i.e. sequence of set-point changes) to be followed. After we develop the search coordinates as defined by the released and tight constraints, we can directly identify an operational route between two operating points. Consequently, a sequencing tree can be constructed as shown in Figure 1. Each node F_{ij} represents an intermediate operating point with a set of constraints to be controlled. Moreover, each branch between two nodes uniquely describes the control action (i.e. regulatory and optimizing) taken by the current control structure. We will now summarize the procedure to screen among the alternative routes of the sequencing tree in order to find the most appropriate sequence of set-point changes. The screening strategy uses an imbedded branch and bound approach, at various levels of increased sophistication and complexity (20).

i) First-Level Screening. It deals primarily with the characteristics of the control loops at each node and their effect on the plant operation.

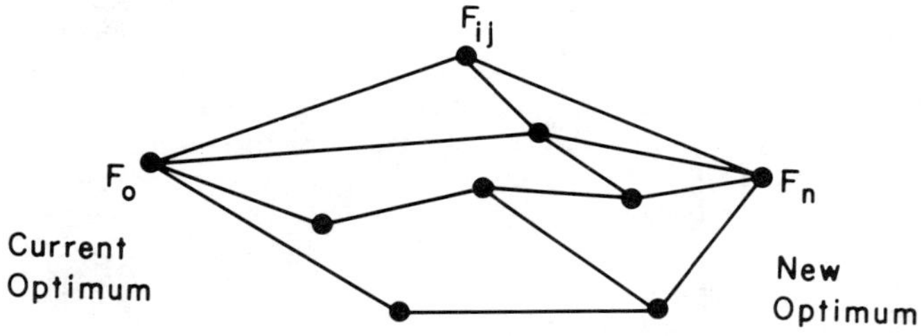

Figure 1. Sequencing tree

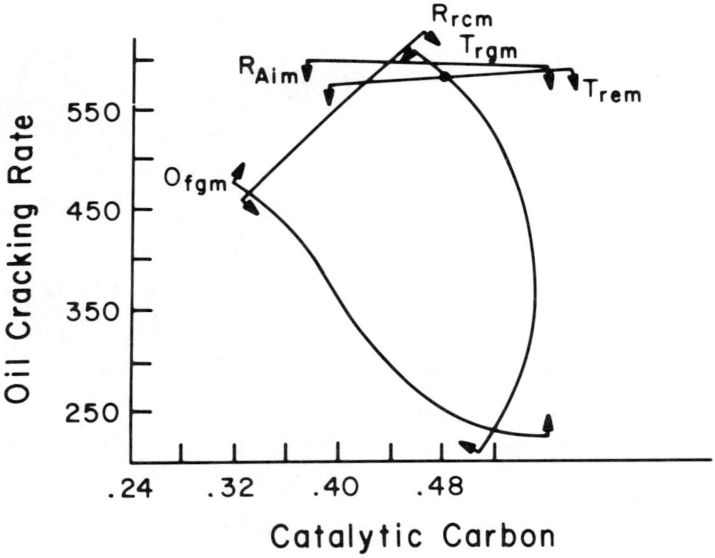

Figure 2. Feasible region for FCC when $F_{cf} = 0.1$

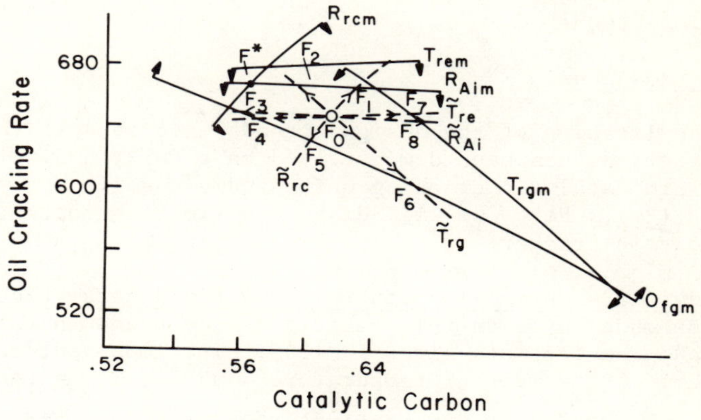

Figure 3. Feasible region for FCC when $F_{cf} = 0.01$

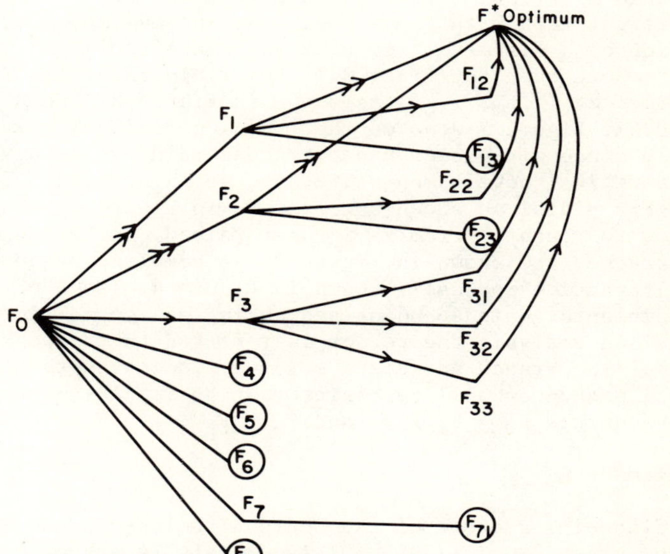

Figure 4. Sequencing tree of alternative routes

ii) Second-Level Screening. It examines the features of a completed sequence of operations in terms of the number of set-point changes and the number of control structure modifications required.

Numerical Examples

The application of the design methods developed in the previous sections has been demonstrated on a CSTR, a distillation column, fluid catalytic cracking units and a gasoline polymerization plant (20). Here, we will discuss optimizing control of the fluid catalytic cracker.

Steady State Optimizing Control of a Fluid Catalytic Cracker. The process model used in this example can be found in (21) while the design parameters are given in (20). The important constraints are T_{re} = reactor temperature $\leq$ 930°F, T_{rg} = regenerator temperature $\leq$ 1200°F, R_{rc} = catalyst recirculation rate $\leq$ 60 tons/minute, O_{fg} = oxygen in flue gas $\leq$ 0.2 mol%, R_{Ai} = Airblower capacity $\leq$ 400 K lb/hr.

Figure 2 denotes these constraints and the location of the optimum, when F_{cf} (coke formation factor of the feed) = 0.1. For a different feed with F_{cf} = 0.01, the optimum moves to the intersection of a different set of constraints (point F^* in Figure 3). Note that for F_{cf} = 0.1, the fluid catalytic cracker is controlled at (T_{rgm}, T_{rem}) as shown in Figure 2. When F_{cf} drops to 0.01, Figure 3 denotes that this control policy will not be feasible since air blower capacity constraint will be violated. As the theoretical developments dictate, we first find a feasible point for F_{cf} = 0.01 as shown by point F_o in Figure 3. The question of how to reach F^* from F_o is next raised. Alternative initial branches are shown in Figure 3. A complete sequencing tree of alternative routes is shown in Figure 4. Each of these routes represents an independent sequencing of control actions. For a detailed analysis the reader is referred to the work of Arkun (20). The branch and bound strategy indicates that only two control sequences will be of interest to retain and examine further: Sequence $F_o \rightarrow F_1 \rightarrow F^*$ and $F_o \rightarrow F_2 \rightarrow F^*$ (Figure 4).

Acknowledgement

The financial support of the National Science Foundation through the grant ENG 75-11165-A01 is gratefully acknowledged.

Literature Cited

1. Buckley, P.S. "Techniques of Process Control", J. Wiley & Sons Inc., New York, N.Y., 1964.

2. Govind, R.; Powers, G.J. Paper presented at the 82nd National AIChE Meeting, Atlantic City, N.J., 1976.

3. Douglas, J.M. Paper presented at the 70th Annual AIChE Meeting, New York, N.Y., 1977.

4. Umeda, T.; Kuriyama, T., Proceedings of IFAC Congress, Helsinki, Finland, 1978.

5. Latour, P.R. AIChE Sympsoium Series, 1976, 72, 159.

6. Ellingsen, W.R. AIChE Symposium Series, 1976, 72, 159.

7. Barkelew, C.H. AIChE Symposium Series, 1976, 72, 159.

8. Lee, W.; Weekman, WV., Jr. AIChE J., 1976, 22.

9. Kaiser, V.A.; Mahoney, J.D., ISA, 1966, 3.

10. Maarleveld, A.; Rijnsdorp, J.E., Automatica, 1970, 6, 51.

11. Rijnsdorp, J.E., Chem. Engr. Progr., 1967, 63 (7), 97.

12. Baxley, R.A., Instrumentation Technology, 1969, 57.

13. Kuehn, D.R.; Davidson, H., Chem. Engr. Progr., 1961, 57 (6), 45.

14. Shah, M.J.; Stillman, R.E., Ind. Eng. Chem., 1970, 62 (12), 59.

15. Ishida, C., Proceedings of Japan-USA Joint Seminar, Kyoto, Japan, 1975.

16. Davis, T.A.; Griffin, D.E.; Webb, P.V., Chem. Engr. Progr., 1974, 70 (11), 53.

17. Duyfjes, G.; Van Der Grinten, P.M.E.M., Automatica, 1973, 9, 537.

18. Webb, P.V.; Lutter, B.E.; Hair, R.L. Paper presented at the 70th Annual AIChE Mtg., New York, N.Y., 1970.

19. Morari, M., "Studies in the Synthesis of Control Structures of Chemical Processes", Ph.D. Thesis, Univ. Minnesota, Minneapolis, MN, 1977.

20. Arkun, Y. "Design of Steady State Optimizing Control Structures for Chemical Processes", Ph.D. Thesis, Univ. Minnesota, Minneapolis, MN, 1979.

21. Kurihara, H. "Optimal Control of Fluid Catalytic Cracking Processes", Sc.D. Thesis, M.I.T., Cambridge, MA, 1967.

RECEIVED November 5, 1979.

11

An Iterative Approach for the Solution of a Class of Stiff ODE Models of Reacting Polydispersed Particles

FERHAN KAYIHAN

Department of Chemical Engineering, Oregon State University, Corvallis, OR 97331

The mathematical models of the reacting polydispersed particles usually have stiff ordinary differential equations. Stiffness arises from the effect of particle sizes on the thermal transients of the particles and from the strong temperature dependence of the reactions like combustion and devolatilization. The computation time for the numerical solution using commercially available stiff ODE solvers may take excessive time for some systems. A model that uses K discrete size cuts and N gas-solid reactions will have K(N + 1) differential equations. As an alternative to the numerical solution of these equations an iterative finite difference method was developed and tested on the pyrolysis model of polydispersed coal particles in a transport reactor. The resulting 160 differential equations were solved in less than 30 seconds on a CDC Cyber 73. This is compared to more than 10 hours on the same machine using a commercially available stiff solver which is based on Gear's method.

In the following sections some background information on stiff ordinary differential equations will be given and the general finite difference approximations for particle temperatures will be derived. Later, the technique will be applied to coal pyrolysis in a transport reactor where the difference equations for reaction kinetics will be discussed and the calculation results will be compared with those obtained by the previously established techniques.

BACKGROUND

Mathematical models that contain ordinary differential equations face an inherent computational difficulty associated with the stiffness of the equations. Stiffness of ordinary differential equations depends on the relative magnitudes of the response modes or the characteristic time constants of the system being modeled. In solid fuel conversion problems where particles of varying sizes are considered the differential equations for the thermal transients of the particles are usually stiff. Estab-

0-8412-0549-3/80/47-124-217$05.00/0

lished numerical techniques, like Gear's method, are available for the solution of these problems. However, this becomes very costly for systems with large number of coupled differential equations. The present work illustrates an alternative approach through finite difference approximations which is shown to be effective for problems dealing with the reaction of polydispersed particles. After a brief description of the stiffness conditions in ordinary differential equations the approximate solution technique will be presented in general terms. Then, a specific application to coal devolatilization will be discussed in detail.

Consider a nonlinear model of n differential equations

$$\frac{dy}{dt} = f(y,t) \ , \ y(t_o) = y_o \tag{1}$$

where $y = (y_1\ y_2\ \ldots\ y_n)^T$. The linearization of this model around (y_o,t_o) gives

$$\frac{dy}{dt} \simeq f(y_o,t_o) + f_y(y_o,t_o)(y-y_o) + f_t(y_o,t_o)(t-t_o) \tag{2}$$

or

$$\frac{dy}{dt} \simeq J_o y + g(t) \tag{3}$$

where J_o is the Jacobian at (y_o,t_o) and $g(t)$ is only a function of the independent variable t. The local solution is

$$y(t) = e^{J_o(t-t_o)} y_o + \int_{t_o}^{t} e^{J_o(t-\tau)} g(\tau)d\tau \tag{4}$$

For a moment, assume that Eq. (3) is correct for all t where Eq. (4) gives the true solution. Then the response modes of the system are given by the eigenvalues (λ_i) of the Jacobian J_o. For a stable system, the eigenvalues have the property that

$$Re(\lambda_i) < 0 \qquad i=1,\ldots,n \tag{5}$$

For the special case of distinct eigenvalues the solution is written in terms of

$$e^{-t/\tau_1},\ e^{-t/\tau_2},\ \ldots,\ e^{-t/\tau_n}$$

where $\tau_i = -1/\lambda_i$ for i=1,...,n are the system time constants.

In a transient problem we are interested in the solution for $t_o < t < t_{max}$ where the solution time $(t_{max}-t_o)$ is a few times larger than the maximum time constant of the system, i.e.,

$$(t_{max} - t_o) \cong 3\tau_{max}$$

where $\tau_{max} = \mathrm{Max}\ \{\tau_1,\ \tau_2,\ \ldots,\ \tau_n\}$. However, the numerical solution techniques require step sizes which are comparable to the smallest time constant of the system to maintain stability and accuracy. Therefore, the difficulty of the numerical solution depends on the ratio $(t_{max}-t_o)/\tau_{min}$ or τ_{max}/τ_{min}. The latter one is usually a good test of the stiffness. An ordinary differential equation system is considered to be stiff if τ_{max}/τ_{min} is greater than 100.

Returning back to the original nonlinear problem we observe that the Jacobian changes with time. Therefore, stiffness is actually a local property. A problem is considered to be stiff even when the ratio of the time contants is moderate in most of the time range but excessively large in a short range.

REACTION MODEL FOR POLYDISPERSED PARTICLES

In the modeling of solid fuel conversion reactors differential equations arise for the description of particle temperatures and gas-solid reactions among others. These equations are coupled and they must be solved simultaneously. Because of the usually wide range of particle sizes the time constants for thermal transients of solids differ considerably. This causes stiffness in the differential equation model. Depending on the type of the gas-solid reactions stiffness may also be introduced by the variation of reaction rates with individual reaction type and with temperature.

For practical purposes, particle size distributions are incorporated into reactor modeling through discrete-cut sizes. Consider K different particle sizes. Then, there must be K differential equations to describe the different temperature-time histories. Of course, here it is assumed that particles are small enough not to have any radial temperature profiles. If there are N reactions that occur on or in a particle which are necessarily dependent on particle temperature, then there must be N differential equations per particle describing reaction kinetics. Thus, the total number of equations for the model is K + KN or K(N + 1) plus other algebraic or differential equations as necessary. For a system like this K(N + 1) also becomes the minimum size of the Jacobian. The established stiff differential equation solving methods (1,2,3) use implicit techniques which require the solution of algebraic equations constructed from the Jacobian. This becomes the main time consuming step in the numerical solution. The better numerical techniques update the information gained by the Jacobian only when necessary. Depending on the local stiffness of the problem the frequency of this algebraic equation solution may vary from every step to every fifth to tenth. But still, the computation time may become excessively large for systems where K(N + 1) is on the order of 100.

As an alternative to the simultaneous solution of stiff differential equations through an implicit technique a method is described here which approximates the solution by successive computations of the corresponding finite difference equations. The successive nature of this method essentially decouples the K(N + 1) equations so that they can be solved on a sequential basis. Furthermore, the finite difference approach provides fast and accurate computations.

THE FINITE DIFFERENCE APPROXIMATION

To develop the general technique consider a reacting solid particle in a medium at temperature T_F. For the simple case of constant physical properties the thermal transient of the particle is given by

$$m_j c \frac{dT_j}{dt} = a_{sj} h_j (T_F - T_j) + \frac{dm_j}{dt} [(C - c)T_j + \Delta h_R] \tag{6}$$

where m_j is the mass of the jth size cut. Here h_j incorporates both the convective and the radiative heat transfer resistances. Let us define

$$a_j = \frac{m_j c}{a_{sj} h_j} \text{ and } b_j = \frac{(C-c)T_j + \Delta h_R}{a_{sj} h_j}$$

Then

$$a_j \frac{dT_j}{dt} + T_j = T_F + b_j \frac{dm_j}{dt} \tag{7}$$

Using the integrating factor $\exp(t/a_j)$ gives

$$d(e^{t/a_j} T_j) = \frac{e^{t/a_j}}{a_j} (T_F + b_j \frac{dm_j}{dt}) dt \tag{8}$$

Over a time increment $\Delta t_n = t_n - t_{n-1}$ Eq. (8) can be approximately integrated to give

$$T_j(t_n) \simeq T_j(t_{n-1}) e^{-\Delta t_n/\tilde{a}_j} + \tilde{T}_F(t_n)[1 - e^{-\Delta t_n/\tilde{a}_j}] + \frac{\tilde{b}_j}{\tilde{a}_j} [m_j(t_n)(1 - \frac{\Delta t_n}{2\tilde{a}_j}) - m_j(t_{n-1})(1 + \frac{\Delta t_n}{2\tilde{a}_j}) e^{-\Delta t_n/\tilde{a}_j}] \tag{9}$$

where the quantities

$$\tilde{a}_j(t_n) = [a_j(t_n) + a_j(t_{n-1})]/2$$

$$\tilde{b}_j(t_n) = [b_j(t_n) + b_j(t_{n-1})]/2$$

$$\tilde{T}_F(t_n) = [T_F(t_n) + T_F(t_{n-1})]/2$$

are assumed to be constant for the interval Δt_n.

The corresponding expression for $m_j(t_n)$ can be derived according to the particular reaction. In general terms the rate of mass loss, $-dm_j/dt$, is a function of particle radius r_j, kinetic rate $\mathcal{R}(T_j, T_F, P)$, and gas phase compositions y_G's. This can be written as

$$- \frac{dm_j}{dt} = f(r_j)\ R\ (T_j,\ T_F,\ P)\ g\ (y_G\text{'s}) \tag{10}$$

Eq. (10) can be approximately integrated through analytical techniques similar to Eq. (7) over the time increment Δt_n either by assuming a constant average particle temperature between $T_j(t_n)$ and $T_j(t_{n-1})$ or by using a linear temperature profile in this range. For combustion reactions Levenspiel (4) gives the constant temperature integration for reaction and gas or ash diffusion controlled cases. The integration of the pyrolysis kinetics will be demonstrated in the following section.

Eq. (9) and the integrated version of Eq. (10) constitute the finite difference approximation for the model. The solution is generated through successive iterations on the complete temperature profiles and other variables. A convenient initial guess can be generated by using $dm_j/dt = 0$ which corresponds to the simple heating of the particles. The details of the computation procedure is described in the following example on coal pyrolysis.

EXAMPLE: PYROLYSIS OF POLYDISPERSED COAL PARTICLES

At low pressures the kinetics of thermal decomposition of pulverized coal particles can be represented by an infinitely large number of parallel first order reactions as shown by Howard, et al. (5,6). For those cases where data on species production rates are not available the devolatilization kinetics can be written in terms of one expression which combines the kinetics of all possible reactions by using a normally distributed activation energy function. However, if data are available on species kinetics, then devolatilization can be characterized by a finite number of first order reactions. For example, Suuberg (7) gives the kinetic parameters for 15 parallel reactions for eight identifiable volatile species in the pyrolysis of Montana lignite. This particular set of data was used in this study to examine the computational problems that arise in modeling of pulverized coal pyrolysis.

For reactor design calculations it is necessary to know the total devolatilization rate as well as the species production rates. Therefore, one needs to include in the reactor model all the reaction rates that are available for the devolatilization of the particular coal. Kayihan and Reklaitis (8) show that the kinetic data provided by Howard, et al. (5,6) can be easily incorporated in the design calculations for fluidized beds where the coal residence times are long. However, if the residence time of pulverized coal in the reactor is short as it is in entrained bed reactors, then the handling of ordinary differential equations arising from the reaction kinetics require excessive machine computation time. This is due to the stiffness of the differential equations. It is found that the model equations cannot be solved by non-stiff numerical techniques and that the Gear's method requires excessively long computation times due to the large number of equations in the model. However, using the method described above, the execution time is significantly reduced while maintaining the same accuracy as the Gear's method.

The Model. The physical system considered is an entrained bed reactor. Pulverized coal, carried by a gas stream, mixes with hot gas at the reactor entrance. As coal particles are carried their temperatures increase and devolatilization takes place. For practical purposes the particle size distribution was approximated by 10 discrete cuts. Since the devolatilization kinetics are highly temperature dependent and the temperature transient of a particle is affected by its size and mass separate account must be taken of each of the 15 reactions in each of the 10 different size particles. Without any detail on the derivation of the model, the equations can be summarized as:

Mixed gas flowrate

$$F = F_o + G_o(1 - e^{-t/\tau_G}) + m_o \sum_i \sum_j p_j v_{ij} \tag{11}$$

Mass flowrate of particles with size j

$$m_j = m_o p_j (1 - \sum_i v_{ij}) \qquad j = 1, \ldots, K \tag{12}$$

Temperature of mixed gas

$$T_F = [G_o(1 - e^{-t/\tau_G}) C T_{G_o} + (F_o C + m_o c) T_{F_o} - C \sum_j m_j T_j - (m_o \sum_i \sum_j p_j v_{ij}) \Delta h_R]/FC \tag{13}$$

Particle temperature

$$\frac{dT_j}{dt} = \{T_F - T_j + \frac{\rho r_j^2}{3k} [(c-C)T_j - \Delta h_R] \times \sum_i k_{oi} (v_i^* - v_{ij}) e^{-E_i/RT_j}\} / \frac{m_j}{m_o p_j} \frac{c \rho r_j^2}{3k}$$

$$j = 1, \ldots, K \tag{14}$$

Devolatilization kinetics

$$\frac{dv_{ij}}{dt} = k_{oi}(v_i^* - v_{ij}) e^{-E_i/RT_j}$$

$$i = 1, \ldots, N$$

$$j = 1, \ldots, K \tag{15}$$

As shown before the total number of differential equations is K(N + 1). In this study, K = 10 and N = 15. The choice of 10 is considered to be a reasonable number for the characterization of particle size fractions for design calculations. Therefore, the number of differential equations, 160, was not artificially reduced by taking only a few discrete cuts. However, the possibility of representing the particle behavior with one average size was explored. Different averages like mean surface, surface mean, volume mean, etc., were tried; but, none proved to be applicable for this problem where heat transfer and devolatilization occur simultaneously.

The stiffness of the system arises partly from the wide range of reaction rates and from the strong temperature dependence of the kinetics given by Eq. (15). The data used in this paper are those obtained by Suuberg (7). As a relative measure of the stiffness of the system the variable time constants of Eq. (15) were calculated at various temperatures between 500°K and 1000°K. The values for

$$(k_{oi} e^{-E_i/RT})^{-1} \qquad i = 1, \ldots, 15$$

are given in Table I. There is significant variation between the smallest and largest value evaluated at each temperature where the ratios of the two (the stiffness ratios) range from 10^7 to 10^{16}. But with a wide range of particle sizes it is possible to have a large particle at 500°K while a small particle being already at 1000°K. Therefore, the actual ratio of largest to smallest time constant can be on the order to $2.319 \times 10^{20}/.9831 \times 10^{-4} \simeq 10^{24}$ which indicates that the eigenvalues of the Jacobian for the system can vary as much as by a factor of 10^{24}.

Table I. The Wide Range of Time Constants of the Pyrolysis Kinetics (Sec):

RXN	500 K	600 K	700 K	800 K	900 K	1000 K
1	.2319E+21	.8665E+14	.2222E+10	.8005E+06	1680E+04	.1210E+02
2	.2647E+09	.5038E+05	.1109E+03	.1126E+01	.3173E-01	.1825E-02
3	.1054E+08	.6440E+04	.3261E+02	.6190E+00	.2835E-01	.2406E-02
4	.2633E+14	.1299E+10	.1090E+07	.5374E+04	.8627E+02	.3164E+01
5	.3916E+16	.2321E+12	.2220E+09	.1207E+07	.2091E+05	.8155E+03
6	.3904E+05	.9359E+02	.1258E+01	.4965E-01	.4019E-02	.5379E-03
7	.1641E+15	.3639E+10	.1724E+07	.5536E+04	.6367E+02	.1789E+01
8	.3165E+12	.2886E+09	.1945E+07	.4573E+05	.2475E+04	.2400E+03
9	.1583E+09	.2915E+05	.6264E+02	.6252E+00	.1737E-01	.9881E-03
10	.2951E+16	.2797E+11	.7221E+07	.1471E+05	.1187E+03	.2512E+01
11	.1719E+13	.6623E+07	.8989E+03	.1130E+01	.6271E-02	.9831E-04
12	.2407E+14	.1022E+10	.7704E+06	.3504E+04	.5284E+02	.1844E+01
13	.1637E+15	.1380E+10	.3279E+06	.6274E+03	.4823E+01	.9818E-01
14	.2305E+05	.4524E+02	.5271E+00	.1869E-01	.1392E-02	.1743E-03
15	.2526E+16	.8953E+10	.1145E+07	.1377E+04	.7379E+01	.1125E+00
			Stiffness Ratios:			
	.1006E+17	.1915E+13	.4215E+10	.6458E+08	.1502E+08	.8294E+07

Another reason for the stiffness of the system is the wide range of thermal transients as indicated before. The time constant for the thermal transient depends on the square of the particle size. Therefore for a system where the ratio of largest to smallest size is 100 the stiffness ratio becomes 10^4.

For computations the physical properties are assumed to be constant. Particle sizes are discretized by 10 equal-weight cuts from the Rosin-Rammler distribution

$$w(x) = \exp\left(-0.0004x^{1.5}\right) \tag{16}$$

The physical properties and the reactor operating conditions are given in Table II.

Table II. Physical properties and parameters used for calculations

m_o	(gm/sec)	:	1.0
F_o	(gm/sec)	:	0.5
G_o	(gm/sec)	:	8.5
T_{Fo}	(°K)	:	400
T_{Go}	(°K)	:	1200
t_G	(sec)	:	0.05
c	(cal/gm°K)	:	0.5
C	(cal/gm°K)	:	0.7
k	(cal/gm sec °K)	:	1.2×10^{-4}
Δh_R	(cal/gm)	:	0
ρ	(gm/cm^3)	:	1.4

Solution by a Stiff Solver. Trial solutions with various single size particles representative of the average size determined from Eq. (16) indicated that for 80 to 90% completion of devolatilization and under typical operating conditions the coal residence time in the reactor must be about one second. To check different computation schemes while considering all particle sizes only a fraction of this estimated residence time was used.

A commercial stiff ordinary differential equation solver subroutine, DVOGER, is available in the IMSL Library (3). This subroutine uses Gear's method for the solution of stiff ODE's with analytic or numerical Jacobians. The pyrolysis model was solved using DVOGER and the analytical Jacobians of Eqs. (14) and (15). For a residence time of 0.0511 in dimensionless time, defined as $t/\bar{t}$ where

$$\bar{t} = \frac{c\rho r_{AV}^2}{3k} \tag{17}$$

and with relative error criterion set at 0.001 the computations took 8.59 minutes on a CDC Cyber 73 computer. With physical properties used in this work and with the particle size distribution given by Eq. (16) the characteristic time constant $\bar{t}$ is 0.1839 seconds. This means that the residence time used was about 1/100 of the expected coal residence time in an actual reactor. Considering the stiffness of the kinetic relations over a wide temperature range and the different temperature time histories of different particle sizes it is reasonable to expect that the computation times for $t/\bar{t} \simeq 5$ or for $t \simeq 1.0$ sec will be proportionately longer.

The main reason for slow computations is the dimensions of the sparse 160 x 160 Jacobian. Although other routines are available through the work of Hindmarsh (1,2) to handle banded Jacobians or Jacobians with certain structures there are no commercial routines that could be used for the particular Jacobian that arises in the modeling of the low pressure pyrolysis of polydispersed coal particles.

The Finite Difference Approximations and the Iterative Technique. The reaction kinetics are dependent on the rest of the equations through particle temperatures. If we have an initial guess on the coal particle temperature history during its total residence in the reactor then the relations given by Eq. (15) are decoupled from the rest and can be solved separately. Consequently, the volatilization estimates, $v_{ij}(t)$'s, can be used to update particle temperatures with Eqs. (11) through (14).

First, consider the approximate analytical integration of Eq. (15). At any time t volatilization in particle size j due to reaction i is given by

$$v_{ij}(t) = v_{ij}^{*}\left[1 - \exp\left(-k_{oi} \int_{o}^{t} e^{-E_i/RT_j} dt\right)\right] \tag{18}$$

Now consider the time increments of Δt and the discrete points $t_o = 0$, $t_1 = \Delta t$, $t_2 = 2\Delta t$, etc. Then, Eq. (18) can be written as

$$v_{ij}(t) = v_{ij}^{*}\left[1 - \exp\left(-k_{oi} \sum_{m=1}^{n} \int_{t_{m-1}}^{t_m} e^{-E_i/RT_j} dt\right)\right] \tag{19}$$

For convenience define

$$\Phi_{ij}(t_m) = \int_{t_{m-1}}^{t_m} e^{-E_i/RT_j} dt \tag{20}$$

It $T_j(t_{m-1}) = T_j(t_m)$, then the integral is simply

$$\Phi_{ij}(t_m) = \Delta t\, e^{-E_i/RT_j(t_m)} \tag{21}$$

However, for any other case where the particle is not isothermal during a time increment an analytical solution is not available. But, if it is assumed that Δt is small enough so that the temperature profiles within each time increment are straight lines, then the following approximate relation for $\Phi_{ij}(t_m)$ can be obtained

$$\Phi_{ij}(t_m) \simeq \frac{R}{E_i s_j(t_m)} [T_j^2(t_m)(1 - 2 \frac{RT_j(t_m)}{E_i}) e^{-E_i/RT_j(t_m)} - T_j^2(t_{m-1})(1 - 2 \frac{RT_j(t_{m-1})}{E_i}) e^{-E_i/RT_j(t_{m-1})}] \quad (22)$$

where

$$s_j(t_m) = \frac{T_j(t_m) - T_j(t_{m-1})}{\Delta t}$$

and $T_j(t_m) \neq T_j(t_{m-1})$.

Now Eq. (19) becomes

$$v_{ij}(t_n) \simeq v_i^* [1 - \exp(-k_{oi} \sum_{m=1}^{n} \Phi_{ij}(t_m))] \quad (23)$$

Assume that all $T_j(t_n)$'s and $v_{ij}(t_n)$'s are available from the previous iteration or from the initial guess. Eq. (14) can be integrated between t_{n-1} and t_n if the following coefficients are assumed to be constant within the time interval as defined below:

$$a_j(t_n) = \frac{[m_j(t_{n-1}) + m_j(t_n)]/2}{m_o p_j} \frac{c\rho r_j^2}{3k} \quad (24)$$

$$b_j(t_n) = \frac{\rho r_j^2}{3k} [(c - C) \frac{T_j(t_{n-1}) + T_j(t_n)}{2} - \Delta h_R] \quad (25)$$

$$\tilde{T}_F(t_n) = \frac{T_F(t_{n-1}) + T_F(t_n)}{2} \quad (26)$$

Eq. (14) gives

$$T_j(t_n) \simeq T_j(t_{n-1}) e^{-\Delta t/a_j(t_n)} + \tilde{T}_F(t_n)(1 - e^{-\Delta t/a_j(t_n)}) + \frac{b_j(t_n)}{a_j(t_n)} [\sum_i v_{ij}(t_n)(1 - \frac{\Delta t}{2a_j(t_n)}) - \sum_i v_{ij}(t_{n-1})(1 + \frac{\Delta t}{2a_j(t_n)}) e^{-\Delta t/a_j(t_n)}] \quad (27)$$

With these finite difference approximations the solution cannot be achieved in one step but the complete temperature and devolatilization profiles must be calculated repetitively. As indicated in Eqs. (22), (23) and (27) the right hand sides have variables which must be evaluated at the same time mode as the left hand side. Instead of a simultaneous solution an iterative approach was taken where the right hand sides can be evaluated by the values obtained in the most recent iteration. The complete calculation procedure can be summarized as follows:

Step 1. Divide total time into NP equal increments, guess a temperature profile for each particle size, and set $v_{ij} = 0$.

A good initial guess is the temperature profile that would occur if there were no devolatilization reactions, but only heat transfer effects on particles.

Step 2. Evaluate Eqs. (11), (12) and (13) for all time increments.

Step 3. Evaluate Eq. (23) for all particle sizes, reactions and time increments.

Step 4. Update temperature profiles through Eq. (27).

Step 5. Stop if convergence is achieved; otherwise, go to Step 2.

This procedure had converged in 4 or 5 iterations to four significant figures for all cases tried in this study. The accuracy of the calculations depends on the time increment Δt because the finite difference approximations become more accurate as Δt gets smaller. A summary of some iteration results and a comparison between this technique and the numerical integration with Gear's method will be presented after the following discussion on the stability of the temperature equation.

Stability of the Temperature Equation. Consider Eq. (27) where $T_j(t_n)$ is dependent on $T_j(t_{n-1})$ and $T_j(t_n)$ computed by the previous iteration through the first term on the right and through $b_j(t_n)$. The latter can cause instability in the explicit iterative calculations for some choices of Δt. First, consider an idealized case where $c = C$ and $\Delta h_R = 0$. Then $b_j(t_n) = 0$ and Eq. (27) takes the form

$$T_j(t_n) \simeq T_j(t_{n-1})e^{-\Delta t/a_j(t_n)} + \tilde{T}_F(t_n)(1 - e^{-\Delta t/a_j(t_n)}) \qquad (28)$$

For this ideal case the iterations for $T_j(t_n)$ will be stable for all choices of Δt. Therefore, this approximation is a good start for exploratory reactor design calculations because of its simplicity and stability. Usually, however, the above assumptions will not hold.

To establish a safe Δt for stable calculations observe that $b_j(t_n)$ in Eq. (27) is on the order of

$$\frac{\rho r_j^2}{3k} [c - C)T_j(t_{n-1}) - \Delta h_R]$$

Substituting this into Eq. (27) gives

$$T_j(t_n) \simeq T_j(t_{n-1}) \{e^{-\Delta t/a_j(t_n)} + \frac{\rho r_j^2}{3k} \frac{c-C}{a_j(t_n)} [\sum_i v_{ij}(t_n)(1 - \frac{\Delta t}{2a_j(t_n)}) - \sum_i v_{ij}(t_{n-1})(1 + \frac{\Delta t}{2a_j(t_n)}) e^{-\Delta t/a_n(t_n)}]\} + \tilde{T}_F(t_n)(1 - e^{-\Delta t/a_j(t_n)}) - \frac{\rho r_j^2}{3k} \Delta h_R [\sum_i v_{ij}(t_n)(1- \frac{\Delta t}{2a_j(t_n)}) - \sum_i v_{ij}(t_{n-1})(1 + \frac{\Delta t}{2a_j(t_n)}) e^{-\Delta t/a_j(t_n)} \qquad (29)$$

A safe choice for Δt will be

$$\Delta t \leq 2a_j(t_n) \qquad (30)$$

This would be restricted by the smallest particle size, j = 1, and by the extent of devolatilization. Usually, the volatile matter of coal is not more than 50% by weight. Therefore, choose

$$\Delta t \simeq a_1(t_o) = \frac{c\rho r_1^2}{3k} \qquad (31)$$

With physical properties given in Table I, this becomes

$$\Delta t = 1944\ r_1^2$$

and for r_1 = 10 μm

$$\Delta t = 0.0019 \text{ sec}$$

or NP = 514 points for 1 second integration time.

While, for r_1 = 5 μm

$$\Delta t = 4.9 \times 10^{-4} \text{ sec}$$

and

NP = 2058 points.

Because it is necessary to adjust the time increment according to the smallest particle size the number of discrete points, NP, may be unnecessarily high. This would increase the computation time without significantly improving the accuracy. In this study, instead of adopting the above stability criterion, it was observed that the smaller particles will respond to temperature changes in their environment much faster than the large particles. Therefore, it was assumed that for an arbitrary Δt if $a_j(t_o)$, the initial temperature time constant of the particle, is smaller than Δt, then it will respond to changes in T_F instantaneously, i.e., $T_j(t_n) = T_F(t_n)$ if $a_j(t_o) \leq \Delta t$. This allows Δt to be adjusted until desired accuracy is achieved in the converged iterative results. The procedure described here has provided a stable and an accurate iterative technique without unnecessarily large number of time increments.

Comparison of the Computation Results. As indicated above, Gear's method was used to solve the model equations only for a fraction of the total residence time in the reactor which took 8.59 minutes of machine computation time. The same set of equations was solved by the approximate iterative technique for the same time interval in 5.8 seconds of computer time. As a comparison of the accuracy overall devolatilization $v_T = \Sigma\, \Sigma\, v_{ij}$ as predicted by the two techniques are plotted on a dimensionless scale in Figure 1. The definitions for the dimensionless quantities used are:

Dimensionless time = $t/\bar{t}$

Dimensionless devolatilization = $v_T/v_T{}^*$

where $\bar{t}$ is the time constant for the temperature transient of non-devolatilizing average size particle and $v_T{}^*$ is the mass fraction of total volatile matter in coal.

Typical reactor calculations would take about five times the average particle time constant $\bar{t}$. For these cases the approximate relations were used to predict some of the reactor conditions as a function of time. To give an idea of the dependence of the computation time and accuracy on the time increment, Δt, three different cases are compared in Table III.

These results show that the proposed technique provides a fast and reliable method for the solution of stiff ODE models of reacting polydispersed particles. Recently, Turton (9) applied this method successfully to the modeling of wood char combustion in a transport reactor.

CONCLUSIONS

Models for the reacting polydispersed particles contain stiff ordinary differential equations. The stiffness is due partly to the wide range of thermal time constants of the particles and partly to the high temperature dependence of reactions like combustion and devolatilization. As an alternative to the established solution techniques based on Gear's method an iterative approach is developed which uses the finite difference representations of the differential equations. The finite differences are obtained by

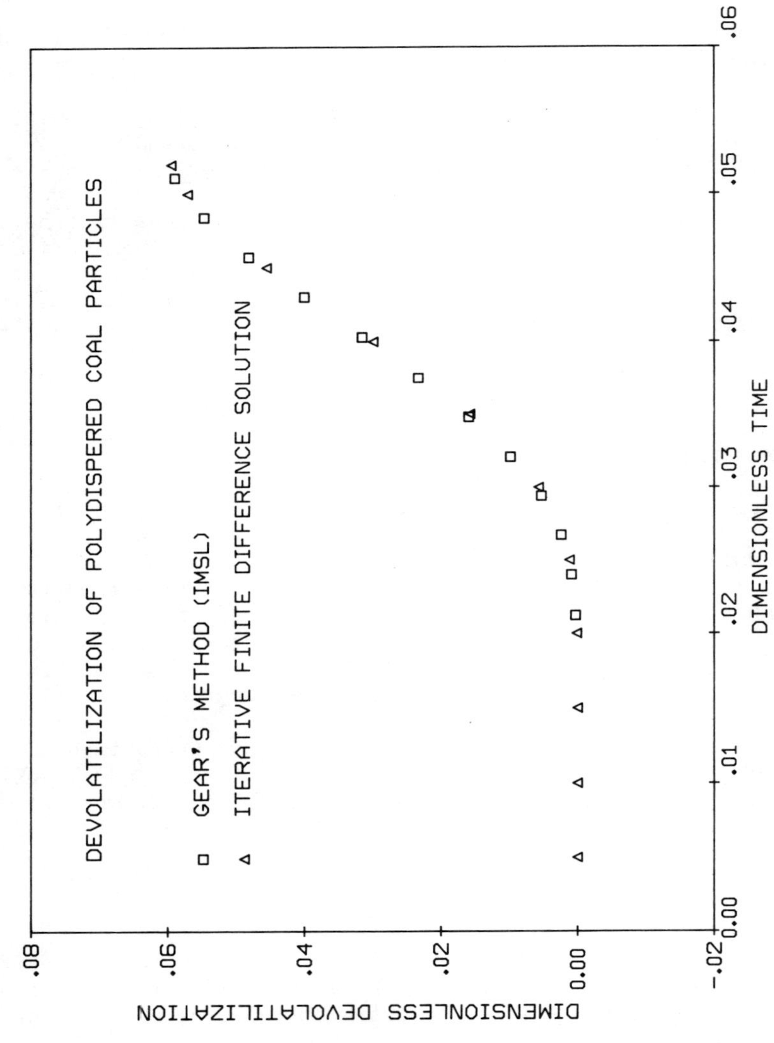

Figure 1. Comparison of the iterative finite difference solution of this work to Gear's method

Table III. Sample calculations for particle temperatures and for total devolatilization showing the effect of Δt on accuracy and execution time.

Number of Points = 50

Time	T4	T7	VT
0.0000	.3333	.3333	.0000
0.5000	.8113	.5563	.3173
1.0000	.9076	.7063	.4808
1.5000	.9105	.8044	.5832
2.0000	.9077	.8579	.6600
2.5000	.9056	.8834	.7096
3.0000	.9041	.8947	.7536
3.5000	.9032	.8994	.7794
4.0000	.9024	.9011	.8060
4.5000	.9018	.9016	.8369
5.0000	.9014	.9015	.8564

11.879 CP seconds execution time

Number of Points = 100

Time	T4	T7	VT
0.0000	.3333	.3333	.0000
0.5000	.8252	.5660	.3173
1.0000	.9085	.7123	.4816
1.5000	.9104	.8083	.5838
2.0000	.9075	.8598	.6605
2.5000	.9055	.8843	.7112
3.0000	.9041	.8951	.7541
3.5000	.9031	.8996	.7799
4.0000	.9024	.9012	.8069
4.5000	.9018	.9016	.8380
5.0000	.9014	.9015	.8566

23.787 CP seconds execution time

Number of Points = 150

Time	T4	T7	VT
0.0000	.3333	.3333	.0000
0.5000	.8292	.5692	.3156
1.0000	.9087	.7143	.4814
1.5000	.9103	.8095	.5838
2.0000	.9075	.8604	.6604
2.5000	.9055	.8845	.7116
3.0000	.9041	.8952	.7541
3.5000	.9031	.8996	.7800
4.0000	.9024	.9012	.8072
4.5000	.9018	.9016	.8382
5.0000	.9013	.9014	.8566

37.150 CP seconds execution time

Time = $t/\bar{t}$

T4 = T_4/T_{Go}

T7 = T_7/T_{Go}

VT = $V_T/V_T{}^*$

approximate analytical integrations through small time increments. This technique allows successive iterations on the complete solution through sequential evaluations of the equations rather than a simultaneous approach. Application to the modeling of polydispersed coal pyrolysis at low pressures gives encouraging results. For this case the direct solution of the model equations with a stiff solver like DVOGER in the IMSL Library is expected to take more than 10 hours on a CDC Cyber 73. Other stiff ODE solvers that can handle particular Jacobian structures are not applicable to this problem. The method proposed here reduces the computation time to less than 30 seconds with three digit accuracy in the computed variables. This iterative technique would be practical for reactor design and optimization studies.

NOMENCLATURE

a_{sj}	surface area of particle, cm^2
c	specific heat of coal, cal/gm°K
C	specific heat of gas, cal/gm°K
E_i	activation energy of ith reaction, cal/gmol
F	mass flowrate of mixed gas, gm/sec
F_o	initial mass flowrate of coal carrier gas, gm/sec
G_o	initial mass flowrate of hot gas, gm/sec
k	thermal conductivity of gas, cal/gm sec °K
k_{oi}	pre-exponential factor of ith reaction, 1/sec
K	number of discrete particle size cuts
m_o	coal mass flow rate at reactor entrance, gm/sec
m_j	mass flowrate of char particles in the jth size cut, gm/sec
N	number of gas-solid reactions
p_j	mass fraction of m_o that is in size cut j
r_j	particle radius of size cut j, cm
r_{AV}	root-mean-square average radius of particle size distribution, cm
R	gas constant, cal/gmol °K
t	time, sec
$\bar{t}$	time constant for the temperature transient of size r_{AV}, defined in Eq. (17), sec
T_F	temperature of gas stream F, °K
T_{F_o}	temperature of gas stream F_o, °K

T_{G_o} temperature of gas stream G_o, °K

T_j temperature of particles of size j, °K

v_i^* maximum volatilization due to reaction i, gm of volatiles/gm of original coal

v_{ij} volatiles produced in size j through reaction i, gm/gm

w Rosin-Rammler mass fraction size distribution function

x particle diameter in w(x), cm

Δh_R heat of reaction, cal/gm

Δt time increment for discrete solution, sec

ρ density of coal, gm/cm^3

τ_G characteristic mixing time of hot and cold gas streams, sec

ACKNOWLEDGEMENT

The computer time was provided by the Oregon State University Milne Computer Center.

LITERATURE CITED

1. Hindmarsh, A.C., "GEARB: Solution of Ordinary Differential Equations Having Banded Jacobians," Lawrence Livermore Laboratory Report UCID-30059, June 1977, Rev. 2.

2. Hindmarsh, A.C., "Preliminary Documentation of GEARBI: Solution of ODE Systems with Block Iterative Treatment of the Jacobian," Lawrence Livermore Laboratory Report UCID-30149, December 1976.

3. The IMSL Library, International Mathematical and Statistical Libraries, Inc., July 1977, 6th Ed.

4. Levenspiel, O., "Chemical Reaction Engineering," 1972, Second Edition, Wiley.

5. Anthony, D.B., Howard, J.B., AIChE J., 1976, 22, 625.

6. Suuberg, E.M., Peters, W.A., Howard, J.B., I&EC Process Design and Development, 1978, 17, 37.

7. Suuberg, E.M., Sc.D. Thesis, Dept. of Chemical Engineering, Massachusetts Institute of Technology, Cambridge, MA, 1977.

8. Kayihan, F., Reklaitis, G.V., "Modelling of Staged Fluidized Bed Coal Pyrolysis Reactors," Paper presented at 85th National AIChE Meeting, June 1978.

9. Turton, R., "Combustion of Wood Char in a Transport Reactor," M.S. Thesis, Dept. of Chemical Engineering, Oregon State University, Corvallis, OR, 1979.

RECEIVED November 5, 1979.

12

Reaction Rate Equations from Integral Reactor Experiments

CHARLES H. WARE, JR.

Commercialization Insights, 33 Sandi Drive, Poughkeepsie, NY 12603

A general method is presented for the development of chemical reaction rate equations from integral reactor and single-sample batch reactor data such as are obtained in process development studies. Following the scope of the method, three earlier foundation stones upon which the method rests, the method itself, and a simple illustration are presented.

Scope

The methodology of this paper is intended for use in chemical process research and development studies. It is applicable when the experimenter wishes to develop a set of reaction rate equations which represents the system performance. The objective may be to gain sufficient understanding of how the various rates are affected by reaction conditions so that innovative approaches are triggered. For example, an immediate question might be whether the loss in selectivity that is observed at high conversions is due to chemical (differences in intrinsic reaction rates) or physical (diffusional or distributional) phenomena. Alternatively, the objective might be to predict the reaction rates at a variety of reaction conditions with sufficient precision to permit a reliable process design optimization.

Before the methodology can be implemented, some data must be available. There must be replicate data sets from which experimental errors may be estimated. If the reactions are catalytic, it is highly recommended that the data include at least three different batches of catalyst (not 3 samples of the same batch). If the catalyst is experimental (either proprietary or from a vendor), it is also highly recommended that the reference data set include six different batches of catalyst. This is to identify unacceptably large variations in the catalyst and to reduce the possibility of formulating a set of rate equations which do not represent the catalyst batches which can be reproduced. The reference data set must also include results at two or more temperatures.

0-8412-0549-3/80/47-124-235$05.00/0

The data will be used to formulate a tentative model to be used in the design of the experiment. It will consist of a simple power-law expression for each step with each of the parameters estimated from the reference data. For the reaction scheme:

$$A + B \longrightarrow C$$
$$C + B \longrightarrow D$$
$$B + B \longrightarrow E$$

the tentative model might be:

$$r_1 = k_1 C_a C_b \qquad (1)$$

$$r_2 = k_2 C_b C_c \qquad (2)$$

$$r_3 = k_3 C_b^2 \qquad (3)$$

$$k_1 = k_{1_o} \exp(-E_1/R(1/T-\overline{1/T}))$$

$$k_2 = k_{2_o} \exp(-E_2/R(1/T-\overline{1/T}))$$

$$k_3 = k_{3_o} \exp(-E_3/R(1/T-\overline{1/T}))$$

where k_{1_o} = 0.1 liter/g-mole-hr

k_{2_o} = 0.1 liter/g-mole-hr

k_{3_o} = 0.02 liter/g-mole-hr

E_1 = 12000 cal/g-mole

E_2 = 12000 cal/g-mole

E_3 = 21000 cal/g-mole

R = 1.987 cal/g-mole-^oK

T = temperature, ^oK

C_a, C_b, C_c = concentration, g-mole/liter

$\overline{}$ denotes average value to be used in experiment

For each reaction, r is the rate of formation of product, g-moles/liter-hr. The numerical values of the parameters would be obtained by fitting the tentative model to the existing data.

Fitting Reaction Rate Models to Rate Data

When reaction rate data can be obtained, results from factorial experiments may be analyzed to deduce acceptable forms of the chemical reaction.(1) Consider the following single-step reaction:

$$A + B \longrightarrow C$$

Suppose that reaction rates are measured in a factorial experiment at each of the conditions shown in Figure 1. The data are then correlated by means of the following equation:

$$\begin{aligned} \ln r = \overline{\ln r} &+ a_1(1/T-\overline{1/T}) + a_2(\ln C_a-\overline{\ln C_a}) + a_3(\ln C_b-\overline{\ln C_b}) \\ &+ a_{12}(1/T-\overline{1/T})(\ln C_a-\overline{\ln C_a}) + a_{13}(1/T-\overline{1/T})(\ln C_b-\overline{\ln C_b}) \\ &+ a_{23}(\ln C_a-\overline{\ln C_a})(\ln C_b-\overline{\ln C_b}) \end{aligned} \qquad (4)$$

By virtue of the fact that a factorial design has been used to generate the data, the estimates of the a's are uncorrelated. This means that the numerical values of the estimates of the a terms will not change if one or more of the other terms is deleted from the correlating equation.

The analysis uses the fact that equation 4 is the Taylor expansion of ln r as a function if 1/T, $\ln C_a$ and $\ln C_b$. The values and the 95% confidence intervals of the a terms have been estimated. The true reaction rate equation, the form which is not yet known, also has a Taylor expansion of ln r as a function of these three variables. We will never know the true equation, but we can deduce the approximate correct form of the equation and estimate its parameters as follows:

1. Propose a tentative model
2. Infer the permissible range for each value of a
3. Compare those required by the true model with the ranges estimated from the data
4. If all the requirements of the true model are met, accept the tentative model and propose alternatives
5. If one or more of the requirements of the true model are not met, reject the tentative model and propose alternatives.

Continue proposing alternative models until all the alternatives of a given level of complexity (all 3-parameter models, for example) have been explored. If one or more of these is acceptable, quit. If none is, consider all the alternatives of the next higher degree of complexity.

For simple models, the use of the Taylor expansion of ln r provides the following physical interpretation for a values:

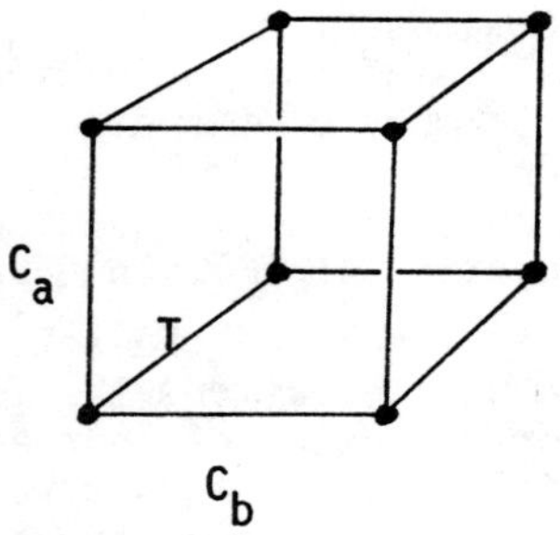

Figure 1. Factorial experiment

$a_1 = -E/R^*$

$a_2 = \alpha$, the order of reaction with respect to C_a^*

$a_3 = \beta$, the order of reaction with respect to C_b^*

$$a_{12} = \frac{\partial - E/R}{\partial \ln C_a} \quad \text{and} \quad \frac{\partial \alpha}{\partial 1/T}$$

$$a_{13} = \frac{\partial - E/R}{\partial \ln C_b} \quad \text{and} \quad \frac{\partial \beta}{\partial 1/T}$$

$$a_{23} = \frac{\partial \alpha}{\partial \ln C_b} \quad \text{and} \quad \frac{\partial \beta}{\partial \ln C_a}$$

* If the parameter is not constant, this is true only at the center of expansion: where $1/T = \overline{1/T}$; $\ln C_a - \overline{\ln C_a}$; and $\ln C_b = \overline{\ln C_b}$

For example, suppose $a_2 = 0.8 \pm 0.1$ $a_{12} = -1000 \pm 500$ and $a_{23} = 0.20 \pm 0.10$. The hypothesis that $\alpha = 1.0$ must be rejected because the 95% confidence interval is 0.7 to 0.9 and it is unlikely that it is really equal to 1.0. A power law model $r \sim C_a^{0.8}$ must also be rejected because it would mean that α = a constant and a_{12} and a_{23} must be zero. But the results are that they are both nonzero. We must now propose some more complex model. Depending upon the system, it might be based upon a free-radical mechanism, Langmuir Hinshelwood rate equations, or other. In any case, the numerical values of the coefficients in equation 4 and their confidence intervals are available to test every model which is proposed.

Model-building in batch reactors

Box and Hunter described a method which applies to batch reactors with intermittent sampling — product concentrations are required at several reaction times.(2) Their analysis is as follows:

1. write tentative reaction rate equations for the system with unknown rate constants, k_i
2. integrate the equations, and for each experimental run, fit the equations to the data by estimating best values of the rate constants; there will be one set of rate

constants for each run

3. if the rate constants do not vary from run to run, conclude that the rate equations are adequate
4. if the rate equations do vary, fit ln k_1 to experimental conditions - temperature, initial concentrations, catalyst dosage, etc. - to determine how they vary with the independent variables
5. modify the rate equations to eliminate the functional dependence of the k_is.

Suppose, for example, the true model is $k_1C_a^2$ but the tentative model is $k_1'C_a$. Then, when the initial concentration of A (C_{a_o}) is varied, k_1' will vary with C_{a_o} : $k_1' = k_{1_o}C_{a_o}$, approximately. From this the experimenter will infer that the tentative model is incorrect and some higher-order dependence is indicated. He might choose to test the hypotheses $r=k_1C_a^2$ or determine the order by the direct estimation of this non-linear parameter.

The Box-Hunter method permits the experimental observations of the product concentrations to be transformed into a set of "observations" of the k_is. Then these k_is or ln k_is can be expanded in a Taylor series as functions of experimental conditions. This is exactly analogous to the analysis of ln r described previously except that, by means of a tentative model, the primary reaction rate dependence on concentrations, temperature, and other experimental factors has been eliminated. This permits the rate equations to be integrated approximately correctly.

Modifications to the Box-Hunter method

To extend their analysis to integral reactors and single-sample batch reactors, three modifications are appropriate. First, use a more appropriate reactor concentration to analyze the data than the initial concentration. The simplest, and an effective one is

$$\overline{C}_a = 1/t \int_0^t C_a dt \qquad (5)$$

The $C_a(t)$ can be obtained from the tentative model with the preliminary estimate of the parameters. Second, to improve precision of the estimates of the k_is and therefore overcome the shortcoming of single samples, use all reactor concentrations to estimate the k_is, not just one as the authors did. Finally, use optimum conversion levels to minimize the variance of (ln k)s that are estimated from the data.(3,4) For single step reactions, the reaction time which will minimize the variance is

$$t_{opt} = 1/kC_{a_o}^{\alpha-1} \tag{6}$$

and the optimum outlet concentration is

$$(C_a)_{opt} = C_{a_o}\alpha^{-\frac{1}{\alpha-1}} \tag{7}$$

where α = reaction order with respect to C_a

Some typical results for k = 1

α	t_{opt}	$(C_a)_{opt}$
1	1	0.368 C_{a_o}
2	$1/C_{a_o}$	0.5 C_{a_o}

The values of k for each run are estimated from the preliminary data. For multi-step reactions, the method of Box and Lucas(3) may be used. Alternatively, we can make use of the fact that the model, its parameters, and the experimental error are "known". Then we may calculate the standard errors of ln k, $s_{\ln k}$, for any experimental design we propose.

Design of integral reactor experiments

A difficulty which we have just created by using ln $\overline{C}_a$ in the Taylor expansion is that of obtaining an efficient experimental design. The classical factorial designs depend upon the experimenter's ability to choose levels of his independent variables to hold them at these prescribed values. By defining the independent variables to be ln $\overline{C}_i$ with the intention of calculating from $\overline{C}_i$ from the data after the experiment is over, we have sacrificed our ability to satisfy the requirements of the factorial designs. How can we get around this difficulty? We have already hypothesized a mathematical model for the process under study in order to define some reaction rate constants into which we will transform the observations. If this model is correct then the only additional information we need in order to predict the values of $\overline{C}_i$ in the experiment about to be run is the numerical values of the reaction rate constants. With them we would be able to formulate a grand null hypothesis comprised of the mathematical model and the numerical values of the parameters. With this we could calculate the outcome of every planned experiment, and in particular the $\overline{C}_i$s. These effective average concentrations would necessarily be functions of the initial concentrations. As the latter were varied, the calculated values of the former would change and by adjusting the initial concentrations we could obtain the desired levels of the averages. Then if the null hypothesis were true, the $\overline{C}_a$ would correspond to the factorial design. To the extent that the null hypothesis is incorrect, when the experimental values of $\overline{C}_i$ and $\ln \overline{C}_i$ are calculated they will be found not to coincide with the exact levels of the factorial design. In this case, however, linear regression will still yield the desired values of the coefficients in the Taylor expansion of ln k and the analysis may proceed.

If we employ optimum conversion levels in all of the runs, we may calculate $\overline{C}_a$ in advance of the experiment. Using the results shown above, we obtain, for optimal experiments:

$$\alpha = 1 \qquad \overline{C}_a/C_{a_o} = 1 - 1/e = 0.632 \tag{8a}$$

$$\alpha = 2 \qquad \overline{C}_a/C_{a_o} = \ln 2 = 0.693 \tag{8b}$$

$$\alpha \neq 1, \alpha \neq 2 \qquad \overline{C}_a/C_{a_o} = (\alpha^{\frac{(\alpha-2)}{(\alpha-1)}} - 1)/(\alpha - 2) \tag{8c}$$

We will correlate the data using a Taylor expansion of ln k as a function of $\ln \overline{C}_a - \overline{\ln \overline{C}_a}$ where the overbar denotes the grand average of all the $\ln \overline{C}_a$. For any given initial conditions for the high level of concentration, what are the lower initial concentrations to be? It is known that as the levels of the independent variables are made more widely separated, the precision of the estimate of the coefficients in the regression equation increases. It can also be shown, from propagation of error analysis that as the lower concentration level is reduced, the precision of ln k that is extracted from the experimental result is reduced.

$$s_{\ln k} = s_k/k = \frac{1}{k}\left|\frac{\partial k}{\partial C_a}\right| s_{c_a} \tag{9}$$

$$= \alpha^{\frac{\alpha}{\alpha-1}} s_{C_a}/C_{a_o} \qquad \alpha \neq 1 \qquad (10a)$$

$$= e\, s_{C_a}/C_{a_o} \qquad \alpha = 1 \qquad (10b)$$

The coefficient of $\ln \overline{C}_a - \overline{\ln \overline{C}_a}$ has a standard error, denoted by s_{a_2}, that depends on both the range and the precision of the observation. For constant $s_{\ln k}$:

$$s_{a_2} = s_{\ln k} \Big/ \sqrt{\sum_{v=1}^{v=n} \{(\ln \overline{C}_a)_v - \overline{\ln \overline{C}_a}\}^2} \qquad (11)$$

Selecting the number of runs.

The design of the experiment consists of determining the number of runs to be made at the low level of C_{a_o} for each run made at high level (as in any factorial design). Let n' and n" denote the number of runs at the high and low levels of C_{a_o}, respectively. Because the standard deviation of the average value of a set of "observations", $\overline{\ln k}$ in this case, is inversely proportional to $\sqrt{n}$, in order to obtain equal values of $s_{\ln k}$ at the high and low levels of C_{a_o}:

$$\frac{1}{\sqrt{n'}\, C'_{a_o}} = \frac{1}{\sqrt{n''}\, C''_{a_o}} \qquad (12)$$

$$\text{or} \qquad \frac{C'_{a_o}}{C''_{a_o}} = \sqrt{\frac{n''}{n'}} \qquad (13)$$

For example, if $C''_{a_o} = 0.5\, C'_{a_o}$, then n" = 4n' will produce equal standard deviations.

$$\overline{\ln \bar{C}_a} \equiv \frac{n' \ln \bar{C}'_a + n'' \ln \bar{C}''_a}{n' + n''} \quad (14)$$

For optimum conversions, equation 8 yields

$$\ln \bar{C}_a = \ln \text{Constant} + \ln C_{a_o} \quad (15)$$

Substituting this into equation 14 and simplifying

$$\ln \bar{C}'_a - \overline{\ln \bar{C}_a} = \frac{n''}{n'+n''}\{\ln C'_{a_o} - \ln C''_{a_o}\} \quad (16)$$

$$\ln \bar{C}''_a - \overline{\ln \bar{C}_a} = \frac{n'}{n'+n''}\{\ln C'_{a_o} - \ln C''_{a_o}\} \quad (17)$$

$$\frac{1}{\sqrt{\sum_v^n(\ln \bar{C}_{a_v} - \overline{\ln \bar{C}_a})^2}} = 2\sqrt{\frac{n''+n'}{n''\,n'}}\left(\ln \frac{n''}{n'}\right)^{-1} \quad (18)$$

Equation 18 permits the calculation of the factor by which $s_{\ln k}$ is multiplied to yield s_{a_2}. It is obviously a function of n' and n". For fixed n', the factor is simply a function of n"/n' and the design of experiment can be broken into two steps:

1. choose n' for the desired design
   a. for a 2^n factorial, $n' = 2^{n-1}$
   b. for a 2^{n-p} fractional factorial, $n' = 2^{n-p-1}$.

2. for the fixed value of n', choose n"/n' to obtain the desired precision for a_2 range for $\ln \bar{C}_a$; typically n" = 2

The table below provides the basis for step 2. It is reasonable to wonder whether it might not be better to use addition runs at the high level of C_{a_o} before reaching such high values of n"/n'. (s_{a_2} depends on the standard deviation of ln k at the high levels of C_{a_o} and that can be reduced by increasing n'). This can be determined for a fixed value of n' + n" by differentiating equation 15 with respect to n", setting the result equal to zero, and

n''/n'	$\dfrac{1}{\sqrt{\sum_{v}^{n}(\ln \bar{C}_{a_v} - \overline{\ln \bar{C}_a})^2}}$	C''_{a_o}/C'_{a_o}
1	∞	1
2	3.54	0.71
3	2.10	0.58
4	1.61	0.50
5	1.36	0.45
6	1.20	0.41
7	1.09	0.38
8	1.02	0.35
9	0.96	0.33
10	0.91	0.32

solving for n''/n'. The optimum ratio is 11:1. This means that for all practical cases the design will consist of n' runs at the high level of C_{a_o} (step 1, previous page) and n''/n' runs at the low level of C_{a_o} which corresponds to a factor in the above table which is small enough to produce the desired precision in a_2. The factors shown there are being reduced more rapidly than $\{(n'+n'')/(n'+n''+1)\}^{\frac{1}{2}}$ which is the rate expected for replication. Suppose we wish to design a factorial experiment in three variables with

$$\frac{1}{\sqrt{\sum_{v}^{n}(\ln \bar{C}_{a_v} - \overline{\ln \bar{C}_a})^2}} \doteq 3$$

$$n' = 4$$

$$n''/n' = 3$$

The design will be as shown in Figure 2 with each dot representing one run.

Illustration

The reaction is:

$$A \longrightarrow B$$
$$A \longrightarrow C$$

At 127°C, the following results have been obtained. Reaction time, 0.5 hr, $C_{a_o} = 1.0$, $C_{b_o} = 0.0$, $C_{c_o} = 0.0$ $C_a = 0.286$ $C_b = 0.428$ $C_c = 0.286$. First order rate constants were calculated: $k_1 =$

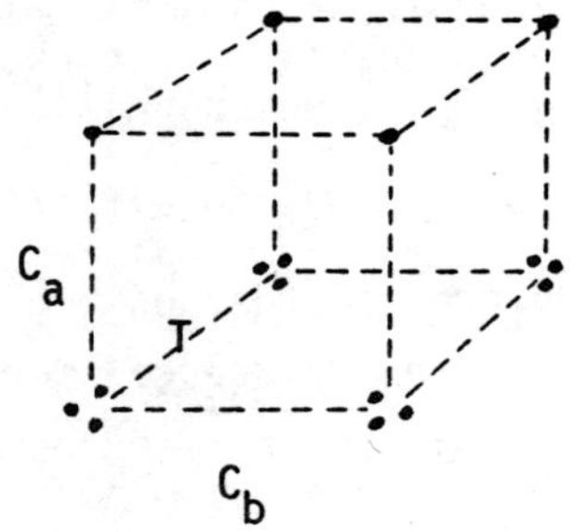

Figure 2. Factorial experiment. Each dot represents one run.

$1.5\ hr^{-1}$; $k_2 = 1.0\ hr^{-1}$. From similar data at higher temperatures the activation energies were calculated: $E_1 = 17.6$ kcal/g-mole and $E_2 = 26.7$ kcal/g-mole. From previous data, s_{C_a}, s_{C_b}, equal 0.01. We wish to design an experiment to obtain data and fit an equation like Equation 4 to them, except using ln k and $\overline{\ln k}$. We wish to achieve the following precision levels: $s_{a_1} = 500$ cal/g-mole; $s_{a_2} = s_{a_3} = 0.1$.

The ks are given by:

$$k_1 = \frac{1}{t} \ln \frac{C_{a_o}}{C_a} \frac{C_b - C_{b_o}}{C_{a_o} - C_a} \tag{19}$$

$$k_2 = \frac{1}{t} \ln \frac{C_{a_o}}{C_a} \frac{C_c - C_{c_o}}{C_{a_o} - C_a} \tag{20}$$

For optimum conversion of A ($\ln C_{a_o}/C_a = 1$) and equal standard deviations for the C's, the propagation of error formula yields:

$$s_{\ln k_i} = (e\, s_{C_a}/C_{a_o})[(\frac{e}{e-1})^2 + (\frac{k_1+k_2}{k_i})^2]^{\frac{1}{2}} \qquad i=1,2 \tag{21}$$

The term in brackets is the factor by which the errors are inflated due to the fact that the ks must be calculated from two measurements. For k_1 this factor is 2.30 while for k_2 it is 2.96; they vary at most 10% over the planned temperature range.

The experiment will be designed to match the precision requirements for $\ln k_2$; for $\ln k_1$, they will be exceeded. We will use a two-level factorial design in 3 variables; from Equation 12 $s_{\ln k} = 0.0805/C_{a_o}$. For n' = 4, $s_{a_2} = 0.0402/C_{a_o} \sum(\ln \overline{C}_{a_v} - \ln \overline{C}_{a_v})^2$ with the summation over n"/n'+1 terms. If $C_{a_o}=1$, and the reciprocal of the denominator is no greater than 2.48, the precision requirements will be met. For n"/n'=3, the factor will be 2.1 which is acceptable and $C''_{a_o}/C'_{a_o}=0.58$. The values for $\overline{C}_b$ are given by:

$$\overline{C}_b = C_{b_o} + \frac{1}{t_f}\int_o^{t_f} \frac{k_1 C_{a_o}}{k_1+k_2}(1 - e^{-(k_1+k_2)t})dt$$

$$= C_{b_o} + \frac{k_1}{k_1+k_2}(C_{a_o} - \overline{C}_a) \qquad (22)$$

Since $\overline{C}_a = 0.632C_{a_o}$, $\overline{C}_b = C_{b_o} + 0.368\ C_{a_o} k_1/(k_1+k_2)$. For any values of C_{a_o}, k_1 and k_2, C_{b_o} may be chosen to achieve a required level of $\overline{C}_b$. Since it will be calculated by difference, it is possible that C_{b_o} will be negative in which case it will be necessary to raise the design level for $\overline{C}_b$. We are still free to choose C_{b_o} at the lower level of $\overline{C}_b$. Two possibilities are : $C_{b_o} = 0.0$ (for convenience or at the level anticipated to be the mixed feed concentration during recycle, say 0.1. For illustration, we will choose $C_{b_o} = 0.1$ to correspond with the lower level of $\overline{C}_b$. At this lower level C_{a_o} will be at two levels (as in any factorial design). In order to calculate the actual values, the ks must be known and, therefore, we must determine the temperature levels. From the design specification:

$$s_{a_1} = 500 = \frac{s_{\ln k}}{\sqrt{\sum_{v=1}^{v=n}(1/T_v - \overline{1/T})^2}} \qquad (23)$$

There will be 8 identical values in the sum, $s_{\ln k} = 0.080$ and we elect to put the center of the design at 400°K:

$$1/T_v = 0.0025 \pm 0.000057$$

$$T_v = 391,\ 409^oK\ (118,\ 136^oC)$$

From the activation energies, we calculate for 118°C $k_1 = 1.2$ $k_2 = 0.715$; for 136°C, $k_1 = 1.87$, $k_2 = 1.40$. Then, the complete design is:

Run No.	1*	2*	3	4	5*	6*	7	8
T_1 °C	118	118	118	118	136	136	136	136
C_{a_o}	0.58	0.58	1.0	1.0	0.58	0.58	1.0	1.0
C_{b_o}	0.203	0.453	0.100	0.350	0.245	0.495	0.172	0.422
t, hr	0.74	0.74	0.74	0.74	0.21	0.21	0.21	0.21

* run in triplicate

The levels of $\overline{C}_b$ are 0.345 and 0.595. Their ratio is 0.58, the same as for $\overline{C}_a$, since the precision requirements of their coefficients are the same. The times are those required to make $(k_1+k_2)t = 1.0$. The design is complete.

To illustrate the efficacy of the design, the following results were calculated:

Run No.	1*	2*	3	4	5*	6*	7	8
C_a	0.198	0.212	0.369	0.407	0.202	0.221	0.396	0.396
C_b	0.479	0.685	0.484	0.707	0.463	0.669	0.494	0.707
k_1	1.049	0.857	0.819	0.731	2.896	2.225	2.350	2.081
k_2	0.403	0.500	0.528	0.484	2.126	2.365	2.060	2.330

* Average of 3 results

The correlation coefficients between pairs of independent variables in Equation 4 are: $1/T - \ln C_a$: -0.015; $1/T - \ln C_b$: -0.095; $\ln C_a - \ln C_b$: -0.071. These low values indicate that the estimates have been uncoupled (0.0 indicates complete independence and 1.0 indicates complete correlation). The calculated values of the coefficients are:

i	$\overline{\ln k_i}$	$a_1 = -E/R$	$a_2 = \alpha$	$a_3 = \beta$
1	Estimate	8908	-0.276	-0.101
	True	8857	-0.377	-0.146
	Std Dev	388	0.076	0.088
2	Estimate	13488	0.135	0.106
	True	13437	0	0
	Std Dev	499	0.098	0.113

The results are satisfactory. The estimates are all within 1½ standard deviations of the values used to generate the results.

Conclusion

In integral reactor experiments, the experimenter is limited in extending the range of his observations by the fact that there is a loss of precision when conversions are too high or too low. A method has been described which utilizes minimum-variance conditions and which leads to almost complete uncoupling of kinetic variables which further reduces the uncertainty in the estimates of the kinetic parameters.

Literature Cited

1. Ware, Charles H., Jr. Proceedings of the 1975 Summer Computer Simulation Conference, 1975, II, 368.
2. Box, G.E.P.; Hunter, William G. Technometrics, 1962, 4, 301.
3. Box, G.E.P.; Lucas, H.L. "Design of Experiments in Non-Linear Situations", Biometrika, 1959, 46, 77.
4. Ware, Charles H., Jr. "Some Modifications of the Box-Hunter Model Building Method". Paper presented at Princeton Symposium on Statistics, 1966.

Received November 5, 1979.

Modeling and Control of Continuous Industrial Polymerization Reactors

JOHN F. MACGREGOR

Department of Chemical Engineering, McMaster University, Hamilton, Ontario L8S 4L7, Canada

PAUL W. TIDWELL

Monsanto Company, Decatur, AL 35601

In this paper we will primarily deal with the analysis and control of continuous flow polymerization reactors although much of the methodology will be relevant to other continuous processes and some batch processes as well. Consider the situation depicted in Figure 1 where there is a single or a sequence of flow reactors and one is interested in maintaining good control over polymer properties such as viscosity, MWD, copolymer composition, particle size distribution, polymer end group composition (dye sites), etc.. This can be done by adjusting the set points of primary controllers for one or more adjustable input variables such as initiator flow, catalyst flow, monomer feed ratios and reactor temperature or pressure. It is this secondary level of control over the polymer properties with which we shall be concerned here rather than the primary level of control over the individual manipulated variable themselves; that is, over the reactor temperature or pressure, or the feedrates, etc.. These latter variables often can be measured accurately and continuously and therefore usually can be well controlled by conventional continuous pneumatic or electronic proportional-integral-derivative (PID) type controllers. The control algorithms for the control of the polymer properties will usually act by readjusting the set-points of these primary controllers.

Measurement Problems in the Control of Polymer Reactors.

Before proceeding it is worthwhile discussing in more detail some aspects of measurement problems which present additional difficulties in formulating control strategies for polymerization reactors.

(i) Discrete data: The observations of the output polymer properties are often available only at discrete equispaced intervals of time. This, of course, rules out the use of continuous pneumatic or electronic controllers.

(ii) Off-line analyses: In spite of rapid development in the area of on-line detectors a large number of polymer proper-

0-8412-0549-3/80/47-124-251$05.00/0

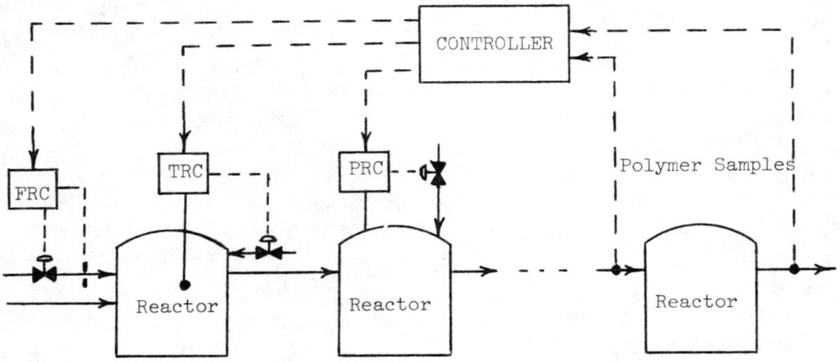

Figure 1. Reactor control configuration

ties (e.g. conversion, MWD, copolymer composition, end group compositions, etc.) are still time consuming measurements made off-line in the quality control laboratory of most plants. This usually results in introducing a considerable time lag into the feedback loop of any control scheme. Although analytical equipment is available for on-line analysis of some of these properties, it is a question of providing economic justification for them and of making them rugged and reliable enough for in-plant use.

(iii) Long sampling intervals: Compounding the above problems, it is not uncommon for many of these properties to be analyzed very infrequently, perhaps only once per 8-hour shift or once per day. This would appear to make the control problem almost impossible but, in fact, such sampling intervals are not inconsistent with the requirement of eliminating many of the major stochastic disturbances in the system. Recent work by Box and Jenkins (4) and MacGregor (14), has emphasized that in the control of processes subject to stochastic disturbances it is only important to sample fast enough to pick up changes in the major forecastable disturbances in the system. In the authors' experience the stochastic disturbances typically affecting polymer reactors persist over periods of days and even weeks, thereby making an 8-hour sampling interval quite consistent with the achievement of good control. In fact, the contrary situation has sometimes been the case, where deterioration in the quality of the control has resulted from too frequent control action, particularly when the measurement noise was large and the control strategy not developed to account for it.

(iv) Large measurement error: The analysis procedures for many polymer properties such as copolymer composition are often involved and difficult to carry out. As a result the measurement error variance is quite large. In fact, it is not uncommon for the two standard deviation limits of these measurements to approach the desired specification limits for grading of the polymer.

Any control procedure for polymer reactors must recognize and deal with these measurement problems. This rules out the use of conventional continuous control ideas taught to engineering students at most universities. These latter ideas are based on continuous measurements, in the presence of little or no measurement noise, and are applicable only to systems with rather short time delays in the feedback loop.

Process Modelling

In order to analyze for process problems or design stochastic controllers one usually needs to run full scale plant tests and develop process dynamic models from the resulting data. This section discusses several aspects related to the modelling of polymer reactors for these purposes.

Type of Model. Depending upon the type of problem being investigated, one can employ models ranging from very complex nonlinear partial differential equations based on theoretical polymerization kinetics and mixing theory to simple linear empirical differential or difference equations built from process data only. Both these types of models have their place in developing process control strategies and in analyzing for process problems.

(i) Mechanistic (nonlinear) Models: These tend to be relatively complex models and usually require a considerable development effort to arrive at an adequate form. Their primary use has been, and probably will continue to be, in the design and simulation of polymer reactors. It is usually the steady-state versions of these models that are used to size new reactor systems or to simulate operation at different conditions of throughput, temperature, feed ratios, etc.. These steady-state models are also useful for supervisory control where they can be used in an on- or off-line mode to optimize the operation conditions of the reactors at various times and to readjust the set-points of the basic temperature or flow controllers accordingly. Such use depends very much on the specific system and objectives.
Dynamic or unsteady-state forms of these mechanistic models may also be useful in simulation or on-line control. They can be used to design reactors or find operating conditions which are less sensitive to specific types of disturbances, or to develop start-up procedures for a sequence of reactors. In the control of batch polymerizers they are potentially useful in controlling the polymerization reaction to follow some desired reaction path or trajectory. For the control of continuous reactors sampled at equi-spaced intervals of time (the main topic of this presentation), they can be used in a feedback control scheme with a process control computer to compute the control action needed to bring the reactor to a desired state in some specified optimal manner, providing the control or sampling interval is long compared to the time necessary to make the necessary computations (see for example Kiparissides et al. (13)). Unless one is interested in operating the reactors over a wide range of conditions the simple empirical models discussed below are usually better suited to this latter feedback control situation.

(ii) Empirical (linear) Dynamic Models: These are usually discrete linear transfer function (difference equation) type models which provide a representation of the dynamic behaviour of the process at the discrete sampling instants. They are in general much simpler to develop than theoretical models, both their structure and parameters being empirically identified from plant data. They are very useful for developing controllers to maintain the polymer properties constant at desired target values. These control schemes can often be made simple enough (at least the univariate ones) that the operators can implement them using tables or a programmable calculator. More powerful schemes are possible if a process minicomputer is available. One of the major drawbacks of these empirical linear models is their limited range of applicability. They hold quite well in the regions of operation for which they were developed, but they extrapolate very poorly due to the nonlinearities present in polymer reactor systems. One method of overcoming this difficulty, under changing conditions, by simultaneously tracking the changing parameters and controlling will be discussed later.

These two general types of models represent the extremes and various levels of sophistication lying somewhere between these are possible. Given the time and manpower it is wise to develop both types of model and compare them with one another and with plant data. The authors are aware of several situations where the simple empirically identified models have revealed important effects not accounted for in the theoretical models, and thereby led to a considerably better understanding of the system.

The remainder of this manuscript will deal with the identification of linear empirical models and their use in developing control strategies.

Plant Experimentation. In order to evaluate process problems or develop simple models for a process, it is often necessary to design and run full scale plant experiments under production conditions. Due to long response times, long sampling intervals and the noisy nature of the measurements, these plant tests usually last from a few days to a month or more. Therefore, they must not drive the polymer properties out of specification for any unusual period of time nor upset any other mode of operation of the plant. There is no more certain way of ensuring termination of the test than trying to obtain too much information too quickly. The test must, therefore, be planned carefully and then performed by the engineers responsible for production. Maximum use of polymer blending facilities or running on only one of many parallel reactor trains can effectively be used to minimize the larger variations in polymer properties which result during the test. Even if only steady-stage information is ultimately desired on the process, it is often necessary to design and analyze the experiments in a manner which accounts for the dynamic or transient effect and then obtain the steady-state information by the appropriate simplification of the resulting dynamic models. The main reason behind this statement is that it is physically difficult or impossible in practice to hold the process steady enough at a number of marginally different fixed levels of operation in the face of major stochastic disturbances in the system, and then to be able to directly estimate meaningful steady-state effects from such data in the presence of the large measurement errors. Furthermore if the major disturbances are of a nonstationary nature, the consequences of trying to estimate steady-state effects by running open-loop factorial type experiments should be obvious. The results obtained by performing the usual factorial regression analysis (which assumes independent, zero mean, random errors) will be incorrect and the confidence limits on the estimated effects will be misleading (Box and Newbold (8)). Furthermore, different results will be obtained upon any replication of the experiment. Replication and randomization procedures have been used successfully in some EVOP and response surface studies on plants subject to nonstationary disturbances. These procedures do not protect us entirely against making incorrect conclusions under those circumstances, and the cost in terms of plant experimentation time for what protection they do afford is prohibitive when compared to the dynamic experimentation route outlined below. For example, one of the authors became involved in experimentation in a petrochemical plant where factorial experiments had been done repeatedly for years in an effort to solve some severe yield problems in part of the plant. These experiments, it turned out, were done with a nonstationary disturbance process operating in the system and hence they yielded silly results. A simple 50-hour experiment and analysis of the data with time series methods solved a problem that had existed and defied solution for about 10 years.

Consider the process configuration in Figure 1. An experimental program consists of introducing designed variations into the process inputs or manipulated variables (up to about 4 of these variables can often be studied simultaneously), and then sampling and analyzing the polymer being produced at the desired locations along the chain of reactors. Such data can sometimes be collected under open-loop conditions, that is, under conditions where there is no feedback control being exercised over the output polymer properties through manipulating some of these input variables. Processes on which this open-loop type of experimentation can be performed are those which are subject to fairly stationary stochastic disturbances over the time period of the experiment and whose output polymer properties will remain reasonably well within defined specification limits if all the input variables are held fixed at suitable mean levels. In this case the design of the experiment consists of choosing when and how to impose deviations in these input variables from their mean levels. The situation is depicted in Figure 2. Various types of input sequences can be used and some research has been done on the optimal design of these input sequences to maximize the amount of information obtained on the process (Box and Jenkins (4), and Mehra (16)). These optimal designs rely upon one's imprecise prior knowledge of the process and disturbances, and so are rarely directly useful. However, they do point out the importance of properly choosing certain characteristics of the input sequences. The most important requirements are as follows:

(i) The input sequences should be uncorrelated with one another in order to enable one to separate out the separate effects of each input in later analysis. This can be guaranteed by using independent random number sequences in generating the inputs.

(ii) The input variations must have the proper frequency content in order to obtain the maximum amount of information on the process dynamics from the small perturbations used. Too high a frequency relative to the time constants of the system will result in most of the variations being damped out and hence practically unobservable in the outputs. Too low a frequency will result in obtaining little information on the important transient effects. The proper frequency must be related to the estimated time constants of the process.

(iii) The magnitude of variation of the inputs must be carefully chosen. They must be varied enough to reveal the important effects in the outputs, but not too much to push the polymer properties out of specification for any inordinate amount of time. This leaves very little room for error and makes the property choice very difficult. One must rely on all prior knowledge of the process and be willing to change the magnitude of the variation as the experimentation proceeds. It is often convenient to design the experimental programs in time blocks with adjustments to the magnitude and frequency content of the input signals being made between blocks. Because the experimentation is being performed on a plant under full production conditions the magnitude of the input variations must often be chosen too small to be able to distinguish the effect of each change on the process outputs from that due to the large measurement errors. As a result a reasonably sophisticated statistical analysis of the data is often necessary.

Some common types of input sequences which can satisfy the above requirements and are commonly used are illustrated in Figure 2. A pseudo-random binary signal is one which takes on values at either of the two levels with the decision to switch from one level to the other being made at each time interval by a random process. This type of signal is quite satisfactory if the switching frequency is chosen properly. For several inputs, uncorrelated pseudo-random binary signals switching according to different random number sequences can be used. An alternative to this is to make use of the orthogonality property of factorial designs and vary different inputs according to the pattern of these designs. This has the added property that if the time spent at each condition is reasonably long in comparison to the process time constants and the disturbance process is stationary, a pseudo-steady-state analysis can be attempted by classical factorial design and analysis methods. A third type of very useful input sequence is the class of stochastic time series inputs generated by sequences of random Normal deviates. These will be discussed in more detail later. In all open loop experimental programs, considerable care must be taken to guard against a lurking controller, a zealous operator who manipulates a variable not included in the program to

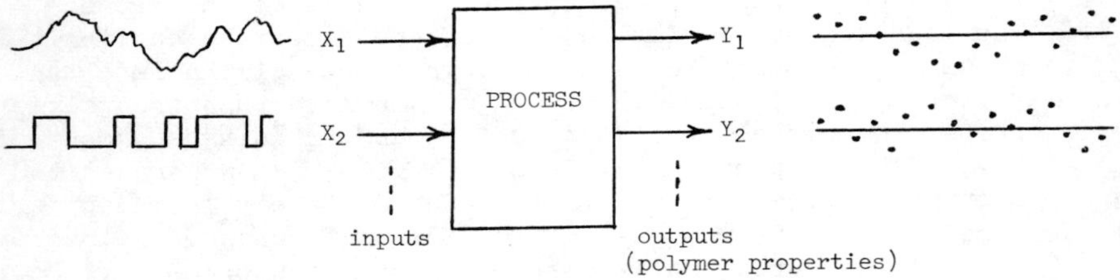

Figure 2. Open-loop experimental configuration

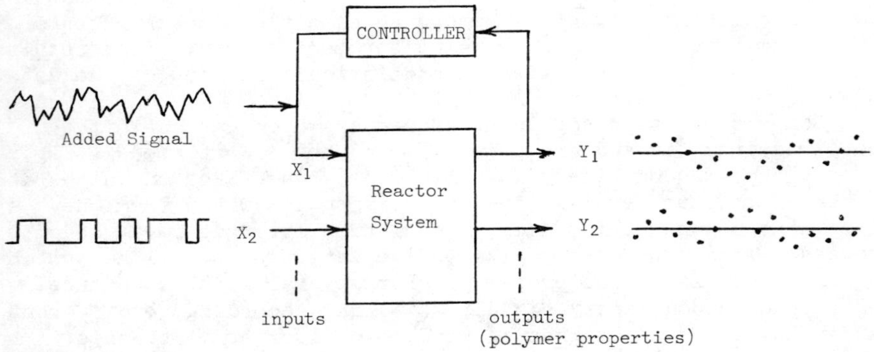

Figure 3. Closed-loop experimental configuration

maintain the response about the target.

In some situations open-loop experimentation cannot be performed due to the presence of large or drifting disturbances in the system which necessitate that some polymer properties be maintained under control throughout the course of experimentation by manipulating one or more inputs in a feedback manner. Such a closed-loop situation is depicted in Figure 3. It is important, in this situation, that to identify the causal relationships between these variables one add a designed input variation to the feedback signals and not try to identify these relationships using only the inherent process input variations resulting from the feedback control action. If such a signal were omitted, any attempt at process model identification could lead not to a model of the process but rather to a model of the feedback controller (Box and MacGregor (7)).

Empirical Model Identification. In this section we consider linear difference equation models for characterizing both the process dynamics and the stochastic disturbances inherent in the process. We shall discuss how to specify the model structure, how to estimate its parameters, and how to check its adequacy. Although discussion will be limited to single-input, single-output processes, the ideas are directly extendable to multiple-input, multiple-output processes.

Process Transfer Function Models: In continuous time, the dynamic behaviour of an ideal continuous flow stirred-tank reactor can be modelled (after linearization of any nonlinear kinetic expressions about a steady-state) by a first order ordinary differential equation of the form

$$\tau \frac{dy}{dt} + y = g X \tag{1}$$

where τ is the process time constant and g is its gain, and X and y are deviations about the steady-state. If this first order process is sampled at discrete equispaced intervals of time $(\ldots(T-1)h, Th, (T+1)h, \ldots)$ and the input (manipulated) variable changed at the sampling instants, but held constant over the following interval (the usual situation in sampled-data control schemes), then its dynamic behaviour at the sampling intervals is given exactly by the first order difference equation

$$y_t = \delta y_{t-1} + g(1-\delta)\, X_{t-1} \tag{2}$$

where $\delta = e^{-h/\tau}$ and the subscript t refers to the discrete time instant. Introducing the backward shift operator z^{-1} (where $z^{-1} y_t = y_{t-1}$ and $z^{-k} y_t = y_{t-k}$) the difference equation (2) can be written as

$$y_t = \frac{g(1 - \delta)z^{-1}}{1 - \delta z^{-1}} X_t \tag{3}$$

This is a first order discrete transfer function of the sampled process. In general, the dynamic behaviour between a single input variable X and an output variable y in most polymer reactor systems can be adequately modelled at the sampling instants by a difference equation model of the form $y_t = \delta_1 y_{t-1} + \delta_2 y_{t-2} + \ldots + \delta_r y_{t-r} + \omega_0 X_{t-b} - \omega_1 X_{t-b-1} - \ldots - \omega_s X_{t-b-s}$ where y_t and X_t are deviation variables about steady-state operating condition. In the backward shift operator notation

$$y_t = \frac{\omega_0 - \omega_1 z^{-1} - \ldots - \omega_s z^{-s}}{1 - \delta_1 z^{-1} - \ldots - \delta_r z^{-r}} X_{t-b} \tag{4}$$

where b is the number of whole periods of time delay between making a change in the input X and first observing its effect on the output. The form or order of the discrete transfer function for any process can be identified directly from plant data.

In practice, the system will be subject to stochastic disturbances or noise and the observed output response deviation Y (polymer properties) will be the sum effect of that induced by the manipulated input X and that induced by all other disturbances, that is

$$Y_t = \frac{\omega(z^{-1})}{\delta(z^{-1})} X_{t-b} + N_t \tag{5}$$

This process disturbance N_t measures the total effect on the output property Y of all unobserved process and measurement disturbances, and it represents the output deviation from target that would occur if no control action were taken. In polymer reactor systems where the process is often subject to large drifting types of disturbances and large measurement errors this term can be very large and often greater in magnitude than the effect of the manipulated variable changes X. Therefore, it is important that a model for these stochastic disturbances also be simultaneously identified from the plant data, firstly to enable one to be able to obtain a valid transfer function (or a mechanistic) model of the process in the presence of such large disturbances, and secondly to enable one eventually to design a stochastic control scheme which can cancel out the effect of these disturbances.

Stochastic Models for the Disturbances: The type of stochastic process disturbances N_t occurring in practice can usually be modelled quite conveniently by statistical time series models (Box and Jenkins (4)). These models are once again simple linear difference equation models in which the input is a sequence of uncorrelated random Normal deviates (a_t) (a white noise sequence) with zero mean and variance σ_a^2. One of the simplest such models is the first order autoregressive model

$$N_t = \phi N_{t-1} + a_t \qquad (6)$$

or as expressed in the backward shift operator notation

$$N_t = \frac{1}{1 - \phi z^{-1}} a_t \qquad (7)$$

Comparing this with equation (3) shows that this can be considered as the output of a first order transfer function in response to a random input sequence. More generally, most stochastic disturbances can be modelled by a general autoregressive-integrated moving-average (ARIMA) time series model of order (p,d,q), that is,

$$N_t = \Phi_1 N_{t-1} + .. + \Phi_{p+d} N_{t+p+d} + a_t - \theta_1 a_{t-1} - \cdots - \phi_q a_{t-q} \qquad (8)$$

or

$$N_t = \frac{(1 - \theta_1 z^{-1} - \ldots - \theta_q z^{-q})}{(1 - \phi_1 z^{-1} - \ldots - \phi_p z^{-p})(1 - z^{-1})^d} a_t \qquad (9)$$

where the autoregressive operator in equations (8) and (9) is seen to contain d roots equal to unity (that is, d difference operators). Usually d = 0 or 1. The presence of a non-zero d permits one to model the drifting type of non-stationary disturbances often seen in practice, and particularly in situations where control is necessary. It will be seen later that only in the presence of non-stationary disturbances ($d > 0$) will the optimal controller contain integral action. The orders p,d, and q, of the autoregressive, differencing, and moving average operators respectively can be identified from plant data.

Identification from Plant Data: Assuming that one is starting with open-loop data collected from the process, Box and Jenkins (5, 4) suggest the following very effective iterative approach to identifying the structure and parameters of the combined dynamic-stochastic model of equations (5) and (9). (A modification of this procedure for use when closed-loop data has been collected is presented by Box and MacGregor (7)). First, the orders of the transfer function and time series models for the process and the disturbances are identified using simple cross-correlation and autocorrelation techniques, plus any available knowledge of the system. The tentatively selected model form is then fitted to the plant data and its parameters estimated. This involves use of any commonly available nonlinear regression program. Diagnostic checks are then applied to the model residuals with the object of uncovering any possible lack of fit and diagnosing its cause.

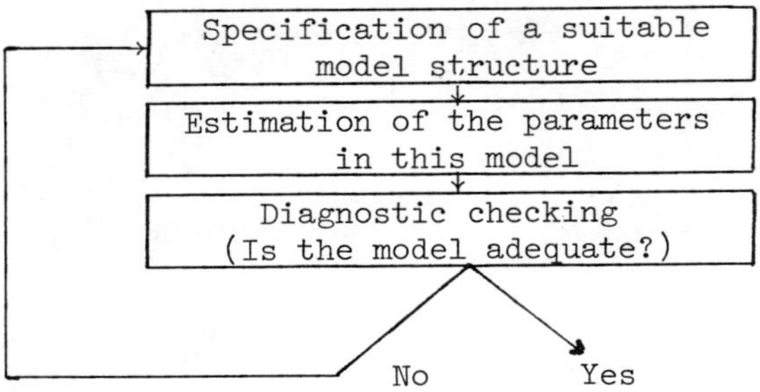

If inadequacy is found, the iterative cycle of identification, estimation, and diagnostic checking is repeated until a suitable representation is found.

Having developed a linear dynamic-stochastic model for the process, one is then in a position to better understand the dynamic or transient behaviour of the process in response to changes in the manipulated variables and to have some appreciation for the characteristics of the disturbances in the system. Before proceeding with a discussion of the theory and application of stochastic control techniques to polymer reactor systems based on these models, it should be stressed that in many instances the improved understanding of the system that has resulted from the development of these models has eliminated the need to implement such control strategies. In some instances, this has come about through simply changing the operating procedures for the process so that desired processing changes are affected by changing variables which have more desirable transient effects on the polymer properties than the variables traditionally used. In other instances, by studying the nature of the disturbance structure at various measurement points in the process, it was possible to track down and eliminate some of these major disturbances at their sources.

Control of Polymer Reactor Systems

The control schemes to be discussed in this section are based on the discrete time empirical models of the previous section. The major emphasis will be on the minimum variance stochastic control schemes of Box and Jenkins (4) and Aström (1), and on modifications of them. These schemes have seen successful application in the polymer industry, and they are intuitively appealing and yet simple enough to be implemented by the plant operators using either a programmable hand calculator or control charts and tables. More powerful adaptive versions can be implemented if a small online mini-computer is available.

Minimum Variance Control: Given that the objective of most continuous flow polymer reactor systems is to maintain the output polymer properties as close as possible to some target or specification values in the face of stochastic disturbances, it would seem to be sensible to design control schemes which would minimize the variance of the output deviation from their target values. For the general theory the reader is referred to the indicated references. In this section the methodology will be illustrated by way of an application to the control of viscosity in a condensation polymerization reactor.

In the production of a condensation polymer the melt viscosity of the final polymer (Y) was measured every 15 minutes by an on-line viscometer. This viscosity could be kept near its target value by manipulating a variable (X). By performing a very simple plant test in which the input (manipulated) variable was perturbed in a pseudo-random binary manner and the resulting viscosity recorded every 15 minutes by the on-line viscometer, the following dynamic-stochastic model was identified for the process:

$$Y_t = \frac{-0.076}{1 - 0.46z^{-1}} X_{t-2} + N_t \tag{10}$$

where Y_t and X_t are deviations from mean values, and the disturbance N_t in the melt viscosity was identified as

$$N_t = \frac{1 - 0.76z^{-1}}{1 - z^{-1}} a_t \tag{11}$$

with variance of the white noise sequence a_t given by $\sigma_a^2 \doteq 15$. The aim of the feedback controller is to compensate for this disturbance in the output viscosity by changing the input variable X_t. The total effect of the disturbance on the output at time t would be cancelled if one could set the effect of the compensation equal to the negative of the effect of the disturbances, that is, by setting

$$\frac{-0.076}{1 - 0.46z^{-1}} X_{t-2} = -N_t$$

Thus, the control action at time t should be such that

$$X_t = \frac{1 - 0.46z^{-1}}{0.076} N_{t+2} \tag{12}$$

However, this action is physically unrealizable since it involves future values of the disturbance. The best one can do is to replace the future value N_{t+2} by its two-step ahead minimum variance forecast $\hat{N}_{t+2}$ taken at the present time t. It can be shown that this will provide the control action which will minimize the variance of the deviations of output viscosity from its target value.

For the disturbance model given in equation (11) the minimum variance forecast is given by

$$\hat{N}_{t+2} = \frac{0.24}{1 - z^{-1}} a_t \tag{13}$$

If this forecast is used in place of N_{t+2} in the control scheme in equation (12), the resulting output deviation from target (e_t) at time t will be given by the two-step ahead forecast error of N_t,

$$(Y_t\text{-Target}) = e_t = (1 + 0.24z^{-1})a_t \tag{14}$$

Eliminating the variable a_t in equation (12) by using (14) yields

$$\hat{N}_{t+2} = \frac{0.24}{(1 - z^{-1})(1 + 0.24\ z^{-1})} e_t \tag{15}$$

which upon substituting for N_{t+2} in equation (13) gives the minimum variance control algorithm

$$X_t = \frac{0.24(1 - 0.46z^{-1})}{0.076(1 - z^{-1})(1 + 0.24z^{-1})} e_t \tag{16}$$

Expanding this out into a difference equation form gives

$$\nabla X_t = -0.24\nabla X_{t-1} + 3.16e_t - 1.45e_{t-1} \tag{17}$$

where $\nabla X_t = (1-z^{-1})X_t = X_t - X_{t-1}$ is the change to be made in the manipulated pressure variable at time t.

The final control algorithm (17) is a very simple difference equation containing only the past two output viscosity deviations and the past manipulated variable change. It could easily be implemented manually using a nomogram type chart or a programmable hand calculator. In reality, this particular controller was implemented with a process control computer in 1978 and performed as expected. Figure 4 shows the performance of that control system in April 1979. Included in Figure 4 is the performance of a companion system fed the same raw material, but controlled by an experienced operator. It is evident that the control performance of the reactor under stochastic control is the better of the two ($\sigma_e^2 = 3.2$ versus $\sigma_e^2 = 5.8$). It is also evident that its performance could be further improved by re-estimating the model parameters and altering the controller parameters accordingly. (An on-line alternative to this latter step is discussed later in this paper.). While the original disturbance model for this system was identified as $\nabla N_t = (1-0.76z^{-1})a_t$, the elimination of a major disturbance source has altered the model to approximately ∇N_t =

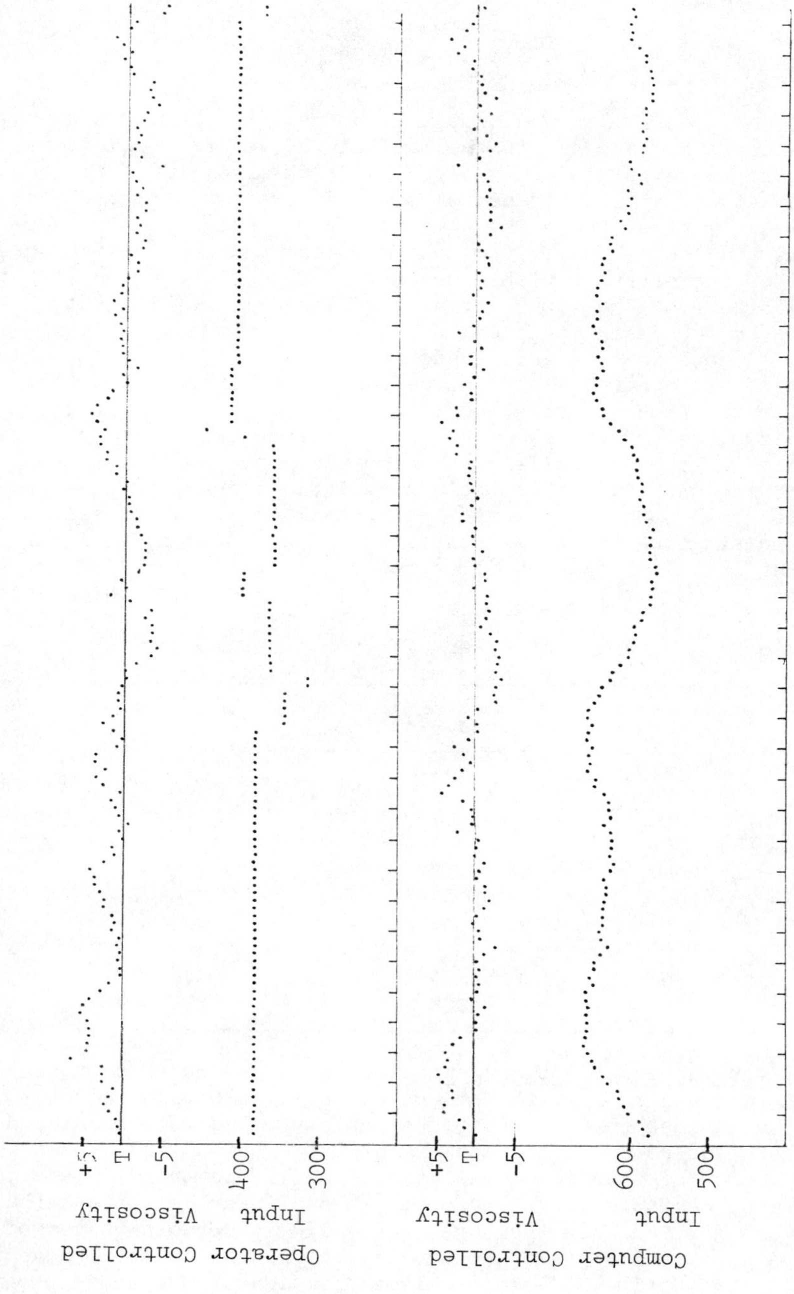

Figure 4. Comparison of operator control and stochastic control of viscosity on two parallel reactor lines

$(1 - 0.3z^{-1})a_t$ with $\sigma_a^2 \simeq 1$. The rederived minimum variance controller would be

$$\nabla X_t = -0.7\nabla X_{t-1} + 9.21e_t - 4.24e_{t-1} \quad (18)$$

However, this controller would require appreciably larger and probably unacceptable changes in the manipulated variable. This is often a problem with minimum variance control algorithms because they effectively attempt to cancel out the forecasted disturbance in the minimum time (the delay time of the process, b). To overcome this the control algorithm can be redesigned to minimize the variance of the output variable deviations subject to a constraint on the variance of the input variable manipulations. The theory is available for designing these constrained control algorithms (17, 11, 15) and in this case leads to the same difference equation form of the control algorithm as that given in equation (17), except for the presence of an additional ∇X_{t-2} term. As more constraint is placed on the allowable manipulated input variance, the performance of the controller, as measured by the variance of the output deviations, will increase. A suitable compromise between the variances of the manipulated and controlled variables can be found by evaluating these variances for a range of values of the constraint parameter (λ), as illustrated in Table I for the current process.

Table I. Variances of e_t and ∇X_t for controllers with various degrees of constraining.

λ	$\mathrm{Var}(e_t)/\sigma_a^2$	$\mathrm{Var}(\nabla X_t)/\sigma_a^2$
0.	1.49	102.8
.01	1.64	20.1
.02	1.73	13.9
.04	1.86	9.4
.06	1.95	7.6
.10	2.09	5.8
.25	2.44	3.6
.50	2.82	2.5
1.0	3.35	1.7

Self-tuning and Adaptive Control: It is obvious from the preceding discussion that the design of full scale plant experiments, the development of suitable dynamic-stochastic models, and finally the design of a stochastic control algorithm requires a certain degree of experience and special skills not readily available in many companies. Furthermore, these linear models and the controllers derived from them are only valid in certain ranges of plant operation and only if the process parameters do not change appreciably with time. These problems have led to the development of what are termed "self-tuning" and "adaptive" versions of these controllers (2,18,10). The basic idea is illustrated below.

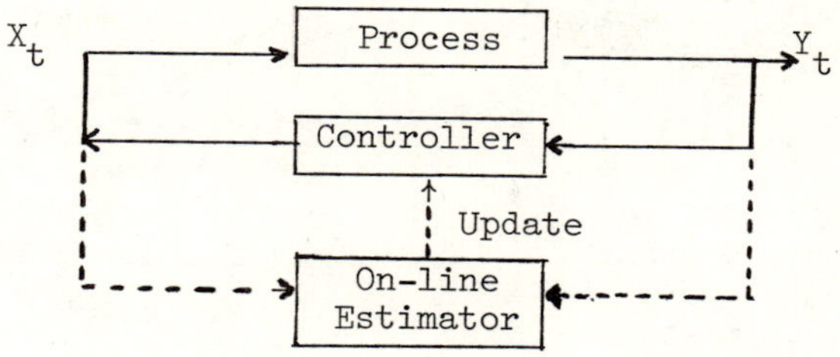

Based on current knowledge of the process and its disturbance characteristics, one may know or choose a reasonable difference equation structure for the controller algorithm. Starting with some assumed initial parameter values in the controller equation, the controller can be implemented on the process as shown. The control algorithm is coupled with an on-line recursive estimation algorithm which receives information on the inputs and outputs at each sampling interval and uses this to recursively estimate the optimal controller parameters on-line and to update the controller accordingly. The idea is to use the data collected from the on-line control manipulations to tune the controller directly. Even starting off with poor initial controller parameter values, the self-tuning algorithm usually converges quite rapidly to a stable controller. After a certain amount of data has been collected one can test whether or not the assumed controller form is optimal (in the sense of the minimum variance) and then change it if necessary.

For processes which operate over a wide range of conditions or production rates it is to be expected that the process and disturbance model parameters will be changing with time. With a very minor change in the recursive estimation algorithm one can use this scheme to track these slowly moving processes and thereby keep the controller well tuned at all times.

Industrial applications of these controllers have been successfully carried out in Sweden on paper machines (9) and on an ore crusher system (3) . A good discussion of the practical aspects of the approach and an application to the control of a packed bed tubular reactor is contained in a paper by Harris et al. (12).

Literature Cited

1. Aström, K.J.; "Introduction to Stochastic Control Theory"; Academic Press, New York, 1970.
2. Aström, K.J.; Wittenmark, B. Automatica, 1973, 9, 185-199.
3. Borisson, U.; Syding R. Automatica, 1976, 12, 1-7.
4. Box, G.E.P.; Jenkins, G.M.; "Time Series Analysis, Forecasting and Control", Holden Day, San Francisco, 1970.

5. Box, G.E.P.; Jenkins, G.M. Applied Statistics, 1968, 17, 91-109.
6. Box, G.E.P.; Jenkins, G.M.; MacGregor, J.F. Applied Statistics, 1974, 17, 158-179.
7. Box, G.E.P.; MacGregor, J.F. Technometrics, 1974, 16, 391-398.
8. Box, G.E.P.; Newbold, P. J. Roy. Statis. Soc., 1971, Series A, 134, 229.
9. Cegrell, T.; Hedqvist, T. Automatica, 1975, 11, 53-59.
10. Clarke, D.W.; Gawthrop, P.J. Proc. Inst. Elec. Eng., 1975, 122, 929-934.
11. Clarke, D.W.; Hastings-James, R. Proc. Inst. Elec. Eng., 1971, 118, 1503-1506.
12. Harris, T.; MacGregor, J.F.; Wright, J.D. Proc. Joint Aut. Control Conf., 1978, Philadelphia, October.
13. Kiparissides, C.; MacGregor, J.F.; Hamielec, A.E. Suboptimal Stochastic Control of a Continuous Latex Reactor, 1979, S.O.C. Report, Faculty of Engineering, McMaster University, Hamilton Canada. (Submitted for publication).
14. MacGregor, J.F. Technometrics, 1976, 18, 151-160.
15. MacGregor, J.F.; Tidwell, P.W. Proc. Inst. Elec. Eng., 1977, 124, 732.
16. Mehra, R.K. IEEE Trans. Aut. Control, 1974, 19, 753-768.
17. Wilson, G.T., 1970, Ph.D. Thesis, University of Lancaster, England. (See also Tech. Rep. No.20, Dept. of Systems Engineering.)
18. Wittenmark, B. "A Self-Tuning Regulator", 1973, Technical Report 7311, Division of Automatic Control, Lund, Inst. of Tech., Lund, Sweden.

RECEIVED November 5, 1979.

14

Minimizing Round-Off Error in Direct Solution of Large Sparse Systems of Linear Equations

M. SHACHAM[1]

Department of Chemical Engineering, Ben Gurion University of the Negev, Beer Sheva, Israel

Many engineering problems require solution of large and sparse systems of linear equations. Typical examples are: modeling and simulation of staged, multicomponent processes, pipeline network design, solution of electrical networks and partial differential equations. Among others, chemical engineers also show a growing interest in the solution method of sparse linear systems (Lin and Mah (7), Gabrielli and Spadoni (2)).

A system of linear equations can be written in the form

$$\underline{A}\underline{x} = \underline{b} \tag{1}$$

where $\underline{A}$ is an nxn matrix, $\underline{x} \varepsilon R^n$ is the unknown vector and $\underline{b} \varepsilon R^n$ is a vector of constants.

Sparse systems encountered in practical problems are characterized by a very large value of n (several hundreds) and the number of non-zero elements in matrix $\underline{A}$ is small, usually less than 5%. When solving such a system, advantage of its sparseness can be taken in two ways: 1. By storage of only the nonzero elements of A and 2. By making arithmetical operations with these elements only.

Several storage schemes can be used for sparse matrices. The most simple scheme is the "random packing" scheme, where $\underline{A}$ is stored in three vectors. One of them (a real array) contains the value of the non-zero elements. Two integer vectors specify their row and column numbers in the corresponding locations. The advantage of this storage scheme is that new nonzero elements (such as the ones generated during the Gauss-elimination process) can be easily added at the end of the list. The major drawback of this scheme is that the scanning of the matrix (both the row-wise and column-wise scanning) is quite complicated and extra computer time is required for it.

---

1 Current address: Department of Chemical Engineering, University of Connecticut, Storrs, Connecticut 06268.

0-8412-0549-3/80/47-124-269$05.00/0

The "systematic packing" scheme is more efficient for such operations. In this scheme, $\underline{A}$ is stored in two vectors. A real vector containing the row number is followed by the value of the nonzero elements in this row, ordered according to increasing column numbers. An integer vector contains the column numbers and also dummy elements to indicate the location of the row numbers in the first vector.

Matrices of special structure like symmetric or band type matrices can be stored even more effectively. In some cases, the matrix $\underline{A}$ must not be stored at all and the elements of a single row or column can be generated during the computation (see the example at the end of this article).

The solution of a sparse system of equations can be carried out in three stages: 1. Partitioning, 2. Reordering or "tearing", and 3. Numerical solution. Stages 1 and 2 contain only logical operations and their objective is to obtain a system which can be solved faster and/or with smaller round-off error propagated. These stages are much more time-consuming than stage 3. (The ratio between the computer time required for stages 1 and 2, to the time required for stage 3 can be up to two orders of magnitude (7)). For this reason, it is recommended to carry out stages 1 and 2 only for linear systems that have to be solved several times with different numerical values.

Partitioning means separation of the matrix into smaller blocks which may be considered consecutively rather than considering the whole matrix simultaneously.

The exact details of the reordering or tearing stage depend on the numerical solution used, so these stages will be discussed together.

Direct or indirect methods are used for the numerical solution of Eq. (1). The direct method involves a fixed number of operations where indirect or iterative methods necessitate the repetition of certain steps until the required accuracy is achieved. All of the direct methods are variants of the Gauss or Gauss-Jordan elimination. The disadvantages of the direct method for a general non-band type sparse system are as follows: 1. During the elimination, the structure of matrix $\underline{A}$ is changed, since new nonzero elements ("fill-ins") are generated. The generation of "fill-ins" increases storage and computation time requirements. 2. Both row and column-wise scanning of the matrix has to be carried out, which makes it necessary to use a larger and more complicated storage scheme.

The amount of "fill-ins" can be minimized by reordering the equations (in stage 2). One of the more recent algorithms for reordering has been published by Lin and Mah (6). The reordering may often be dangerous, since it may prevent "pivoting" which is necessary for reducing round-off error propagation.

Iterative methods (like Gauss-Seidel, Successive over relaxation and conjugate gradient) have often been preferred to the direct method for solving sparse sets of equations for the following reasons: 1. The matrix $\underline{A}$ remains unchanged during the computation. 2. Only row-wise scanning of the matrix is required. These two properties lead to much simpler computer codes and reduction of the overhead time which is necessary for index manipulation. Iterative methods however, cannot be used for all types of problems, since unless the matrix $\underline{A}$ has some special properties, the convergence may be slow or unattainable.

The concept of "tearing" has been developed in connection with the iterative methods. First an output set for the system of equations is chosen. Then one or more tearing variables are selected. These variables are the iterates that need to be chosen to obtain a solution of the system. The number of tearing variables is usually much smaller than the number of the equations. An accepted criteria for selecting tearing variables is the minimum number of such variables which will make it possible to solve the whole system. The ordered set of equations that results is then solved using an iterative method.

The concepts of partitioning and tearing have been discussed recently by Bunch and Rose (1) and Kevorkian (3). Shacham and Kehat (8) introduced a direct method of solution which shares most of the advantages of both the direct and iterative methods. It has been found, however, that this method is quite sensitive to round-off error propagation. This article presents methods both for minimizing round-off error and for iterative refinement of the solution.

Shacham and Kehat's Direct Method

It is assumed that the original system of equations has already been partitioned into the smallest possible blocks and Eq. (1) defines one of these blocks. The linear system given in Eq. (1) is permuted so as to arrive at the following equivalent system:

$$\begin{bmatrix} \underline{T} & \underline{S} \\ \underline{F}_1 & \underline{F}_2 \end{bmatrix} \begin{bmatrix} \underline{x}_1 \\ \underline{x}_2 \end{bmatrix} = \begin{bmatrix} \underline{b}_1 \\ \underline{b}_2 \end{bmatrix} \tag{2}$$

where: $\underline{x}_1 \varepsilon R^m$ and $\underline{x}_2 \varepsilon R^{n-m}$ are subvectors of $\underline{x}$, $\underline{b}_1 \varepsilon R^{n-m}$ and $\underline{b}_2 \varepsilon R^m$ are subvectors of $\underline{b}$, $\underline{S}$ is an (n-m)(n-m) lower triangular submatrix of $\underline{A}$ with non-zero main diagonal $\underline{T}$, $\underline{F}_1$ and $\underline{F}_2$ are (n-m)xm, mxm and mx(n-m) submatrices of $\underline{A}$ respectively.

Let us rewrite Eq. (2) and define $f(\underline{x}_1)$:

$$\underline{x}_2 = \underline{S}^{-1}(\underline{b}_1 - \underline{T}\underline{x}_1) \tag{3a}$$

$$\underline{f}(\underline{x}_1) = \underline{F}_1\underline{x}_1 + \underline{F}_2\underline{x}_2 - \underline{b}_2 \tag{3b}$$

where $\underline{f}(\underline{x}_1)$ is a vector of m functions.

The system consisting of Eqs. (3a) and (3b) can be solved as follows:

1. Assume an initial value for the tear variables ($\underline{x}_1^\circ$);
2. Compute $\underline{x}_2^\circ$ from Eq. (3a), (note that this can be done by substitution since $\underline{S}$ is lower triangular);
3. Compute $\underline{f}(\underline{x}_1^\circ)$ from Eq. (3b);
4. Compute $\underline{\nabla f}(\underline{x}_1^\circ)$ where $\nabla f(\underline{x}_1^\circ)$ is the mxm matrix of partial derivatives of $\underline{f}(\underline{x}_1)$ with respect to $\underline{x}_1$. (The method of calculating this matrix is given below).
5. Use Newton iteration to compute

$$\underline{x}_1^1 = \underline{x}_1^\circ - \nabla f(\underline{x}_1^\circ)^{-1}\, \underline{f}(\underline{x}_1^\circ) \tag{4}$$

6. Compute $\underline{x}_2^1$ from: $x_2^1 = \underline{S}^{-1}\,(\underline{b}_1 - \underline{Tx}_1{}^1)$;
7. The vector $(\underline{x}_1^1, \underline{x}_2^1)$ is the solution of system (1).

The matrix of partial derivatives is calculated numerically column by column by setting $\underline{b} = \underline{0}$; $x_1, x_2 \ldots, x_{j-1}\; x_{j+1}, \ldots x_m = 0$; $x_j \neq 0$, calculating $\underline{f}(\underline{x}_1)$ using equations (3a) and (3b) and then the derivatives from Eq. (5)

$$\frac{\partial f_i}{\partial x_j} = \frac{f_i(\underline{x}_1)}{x_j} \tag{5}$$

The use of this algorithm requires (m+1) times of computation of the system (3a) and (3b) and solution of an (mxm) linear system of equations. (In Eq. (4)). The storage requirements are mainly the sparse storage for the matrix $\underline{A}$ and storage for the mxm dense matrix of the derivatives.

The m variables of $\underline{x}_1$ can be regarded as tearing variables and clearly the smaller m is, the more effective the algorithm. Since computer programs for selection of minimal sets are readily available (Ledet and Himmelblau (5)), this problem will not be discussed.

Shacham and Kehat's direct method has several advantages over similar methods. The matrix $\underline{A}$ remains unchanged during the solution process and only row-wise scanning operations have to be carried out. The "systematic packing" storage scheme can be used and the amount of index manipulation is reduced to minimum. The problem of "fill-ins" does not exist since newly-generated elements are included either in the matrix $\underline{\nabla f}(\underline{x}_1)$ or in the vector $\underline{f}(\underline{x}_1)$.

An example which demonstrates the use and the advantages of this method for a small 6x6 system is given in the Appendix.

Shacham and Kehat (8) showed substantial savings in computer time when using this method. It has been noticed however that for certain systems, round-off error propagation may prevent achievement of correct results. In the next section, the cause of error propagation is investigated.

Error Analysis

For practical problems, m<<n and it can be assumed that error propagation mainly occurs when using equation (3a) for calculating $\underline{x}_2$. Let us denote as $\underline{\delta x}_1$ the error in $\underline{x}_1$ then equation (3a) can be used for obtaining an error bound for $\underline{x}_2$.

$$\|\underline{\delta x}_2\| \leq \|\underline{S}^{-1}\underline{T}\| \, \|\underline{\delta x}_1\| \tag{6}$$

where $\|\cdot\|$ indicates some type of matrix norm. Equation (6) shows that the larger the elements of $\underline{S}^{-1}$, the bigger is the error in the calculated value of $\underline{x}_2$. Since $\underline{S}$ is a triangular matrix, its inverse can be generated using recursive formulas (see p. 243 in (4)). In these formulas, the diagonal elements of $\underline{S}$ are divisors and the off-diagonal elements are multipliers. Obviously both the magnitude of the elements in $\underline{S}^{-1}$ and the round-off error propagation can be minimized by bringing the elements of largest absolute value into the main diagonal of the $\underline{S}$ matrix.

The round-off error in $\underline{f}(\underline{x}_1)$ can be calculated using Eq.(3b):

$$\underline{\delta f}_1 = \underline{F}_1\underline{\delta x}_1 + \underline{F}_2\underline{\delta x}_2 \tag{7}$$

where $\underline{\delta f}_1$ is the error in $\underline{f}(\underline{x}_1)$.

Substituting $\underline{\delta x}_2$ into Eq.(7) gives

$$\underline{\delta f}_1 = (\underline{F}_1 - \underline{F}_2\underline{S}^{-1}\underline{T})\underline{\delta x}_1 \tag{8}$$

We have found in all the practical cases, that when serious error propagation occurs, the term $\underline{F}_1$ is much smaller than $\underline{F}_2\underline{S}^{-1}\underline{T}$ in Eq. (8). Thus a good estimation of $\underline{\delta f}_1$ is obtained from:

$$\|\underline{\delta f}_1\| \simeq \|\underline{\delta x}_2\| \, \|\underline{F}_2\|$$

Using equations (8), (3a) and (3b), it can be shown that the relative errors in the calculated derivative values are equal to the relative error in $\underline{x}_1$.

Introducing (3a) into (3b) gives:

$$\underline{f}(\underline{x}_1) = (\underline{F}_1 - \underline{F}_2\underline{S}^{-1}\underline{T})\underline{x}_1 - \underline{F}_2\underline{S}^{-1}\underline{b}_1 - \underline{b}_2 \tag{9}$$

When calculating the derivatives, $\underline{b}_1$ and $\underline{b}_2$ are set at zero as well as all the elements of $\underline{x}$ except x_j. For that case, equation (9) reduces to:

$$\underline{f}(\underline{x}_1) = (\underline{F}_1 - \underline{F}_2\underline{S}^{-1}\underline{T})\underline{x}_1 \tag{10}$$

and the right-hand side of this equation can be interpreted as a simple vector multiplication by a scalar. This fact can be used to arrive at the following expression:

$$\frac{\left\|\frac{\delta \underline{f}_j}{}\right\|}{\left\|\underline{f}_j\right\|} = \frac{\left\|\delta \underline{x}_j\right\|}{\left\|\underline{x}_j\right\|} \quad ; \text{ for } j=1,2,\ldots m \tag{11}$$

where the subscript j indicates that the calculations have to be made when setting all the elements of $\underline{x}_1$ at zero.

The error in the up-dated value of $\underline{x}_1$ (thus $\underline{x}_1^{\,1}$) can be estimated using equation (4). Keeping in mind that m<<n and assuming that $\underline{\nabla f}(\underline{x}_1)$ is a well-conditioned matrix, leads to the conclusion that the error in the correction term: $\underline{x}_1^{\,1} - \underline{x}_1^{\,o}$ is approximately $\|\underline{F}_2\| \; \|\delta \underline{x}_2\|$. Numerical experiments have verified this result.

For summarizing the practical results of the error analysis, let us assume that the solution of the system of equation started by setting $\underline{x}_1^{\,o} = \underline{1}$. The error in the elements of $\underline{x}_1^{\,o}$ in that case is of the order of 10^{-t} where t is the number of significant decimal digits carried by the computer. The accuracy of the solution: $(\underline{x}_1^{\,1}, \underline{x}_2^{\,1})$ depends mainly on the absolute value of the norm $\|\underline{S}^{-1}\underline{T}\|$. Let us say that this value is 10^k. If k<<t, then the solution will be quite accurate. For $k > \frac{1}{2}t$, the solution will be very inaccurate and some elements of $\underline{x}_2^{\,1}$ will probably be completely incorrect. In such cases, a more accurate solution can be obtained by reordering the matrix $\underline{A}$ so as to bring the elements of largest absolute value into the main diagonal of the submatrix $\underline{S}$.

Iterative Refinement of the Solution

In some cases, the rearrangement of matrix $\underline{A}$ for improving accuracy is undesirable or even impossible, since it increases the number of tearing variables. In such cases, an iterative method can be used to improve the accuracy of the solution.

As we have seen before, even if $\underline{x}_1$ is accurate upto the last digit, the calculated value of $\underline{x}_2$ can still be erroneous, depending on the magnitude of $\|\underline{S}^{-1}\underline{T}\|$. This difficulty can be overcome by updating the correction for $\underline{x}_2$ according to the correction for $\underline{x}_1$, instead of updating $\underline{x}_2$ itself according to $\underline{x}_1$.

Let $\Delta\underline{x}_1^i$ be the correction for $\underline{x}_1$ in iteration number i. Thus: $\Delta\underline{x}_1^i = \underline{x}_1^i - \underline{x}_1^{i-1}$. Substituting $\Delta\underline{x}_1^i$ into equation (4) gives:

$$\Delta\underline{x}_1^i = -\nabla\underline{f}\,(\underline{x}_1^{o})^{-1}\underline{f}(\underline{x}_1^{i-1}) \tag{12}$$

Let $\underline{x}_2^{i+1} = \underline{x}_2^i + \Delta\underline{x}_2^i$. Substitution of $\underline{x}_2^{i+1}$ into Eq. (3) gives:

$$(\underline{x}_2^i + \Delta\underline{x}_2^i) = \underline{S}^{-1}[\underline{b}_1 - \underline{T}(\underline{x}_1^i + \Delta\underline{x}_1^i)] \tag{13}$$

Subtracting $\underline{x}_2^i$ from both sides in Eq. (13) leads to:

$$\Delta\underline{x}_2^i = \underline{S}^{-1}\underline{T}\Delta\underline{x}_1^i \tag{14}$$

Using Eq. (14), the correction term for $\underline{x}_2$ can be calculated only from the correction term for $\underline{x}_1$.
The improved algorithm for solving sparse systems of linear equations follows:

1. Assume an initial value for $\underline{x}_1^o$
2. Compute $\underline{x}_2^o$ from Eq. (3a)
3. Compute $\underline{f}(\underline{x}_1^o)$ from Eq. (3b) and compute $\nabla\underline{f}(\underline{x}_1^o)^{-1}$
4. Calculate $\Delta\underline{x}_1^i$ from Eq. (12) and $\Delta\underline{x}_2^i$ from Eq. (14)
5. Update $\underline{x}_1^i$ and $\underline{x}_2^i$ according to $\underline{x}_1^{i+1} = \underline{x}_1^i + \Delta\underline{x}_1^i$; $\underline{x}_2^{i+1} = \underline{x}_2^i + \Delta\underline{x}_2^i$.
6. Calculate $\underline{f}(\underline{x}_1)$ from Eq. (3b)
7. Test accuracy. If $\|\underline{f}(\underline{x}_1)\|$ is smaller than a prespecified error tolerance, finish, otherwise go back to 4.

The use of this algorithm requires (m+1) times of computation of the system (3a) and (3b) and inversion of an (mxm) dense matrix at the first iteration. Work associated with the subsequent iterations is very small and includes mainly a single evaluation of Eqs. (12), (14) and (3b). It was shown in the previous section that the matrix of the partial derivatives is quite accurate and it must not be updated from iteration to iteration.

Example

A system of linear submatrices obtained by the usual 5 point difference approximation for the Laplace equation in a rectangle Reid (1971) p. 239).
For a kxℓ grid, the matrix A is of the order of n= kxℓ and has an ℓxℓ block form:

$$\underline{A} = \begin{bmatrix} \underline{T} & -\underline{I} & & & & \\ -\underline{I} & \underline{T} & -\underline{I} & & & \\ & -\underline{I} & \underline{T} & -\underline{I} & & \\ & & \cdot & \cdot & \cdot & \\ & & & \cdot & \cdot & \cdot \\ & & & & -\underline{I} & \underline{T} \end{bmatrix} \tag{15}$$

where $\underline{I}$ is the kxk identity matrix and T is the kxk tridiagonal matrix:

$$\underline{T} = \begin{bmatrix} 4 & -1 & 0 & \cdot & \cdot & \cdot & 0 \\ -1 & 4 & -1 & \cdot & \cdot & \cdot & 0 \\ 0 & -1 & 4 & & & & \\ & & \cdot & \cdot & & & \\ & & & \cdot & \cdot & & \\ 0 & & & & -1 & & 4 \end{bmatrix} \tag{16}$$

Obviously, the minimum number of tear streams is the smaller value of either k or ℓ. If k has the smaller value, the first k variables can be used as the tear variables and the system can be transformed into:

$$x_{i+k} = c_1+c_2+c_3+4x_i-b_i \qquad i=1,2,\ldots n-k$$

$$f_j = c_1+c_2+c_3+4x_i-b_i \qquad i=n-k+1,\ldots n,\ j=i+k-n$$

where:

$$c_1 = 0 \text{ for } i \leq k \ ; \quad c_1 = -x_{i-k} \text{ for } i > k$$

$$c_2 = \begin{cases} 0 \text{ for the first row of each block} \\ -x_{i-1} \quad \text{elsewhere} \end{cases}$$

$$c_3 = \begin{cases} 0 \text{ for the last row of each block} \\ -x_{i+1} \quad \text{elsewhere} \end{cases}$$

$$\underline{b} = 1$$

In this problem, the value of $\left\| \underline{S}^{-1}\underline{T} \right\|$ grows very fast when the dimension: n, increases. For example, for k=5 and n=25,50 and 100, the largest elements of the matrix $\underline{S}^{-1}\underline{T}$ are 77, 2.1×10^5 and 5×10^{12} respectively. This problem has been solved for different values of n and k, up to n=500. A CDC Cyber 73 computer with single precision arithmetic (16 decimal digits) has been used. The results are summarized in Table 1.

It can be seen from this Table that up to n=50, quite accurate results are obtained within a single iteration (the error is less than 10^{-7}), but for n=100, the error after the first iteration is of the same order of magnitude as the solution itself.

The use of the proposed iterative method reduced the error below 10^{-10} in all the cases, which means that a solution with at least 10 accurate digits has been obtained. The solution for $\underline{x}$ obtained in this way has been compared to the solution calculated by different methods (i.e. Gauss-Jordan elimination). The results of the comparison verified the above conclusion.

Table 1
Change in the logarithm of the Euclidean norm of $\underline{f}(\underline{x}_1)$ as a function of the iteration number:

Iteration No.	n=50 k=5	n=100 k=5	n=200 k=10	n=500 k=25
1	-7.4	0.35	1.21	2.14
2	-13.3	-5.17	-1.94	0.95
3	-	-10.41	-5.09	-0.2
4	-	-	-7.86	-1.36
5	-	-	-11.03	-2.51
.				.
.				.
.				.
12	-	-	-	-10.67
Computation time (seconds)	0.055	0.105	.38	3.8

The computation time in all the cases tested was below 4 seconds.

Conclusion

The norms $\underline{S}^{-1}\underline{T}$ and $\underline{f}(\underline{x}_1)$ can be used as indicators for the rounding error propagated when using Shacham and Kehat's direct method. The rate of error propagation can be reduced by reordering matrix $\underline{A}$ to bring the elements of largest absolute value to the main diagonal of the submatrix $\underline{S}$.

Iterative refinement of the solution, which requires very small computational effort, leads to accurate solutions even for extremely large problems.

Abstract

The round-off error propagation associated with the use of Shacham and Kehat's direct method for the solution of large sparse systems of linear equations is investigated. A reordering scheme for reducing error propagation is proposed as well as a method for iterative refinement of the solution. Accurate solutions for linear systems, which contain up to 500 equations, have been obtained using the proposed method, in very short computer times.

Literature Cited

1. Bunch, J.R.; Rose, D.J., "Partitioning, Tearing and Modification of Sparse Linear Systems", J. Math. Anal. Appl., 1974, 48(2), 574.

2. Gabbrielli, E.; Spadoni, G., "The Iterative Solution of Large Systems of Linear Equations: Some New Empirical Criteria and Tests", Comp. Chem. Eng., 1977, 1, 113.

3. Kevorkian, A.J., "A Decomposition Algorithm for the Solu-

tion of Large Systems of Linear Algebraic Equations", Proc. of the 1975 IEEE Symp. on Circuits and Systems, Boston, USA, IEEE Cat. No. 75 (II0937-3 CAS, pp. 116-120).

4. Lapidus, L., "Digital Computation for Chemical Engineers", McGraw-Hill Book Co., Inc., New York, 1962.

5. Ledet, W.P.; Himmelblau, D.M., "Decomposition Procedures for Solving of Large Scale Systems", Adv. Chem. Eng., Academic Press, Vol. 8, 1970.

6. Lin, T.D.; Mah, R.S.H., "Hierarchical Partition - A New Optimal Pivoting Algorithm", Math. Prog., 1977, 12, 260.

7. Lin, T.D.; Mah, R.S.H., "A Sparse Computation System for Process Design and Simulation", AIChE J., 1978, 24(5), 830.

8. Shacham, M.; Kehat, E., "A Direct Method for the Solution of Large Sparse Systems of Linear Equations", Comp. J., 1976, 19(4), 353.

APPENDIX

Solution of a small linear system using Shacham and Kehat's method

$$\begin{bmatrix} 4 & -1 & 0 & -1 & 0 & 0 \\ -1 & 4 & -1 & 0 & -1 & 0 \\ 0 & -1 & 4 & 0 & 0 & -1 \\ -1 & 0 & 0 & 4 & -1 & 0 \\ 0 & -1 & 0 & -1 & 4 & -1 \\ 0 & 0 & -1 & 0 & -1 & 4 \end{bmatrix} \begin{bmatrix} x_1 \\ x_2 \\ x_3 \\ x_4 \\ x_5 \\ x_6 \end{bmatrix} = \begin{bmatrix} 100 \\ 0 \\ 0 \\ 100 \\ 0 \\ 0 \end{bmatrix}$$

After rearrangement:

Var. No.: 3 5 6 2 4 1

Eq. No.

$$\begin{array}{c} 6 \\ 3 \\ 5 \\ 4 \\ \\ 1 \\ 2 \end{array} \left[\begin{array}{cc|cccc} -1 & -1 & 4 & 0 & 0 & 0 \\ 4 & 0 & -1 & -1 & 0 & 0 \\ 0 & 4 & -1 & -1 & -1 & 0 \\ 0 & -1 & 0 & 0 & 4 & -1 \\ \hline 0 & 0 & 0 & -1 & -1 & 4 \\ -1 & -1 & 0 & 4 & 0 & -1 \end{array}\right] \left[\begin{array}{c} x_3 \\ x_5 \\ \hline x_6 \\ x_2 \\ x_4 \\ x_1 \end{array}\right] = \left[\begin{array}{c} 0 \\ 0 \\ 0 \\ 100 \\ \hline 100 \\ 0 \end{array}\right]$$

Row-wise "systematic packing" storage

0	6	← Row No.
3	-1	← Coeff.
5	-1	
6	4	
0	3	
3	4	
6	-1	
2	-1	
0	5	
5	4	
6	-1	
2	-1	
4	-1	
0	4	
5	-1	
4	4	
1	-1	
7	-100	
0	1	
2	-1	
4	-1	
1	4	
7	-100	← Variable No.
0	2	← Dummy element
3	-1	
5	-1	
6	4	
1	-1	

Calculation of $f_1(\underline{x}_1^{\circ})$ and $\nabla f(\underline{x}_1)^{\circ}$

The solution

0	6				
3	1/4	1	1	0	4.762
5	1/4	1	0	1	14.286
6	-1	0.5	0.25	0.25	4.762
0	3				
3	4				
6	-1				
2	-1	3.5	3.75	-0.25	14.286
0	5				
5	4				
6	-1				
2	-1				
4	-1	0	-4	4	38.096
0	4				
5	-1				
4	4				
1	-1	-101	-16	-15	38.098
7	-100		($\underline{b}$ = $\underline{0}$)	($\underline{b}$ = $\underline{0}$)	
0	1				
2	-1				
4	-1	$f_1=$	$\frac{\partial f_1}{\partial x_3}=$	$\frac{\partial f_1}{\partial x_5}$	
1	4	-507.5	-63.75	-56.25	
7	-100		($\underline{b}$ = $\underline{0}$)	($\underline{b}$ = $\underline{0}$)	
0	2				
3	-1				
5	-1	$f_2=$	$\frac{\partial f_2}{\partial x_3}=$	$\frac{\partial f_2}{\partial x_5}=$	
6	4	113	30.	-17.	
1	-1				

$$\begin{bmatrix} -63.75 & -56.25 \\ 30. & -17. \end{bmatrix} \begin{bmatrix} \Delta x_3 \\ \Delta x_5 \end{bmatrix} \begin{bmatrix} -507.5 \\ 113. \end{bmatrix}$$

$\Delta x_3 = 3.762 \quad \Delta x_5 = 13.286$

$x_3^1 = 1 + 3.762 = 4.762$

$x_5^1 = 1 + 13.286 = 14.286$

Calculation of $\underline{x}_1^1$

RECEIVED November 5, 1979.

Variable Sample Time Algorithm for Microcomputer Control of a Heat Exchanger

RICHARD R. STEWART and NEIL G. SWEERUS

Department of Chemical Engineering, Northwestern University, Boston, MA 02115

Mutharasan and co-workers have published several papers on the sampled-data control of distributed parameter systems. Mutharasan and Coughanowr (2) developed DDC algorithms for the flow-forced tubular reactor and heat exchanger as well as the wall temperature forced heat exchanger. The non-linear algorithms employed previous values of the manipulated variable to calculate the current inlet state from which the new manipulated variable is calculated. In another paper (3) the same authors showed that sampled data proportional control of the flow-forced tubular reactor results in sustained oscillations for certain step changes in load for a given proportional gain and sampling time. The oscillations were eliminated by the addition of integral action. Mutharasan and Coughanowr (4) also derived minimal prototype algorithms for a linear model of the wall temperature forced heat exchanger. Mutharasan and Luus (5) developed minimal prototype algorithms for the linearized model of a flow-forced heat exchanger. The authors also used an alternate design method which involved a direct search on the parameters of the general linear DDC algorithm.

The purpose of the present study is to present a simply derived and implemented DDC algorithm for the flow-forced heat exchanger which does not require storage of previous values of the manipulated variable or error. The algorithm requires that steady-state exist at the sampling instants which means that the sample time is variable. However, the algorithm is shown to control for the constant sample time case.

Control Algorithm. The equation for the flow-forced heat exchanger is given by

$$\frac{\partial x}{\partial \theta} + (1+a)\frac{\partial x}{\partial \eta} = -\beta\,(1+a)^{b}x \tag{1}$$

where "a" is the change in fluid velocity relative to the original steady-state velocity. For sampled data control of the

0-8412-0549-3/80/47-124-281$05.00/0

exchanger, "a" is constant between sampling instants.

When the steady-state portion of Equation (1) is integrated between inlet and outlet temperatures, x_0 and x_1, the following equation results

$$x_1 = x_0 \exp[-\beta(1+a)^{b-1}] \quad (2)$$

Equation (2) can now be used to calculate the inlet temperature using the measured outlet temperature, x_1, the estimated system parameter, β_{est}, the known system parameter, b, and the current velocity, (1+a).

$$x_{0,calc} = x_1 \exp[\beta_{est}(1+a)^{b-1}] \quad (3)$$

The control algorithm which calculates the new velocity $(1+a)_{new}$ is obtained by, a) in Equation (2), changing x_1 to its set point value, $x_{1,sp}$, changing (1+a) to $(1+a)_{new}$, and changing β to β_{est}, b) substituting the right-hand side of Equation (3) for x_0 in Equation (2), c) solving the resulting equation for $(1+a)_{new}$. The result is

$$(1+a)_{new} = [\ (1+a)^{b-1} - \frac{1}{\beta_{est}} \ln(x_{1,sp}/x_1)]^{\frac{1}{b-1}} \quad (4)$$

Equation (4) is a feedback control algorithm for both setpoint and load changes which computes the new velocity from the current velocity, current and desired outlet temperatures, and estimated and known system parameters. Storage of previous values of the manipulated variable and error are not required for the algorithm of Equation (4).

Because the control algorithm was derived from the steady-state portion of Equation 1 it may be applied, strictly, only after the system has reached steady-state following a change in velocity, that is, after one residence time. This means that the minimum sampling time, T_s, is given by

$$T_s = (1+a)^{-1}_{new} \quad (5)$$

In addition, because the control signal, fluid velocity, varies from one sampling instant to the next, the minimum sampling time is variable.

Mutharasan and Coughanowr (2) in developing their ALG1 algorithm for the case of constant sampling time also used the controlled and manipulated variables to estimate the current inlet state and then compute the new manipulated variable. However, for the constant sampling time case it is necessary to store previous values of the manipulated variables in order to compute the new manipulated variable. A recent paper by Fuhrman, et al (1) presents experimental data using the ALG1 algorithm.

Experimental

The equipment arrangement is shown in Figure 1.

Equipment. A double-pipe heat exchanger consisting of twenty one feet of 3/4-inch Schedule 40 steel pipe within twenty feet of 3-inch Schedule 40 steel pipe was used. The exchanger and steam lines were insulated with 1-inch formed Fiberglas. Outlet water temperature was measured by an industrial type iron-constantan thermocouple and temperature transmitter manufactured by Taylor. The transmitter was spanned to produce a 10 to 50 ma signal over a temperature range of 16°C. The transmitter current signal was passed through a 50 ohm resistor to produce a 0.5 to 2.5 volt signal which was read by the microcomputer A/D converter and recorded by a Brush recorder.

The computer D/A output signal of 0.0 to 2.54 volts was connected in series with a 1.5 volt battery to bias the voltage range to 1.5 to 4.0 volts. A Foxboro Model 62H controller, set for proportional action, was used as a voltage/current converter. The controller output signal in the 10 to 50 ma range was sent to a Taylor current/pressure converter which produced an output signal in the 3 to 15 psig range. This pressure signal operated a 1/2-inch Foxboro control valve with stem positioner, equal percentage trim, and a C_v of 5.0.

Microcomputer. A Cromemco Z-1 microcomputer was used as the controller. I/O devices were a Lear-Siegler CRT and an AR33 Teletype. The microcomputer utilized the Z80 microprocessor at 4 MHz cycle time, 32 K of RAM, and a PROM-based disc operating system. The disc operating system allowed all development work to be performed in FORTRAN IV. The use of the higher level language greatly expedited implementation of the control algorithm. The compiled control algorithm required less than 1K of RAM. The A/D-D/A board contained seven channels each of input and output. Special input and output commands provided control of the A/D and D/A signals.

Results and Discussion

The variable sample time control algorithm was tested experimentally and the results compared with computer simulations. Tests were made with and without modeling error (parameter shift) for set point and load changes.

Set Point Changes, Variable Sample Time. Figure 2 shows the result of the first comparison for a set point change from approximately 55 to 50°C. The upper curves are for the case where

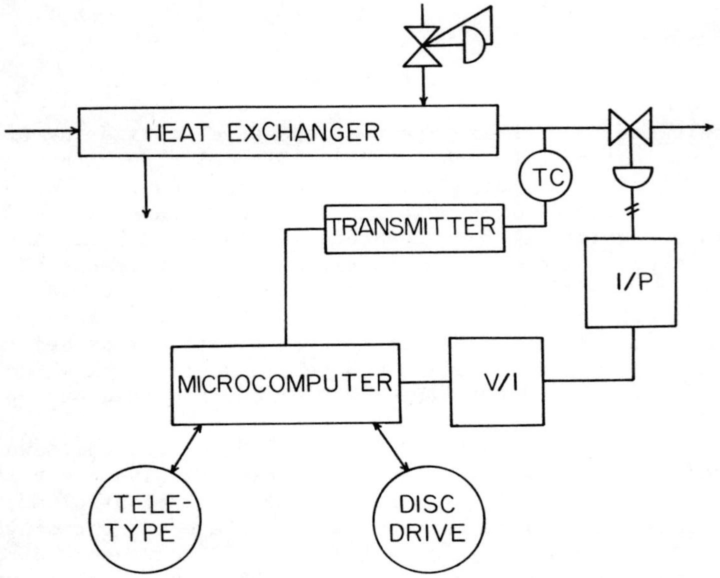

Figure 1. Equipment and arrangement

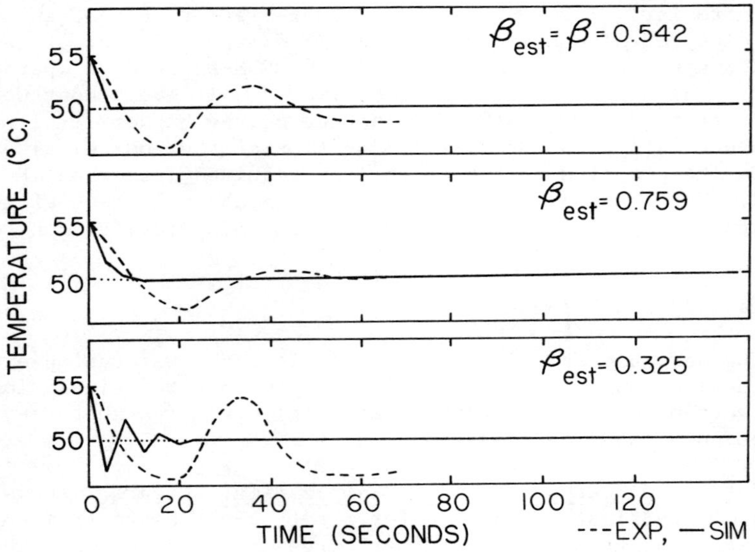

Figure 2. Variable sample time-without thermocouple dynamics in simulation

β_{est} agrees with the experimentally determined value of 0.542. The middle and lower sets of curves are for β_{est} values 40% greater and less than the actual value. Lack of agreement between experimental and simulated results is due primarily to the omission of thermocouple dynamics from the simulations. Measurement lag turned out to affect significantly exchanger dynamics. Because of the measurement lag, the assumption of steady-state at the sampling instants which is required of the variable sample time algorithm, is violated. A secondary reason for disagreement, in the case of β_{est} 40% low, is the omission of control valve saturation from the simulations. The low value of β_{est} results in excessive control action which led to control valve saturation for the set point change employed. Figure 2 shows that as β_{est} is decreased from values above to values below the actual β, the experimental response becomes more oscillatory. The simulated results show the best possible control when β_{est} equals β, overdamped response when β_{est}, exceeds β, and underdamped response when β_{est} is less than β. This is due to calculating exactly, under calculating, and over calculating, respectively, the proper control action.

Figure 3 shows the experimental results of Figure 2 compared with simulations modified to include measurement lag. The experimentally determined thermocouple time constant was four seconds. Inclusion of measurement lag brings the simulations into better agreement with experimental results although for the case of β_{est} 40% low, disagreement, remains due to control valve saturation. As with Figure 2, the sampling instants for Figure 3 are computed solely on the basis of fluid residence time for both experimental and simulated results. Therefore, the requirement of steady-state prior to calculation of the control action is not met due to measurement lag.

In an effort to more closely achieve steady-state at the sampling instants, the variable sample time based on fluid residence time was increased by twenty seconds or five thermocouple time constants. Figure 4 shows close agreement between experimental and simulated results for β_{est} equal to and 40% greater than β. For β_{est} 40% less than β the experimental response exhibits smaller fluctuations about the new set point than does the simulation. This damping is likely due to the effect of the thermal capacitance of the inner pipe wall which was neglected in the simulations but is more pronounced at the low values of β_{est} where the magnitute of the control action is greatest.

Set Point Changes, Constant Sample Time. When the constant twenty seconds was added to the variable residence time, the resulting variable sample times were in the 22-26 second range. Since this variable sample time is similar to a long constant sample time, the performance of the control algorithm was tested with various constant sample times. The constant sample times

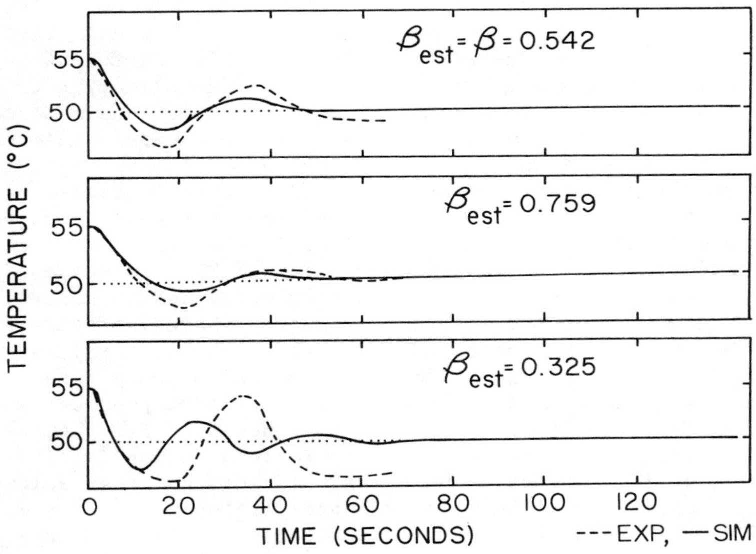

Figure 3. Variable sample time-with thermocouple dynamics in simulation

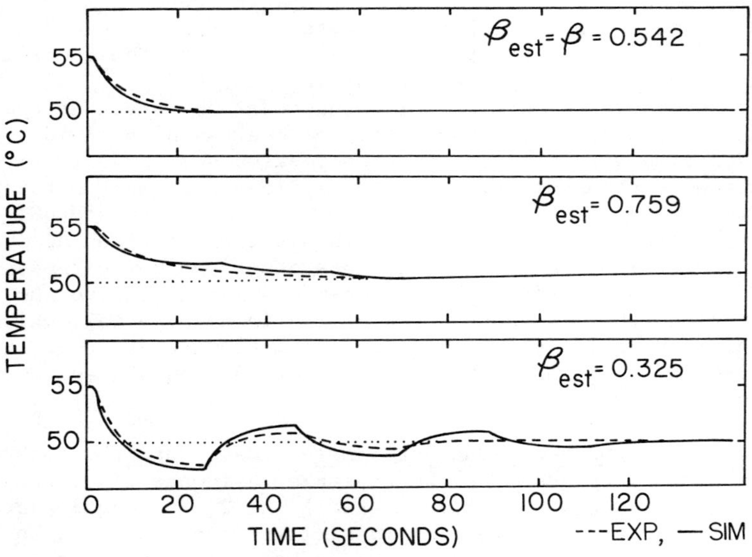

Figure 4. Modified variable sample time

studied ranged from 5 to 25 seconds in 5 second intervals.

The results for the 25 second sample time are not shown since they are nearly indistinguishable for those of Figure 4. Similarly, the results for the 5 second sample time are not shown as they are close to those of Figure 3.

Results for an intermediate constant sample time of 10 seconds are shown in Figure 5. The control in this case and also for the 15 second sample time case, not shown, is improved compared with the constant 25 and 5 second sample times. This suggests that an optimal constant sample time exists for application of the variable sample time algorithm. The optimum constant sample time appears to be on the order of 2 to 3 thermocouple time constants plus the average fluid residence time. This shorter sample time, in addition to giving better set point control, would permit a more rapid response to unexpected load upsets.

Wall Temperature Load Upsets

The dynamic equation from which the variable sample time control algorithm is derived assumes a constant wall temperature. If the wall temperature is changed, the resulting variation in outlet temperature is due to an unmodelled load upset for which the algorithm was not derived. A change in wall temperature can be approximated mathematically through a change in the overall heat transfer coefficient, U_o, with the result that the parameter β assumes a new value. Computer simulations of these load disturbances are performed by changing the value of β. From steady-state experimental data, the new value of β corresponding to a shell-side steam pressure of 40 kPa is 0.481. Experimentally, wall temperature is manipulated by changing the shell-side steam pressure.

The results for a step change of shell-side steam pressure from 110 kPa to 40 kPa and from 40 to 110 kPa are shown in Figures 6 and 7, respectively. For both figures, variable sample times of five thermocouple time constants plus the fluid residence time are used. Despite generally sluggish experimental responses, control is attained with approximately 3 to 5 sample times. The sluggish experimental responses and disagreement with simulations are due to the thermal lag involved in changing the inner pipe wall temperature once the steam pressure is changed. The wall temperature for the simulation, made by changing the value of β, is instantaneous.

Summary. A simply derived and implemented variable sample time control algorithm is shown to control a laboratory flow forced heat exchanger for both set point changes and unmodeled load upsets in the presence of modeling error. The algorithm computes the new control action from the present value of the manipulated variable making storage of prior values of the

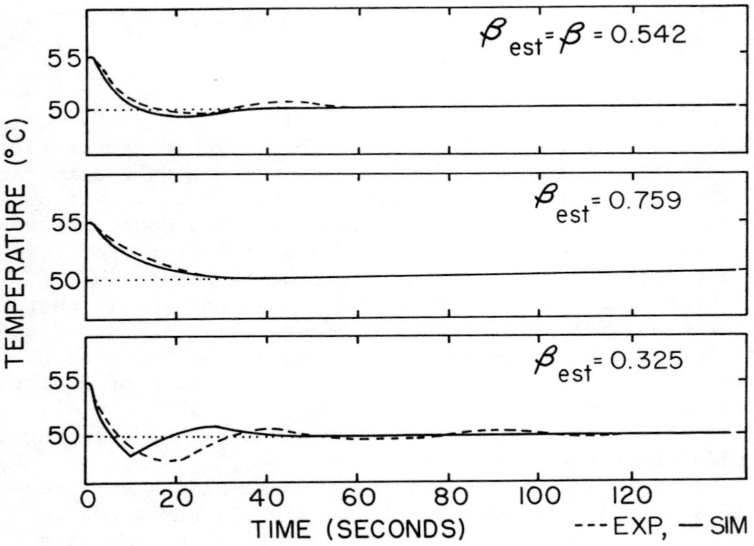

Figure 5. Constant sample time of 10 seconds

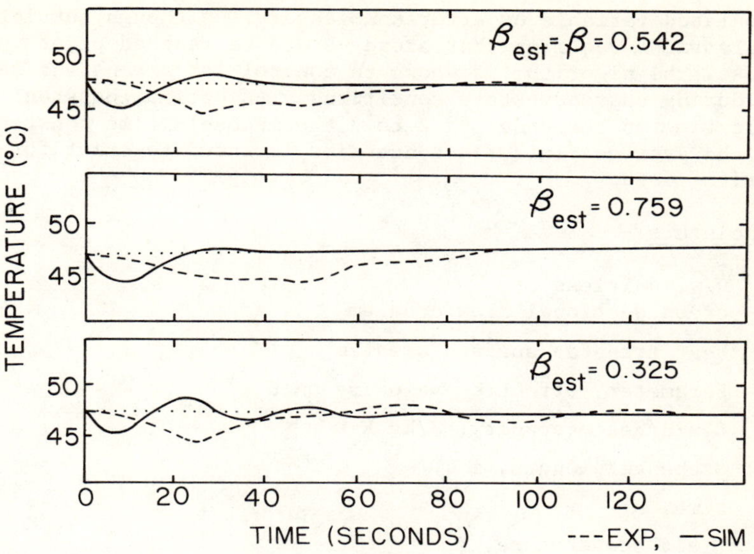

Figure 6. Modified variable sample time for steam pressure change from 110 to 40 kPa

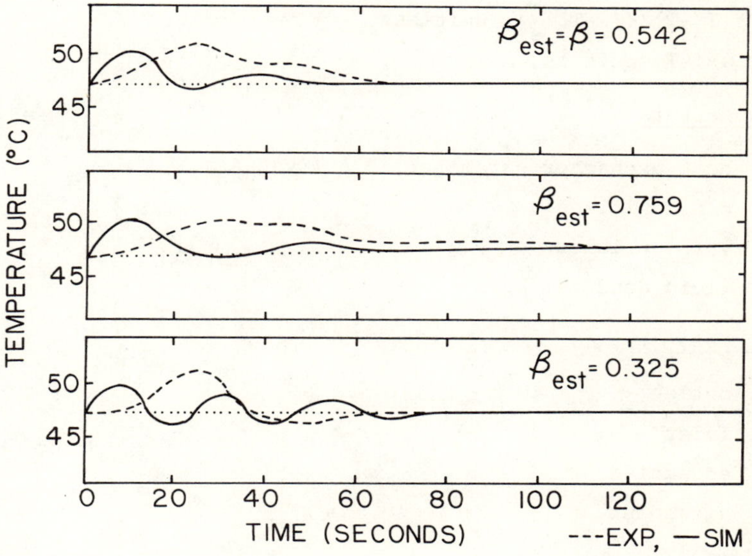

Figure 7. Modified variable sample time for steam pressure change from 40 to 110 kPa

manipulated variable or error unnecessary. Although, strictly, the algorithm requires that steady-state be reached prior to its use, the algorithm is shown to control for a constant sample time during unsteady-state conditions. An optimum constant sample time on the order of 2 to 3 thermocouple time constants plus the average fluid residence time is shown to exist for the algorithm used.

Nomenclature

a	$\tilde{v}/\bar{v}$, unitless
A_c	cross sectional flow area, m^2
A_s	heat transfer surface area, m^2
b	parameter, effect of velocity on U_o
C_p	fluid heat capacity, $J/kg \cdot K$
L	exchanger length, m
t	time, s
T	fluid temperature, K
T_s	sample time, unitless
U_o	overall heat transfer coefficient, $J/m^2 \cdot s \cdot K$
$\bar{v}$	steady-state fluid velocity, m/s
$\tilde{v}$	change in fluid velocity from steady-state, m/s
x	$(T_w - T)/(T_w - T_{0,s})$, unitless
z	axial position, m

Greek Letters

β	heat exchanger parameter, $U_o A_s/(v \rho C_p A_c)$
η	z/L, unitless
θ	$\bar{v}t/L$, unitless
ρ	fluid density, kg/m^3

Subscripts

1	outlet
0	inlet
est	estimated
sp	set point
w	wall
s	steady-state

Literature Cited

1. Fuhrman, J.E., R. Mutharasan, and D.R. Coughanowr, "Computer Control of a Distributed Parameter System", presented at National AIChE Meeting, Houston, Texas, April 1979.

2. Mutharasan, R., and D.R. Coughanowr, IEC Proc. Des. and Dev., 13, 168-176, 1974.

3. Mutharasan, R., and D.R. Coughanowr, IEC Proc. Des. and Dev., 15, 141-144, 1976.

4. Mutharasan, R., and D.R. Coughanowr, IEC Proc. Des. and Dev., 15, 378-381, 1976.

5. Mutharasan, R., and R. Luus, IEC Proc. Des. and Dev., 15, 137-141, 1976.

RECEIVED November 5, 1979.

16

ASPEN: Advanced Capabilities for Modeling and Simulation of Industrial Processes

P. W. GALLIER, L. B. EVANS, H. I. BRITT, J. F. BOSTON, and P. K. GUPTA[1]

Massachusetts Institute of Technology, Cambridge, MA 02139

Abstract

A major development effort has been underway at M.I.T. from 1976 to 1979 to develop a next-generation process simulator and economic evaluation system named ASPEN (Advanced System for Process ENgineering). The 150,000-line computer program will simulate the flowsheet of a proposed or operating plant. In addition to calculating detailed heat and material balances, ASPEN can also provide preliminary estimates of capital and operating costs and economic viability. The project is funded by the Department of Energy which will use ASPEN to evaluate process alternatives for fossil energy conversion.

This paper describes the advanced engineering capabilities of ASPEN and demonstrates their use by means of an example problem. The advanced capabilities include the ability to model unit operations involving solids, the ability to compute the properties of coal and coal-derived materials, and great flexibility for the user to add specialized models or other computations.

In addition to handling the conventional vapor/liquid process operations, the ASPEN library of process models includes solids handling and separation units, a set of generalized reactors, improved flash and distillation unit models and process models from the FLOWTRAN simulator. The user can also include his or her own model or key elements of a model, such as the reaction kinetics, in FORTRAN code.

ASPEN is supported by a versatile set of physical property correlations representing the current state-of-the-art. Physical property monitors control the property calculations in accordance with methods and models specified by the user. The user is allowed to specify different combinations of physical property calculation methods in different parts of the process. For specialized components such as coal or limestone, a collection of non-conventional property models is available. ASPEN includes data banks from which the required physical property constants and correlation parameters can be retrieved automatically at run

[1] Current address: Exxon Corporation, Florham Park, NJ

0-8412-0549-3/80/47-124-293$05.00/0

time. Comprehensive data on more than one hundred coals, covering a wide range of ranks and geographical parameters, is included.

The paper concludes by presenting an example problem showing how ASPEN is used to model a typical industrial process.

Introduction

In 1976, the U. S. Department of Energy (then called ERDA) realized the need for "a rapid, efficient, and consistent means of performing its process evaluation functions" (1). With the large government expenditures for fossil energy process development, it was found to be important to identify problems as soon as possible, before beginning costly construction of pilot and demonstration plants.

ASPEN (Advanced System for Process ENgineering) was funded at that time to provide this technical and economic analysis for the specialized requirements of fossil energy conversion processes. The project was established at the Massachusetts Institute of Technology for $3,285,000 over three years.

The software system, to meet the needs of the 1980's, has wide flexibility and capabilities. For processes such as coal gasification or coal liquefaction, it can be used to perform steady state material and energy balances, calculate sizes of equipment, and carry out economic evaluations. Its flexibility can allow for the handling of coal or other solids in streams and equipment, and its capabilities allow for the simulation of many different types of process equipment and the calculations of physical properties under widely different conditions. Included in this is the ability to analyze conventional chemical and petroleum processes. Another valuable feature is a good preliminary cost estimation capability that permits the comparison of alternative processes on an economically consistent basis at an early stage of development.

The methodology of ASPEN has been to build upon present technology and to engage the cooperation of the entire profession. The full-time staff at MIT has involved about 15 people over the past two years, many of them on loan from industry. Where possible, proven industrial software was acquired to start with the present state-of-the-art and then to build upon it. An advisory committee with representatives from industry, government, and universities has met regularly to provide reviews of progress and to help make certain that ASPEN meets the needs of the ultimate user. Over fifty companies are on the advisory committee representing diverse industries including fossil energy, petroleum, chemicals, construction, pulp and paper, metals, and food.

The ASPEN system is on schedule for a working version to be completed October, 1979. The program system will be comprised of about 150,000 lines of FORTRAN code and data for physical property and cost data banks. Although it will be a working version at that time, ASPEN will need considerable perfecting and user testing before a mature product can be released to the public. A two year testing project has been proposed to the U.S. Department of Energy for these purposes. Some time before the end of that project (October, 1981), ASPEN could be released in source code form for public use.

ASPEN Structures

ASPEN has been designed with the user in mind. Early in the project the advisory committee was involved with the staff in developing design criteria for the system. These design criteria set the premises for the ASPEN structures which included the executive system, the computational architecture, data for streams and equipment models, physical property monitors, and others. Some of these are discussed in condensed form below. More complete descriptions can be found in the ASPEN project quarterly reports (2).

The executive system is a preprocessor type which develops the actual simulation program. An input translator program reads the user input and generates a FORTRAN main program for execution. The executive programs set up the data structures and generate the computing sequence of equipment models. The flow of information in executing an ASPEN calculation is shown in Figure 1. This structure allows: (1) a larger and variable number of model programs to be executed, (2) FORTRAN statements to be inserted for execution, and (3) the need for only a minimum amount of memory for the simulation in executing an ASPEN calculation. The load module thus created is a tailor made simulation for the problem at hand.

The computational architecture is a sequential modular approach with advanced features. To model a process, each equipment module is simulated by a program module. The overall process is simulated by connecting the models together in the same way as the equipment in the flow sheet. When the input streams are known then the outputs can be calculated. The entire flowsheet can be calculated "sequentially" in this manner. Advanced features are discussed below in connection with an example.

For the data of streams and equipment models, ASPEN utilizes a plex data structure of the type proposed by Evans, et al. (3). Information is stored in blocks of contiguous locations known as beads. Beads of any length are created dynamically from a pool of free storage which may be thought of as a lengthy FORTRAN array. The combination of the preprocessor approach and the plex data structure has resulted in the absence of dimensional constraints on the system. There are no maximum numbers of streams, components, models, stages in a column, etc. except as limited by the total memory available.

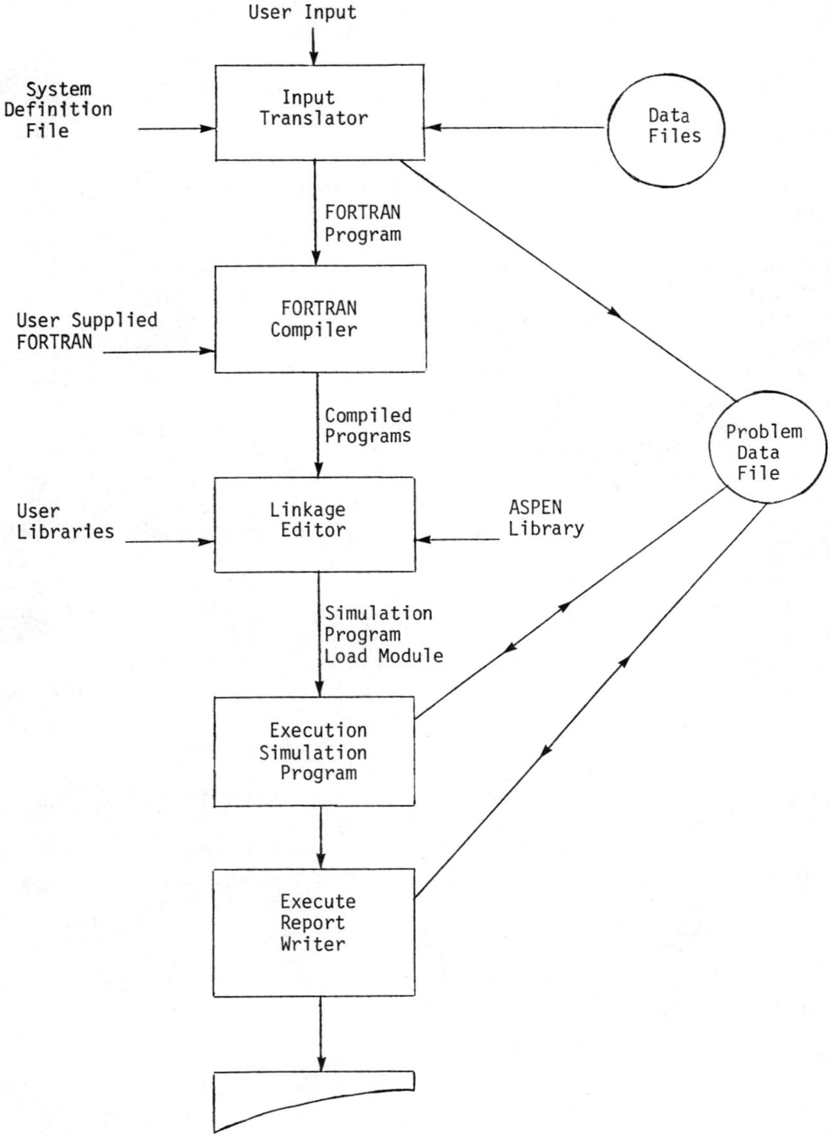

Figure 1. ASPEN flow of information

The Input Translator is completely table driven. This means that all of the information needed to process input statements (such as names of keywords, default values of data items, etc.) is stored in tables in a file called the System Definition File. Therefore, it is easy to add keywords or change defaults by changing entries in the System Definition File. In addition to the Input Language tables, almost any "changeable" information related to Input Translation is stored in the System Definition File. This includes unit conversion tables, attribute descriptions, physical property option models, data structure, unit operation model data, and stream requirements, etc. Thus it is easy to add new system parameters without changing any code in the Input Translator.

Input Language

The ASPEN input language is oriented towards process engineers familiar with chemical engineering calculations, but without extensive knowledge of computer programming. The input can be considered to be made up of paragraphs, sentences, and words. A paragraph begins with a primary keyword and may consist of one or more sentences. Each sentence begins with a secondary keyword that indicates the category of data appearing in the sentence. Tertiary keywords are used to enter data, and their values are the data items.

For example, in the following statement:

```
BLOCK F1     FLASH2
    PARAM         TEMP = 310  PRES = 1(ATM)
```

the word BLOCK is a primary keyword indicating that the paragraph contains block data. The user-specified block identifier is F1 and the unit operation model (FLASH2) is a two phase flash with specified temperature and pressure. PARAM is a secondary keyword indicating that the sentence contains block parameters. TEMP and PRES are tertiary keywords whose values are the temperature and pressure respectively. The value of a tertiary keyword may consist of a single data item or a vector of values.

The internal units for calculation are basic SI units, but the user may specify optional sets of input and output units which include English engineering, metric engineering (a set specific to the company or installation), or SI units. In addition, the user may specify individual units, such as temperature, that override the set.

The input is completely free format, except that primary keywords (and nothing else) must begin in column 1. The order of input language is sorted into a standard order before processing. Although every data item or vector of items has a tertiary keyword, the keyword may be omitted to allow positional input. The default principle is fully exercised and wherever it is meaningful, default values are provided for items the user need not supply.

As an example of using ASPEN, a simplified flowsheet is shown in Figure 2 for making synthetic natural gas (SNG) from a gasifier effluent. Oils, along with the water, are to be dropped out in a quench. Acid gases are to be removed by scrubbing. In the methanation reactor loop, some product gas is recycled back to dilute the reactor feed. Water is removed by cooling and condensation after the reaction.

Figure 3 shows the translation to a block diagram for an ASPEN simulation. Note that control units are shown with the flash blocks in this example to determine what flash temperatures are required to meet some design specifications. Figures 4 and 5 give the input language of the problem and Figure 6 gives an example page of output. Notice an outstanding feature of the input language, that it is self-documenting. The key words in a paragraph, sentence and word structure reveal the meaning and the associated numerical values. In this example, the physical property statements are not shown.

Figure 5 also shows the advanced features for design specifications. The DES-SPEC paragraphs show how variables in the problem can be defined, in a DEFINE statement, to be FORTRAN variables which can be used in any FORTRAN statements to define any arbitrary function. Then, any other problem variable can be varied, in a VARY statement, to drive the function to zero. This flexibility in design specifications is quite powerful.

FORTRAN statements also can be inserted by the user in the paragraph called FORTRAN. Using the DEFINE sentence, as before, any problem variables may be accessed. Using FORTRAN language, any arbitrary transformation of the problem variables may be made and stored. This extremely flexible capability allows the user to (1) modify block calculations, (2) change stream values, (3) insert user FORTRAN blocks, and (4) execute many other powerful, specific functions. A BEFORE or AFTER statement can be used to make the FORTRAN execution before or after any block.

For convergence calculations ASPEN employs some advanced features with the well-proven sequential modular architecture. In many process simulations, the user is responsible for structuring all computations and the computational sequence directly. In ASPEN the system is capable of complete automatic determination of the computational sequence. Alternatively, the user can select certain tear streams and can, in fact, easily specify the entire sequence.

In addition, convergence calculations may be combined simultaneously with design specifications. The usual methods would be to embed the design in a convergence loop and meet the design specification in each recycle calculation. A quasi-Newton method convergence calculation in ASPEN will allow a simultaneous, more efficient solution for the more difficult problems.

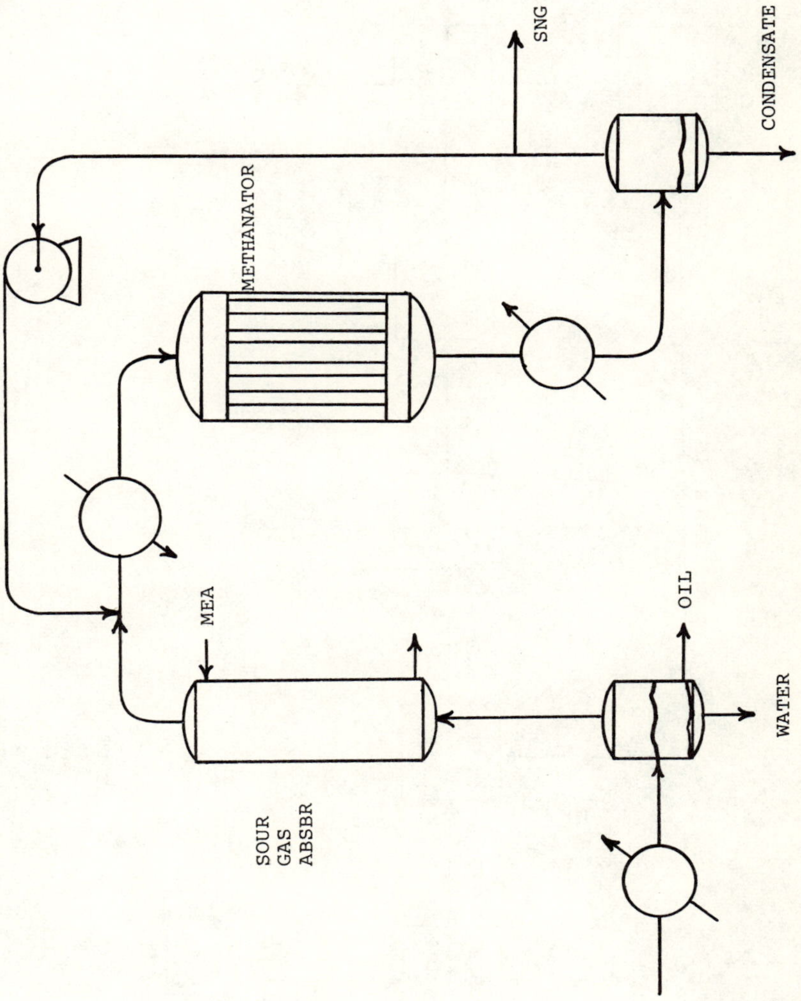

Figure 2. Flowsheet of SNG process

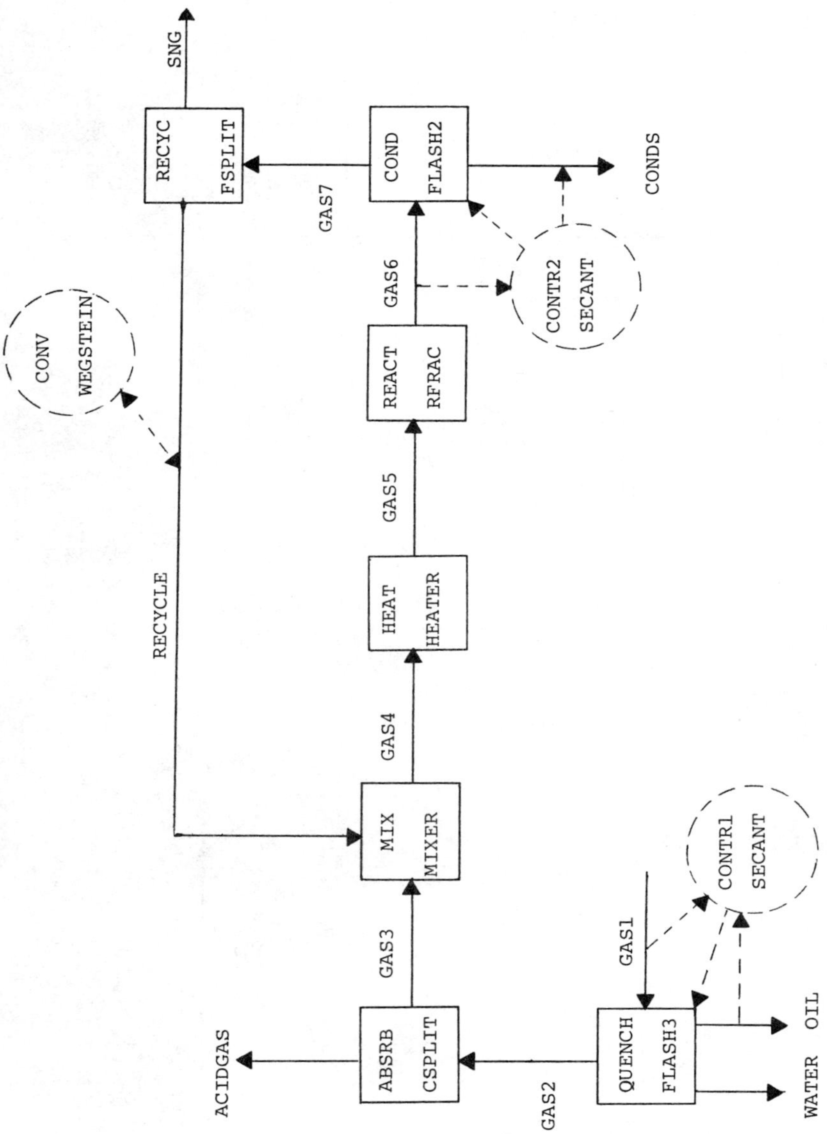

Figure 3. ASPEN block diagram SNG-process example

```
NEW
TITLE       'ASPEN SNG PROCESS EXAMPLE'
COMPONENTS H2 HYDROGEN / CO CARBON-MONOXIDE / CO2 CARBON-DIOXIDE    /
           CH4 METHANE / H2O WATER           / H2S HYDROGEN-SULFIDE /
           C6H14 HEXANE

;          DESCRIBE THE FLOWSHEET CONNECTIVITY

FLOWSHEET
           QUENCH    IN=GAS1              OUT=GAS2     WATER    OIL
           ABSRB     IN=GAS2              OUT=GAS3     ACIDGAS
           MIX       IN=GAS3 RECYCLE      OUT=GAS4
           HEAT      IN=GAS4              OUT=GAS5
           REACT     IN=GAS5              OUT=GAS6
           COND      IN=GAS6              OUT=GAS7     CONDS
           RECYC     IN=GAS7              OUT=RECYCLE SNG

;          DEFINE THE FEED STREAMS

STREAM     GAS1
           SUBSTREAM MIXED T=600 P=1000
           MOLE-FLOW H2   900   / CO      300 / CH4 400 / CO2 700 / H2S 15 /   &
                     H2O 1500 / C6H14 500

;          DEFINE ALL BLOCKS IN THE FLOWSHEET

BLOCK      QUENCH FLASH3
           PARAM   T=150   P=980   KEY=H2O

BLOCK      ABSRB   CSPLIT
           FRAC    SUBS=MIXED   STRM=GAS3                               &
                   COMP=H2 CO CH4 H2O C6H14  CO2  H2S                   &
                   FRAC= 1   1   1   1    1    0    0
           FLASH-SPECS    GAS3  VFRAC=1.0 / ACIDGAS  VFRAC=1.0

BLOCK      MIX     MIXER     PRES=820
           BOPT REST=0

BLOCK      HEAT    HEATER    TEMP=545

BLOCK      REACT   RFRAC
           BOPT    SMLV=5
           PARAM   T=800    P=820   NPKODE=1   KPHASE=1
           STOICHIOMETRY 1 H2 -3 / CO -1 / H2O 1 / CH4 1
           CONVERSION      1 CO   1

BLOCK      COND    FLASH2    TEMP=200    PRES=800    NPKODE=2

BLOCK      RECYC   FSPLIT
           FRAC    RECYCLE 0.25 / SNG 0.75

HISTORY    MSG-LEVEL  PROP=2   SIM=4
```

Figure 4. ASPEN input SNG-process example

```
;           DEFINE THE CONVERGENCE BLOCK

CONVERGENCE CONV WEGSTEIN
            TEAR RECYCLE

;           DEFINE THE COMPUTATIONAL SEQUENCE

SEQUENCE    SEQ     CONTR1 QUENCH (RETURN CONTR1) ABSRB CONV MIX HEAT REACT &
                    CONTR2 COND (RETURN CONTR2) RECYC (RETURN CONV)

REPORT

SIM-OPTIONS RPAS=0
DES-SPEC CONDENST
C       .
C       .  DESIGN SPEC CONDENST.
C       .  CONDENSE OUT 90% OF THE WATER PER PASS THROUGH THE
C       .  CONDENSER BY MANIPULATING THE TEMPERATURE
C       .  BETWEEN 150 AND 250 DEG F.
C       .
 DEFINE CFLOW FMOL CONDS MIXED H2O
 DEFINE X FMOL GAS6 MIXED H2O
F          X = .9*X
 SPEC CFLOW TO X
 TOL-SPEC .00001
 VARY BVAR COND PARAM TEMP
 LIMITS LOWER=150 UPPER=250
F        CALL HURRAY
DES-SPEC OILOUT
C       .
C       .  DESIGN SPEC OILOUT
C       .  MAKE SURE YOU QUENCH THE INPUT STREAM TO A LOW ENOUGH
C       .  TEMPERATURE SO THAT YOU GET 90% OF THE OILS (I.E. HEXANE)
C       .  OUT IN THE OIL PHASE OF THE THREE PHASE FLASH.
C       .  VARY T BETWEEN 100 AND 200 DEG F.
C       .
 DEFINE OILIN FMOL GAS1 MIXED C6H14
C     .
C     .  DROP OUT 90% OF THE OILS IN
C     .
F      X = 0.900 * OILIN
 DEFINE OIL FMOL OIL MIXED C6H14
 SPEC OIL TO X
 TOL-SPEC .00001
 VARY BVAP QUENCH PARAM TEMP
 LIMITS LOWER=100 UPPER=200
F       CALL HURRAY
CONVERGENCE CONTR1 SECANT
 SPEC OILOUT
 PARM MAXIT=15 STEP=-2
CONVERGENCE CONTR2 SECANT
 SPEC CONDENST
 PARM MAXIT=10 STEP=1
```

Figure 5. ASPEN input SNG-process example

```
ASPEN VERSION ONE        SEQUENCE    1                         DATE: 08/10/79  PAGE 6
              ASPEN SNG PROCESS EXAMPLE FOR SIMULATION COURSE
                                STREAM SECTION

STREAM SUMMARY: RECYCLE
    STREAM: RECYCLE     FROM: RECYC      TO: MIX

    COMP.   COMP.       LBS/HR          MOLES/HR        MOLE PC
    NO.     ID
      1     H2           0.12316D+02   0.61093D+01     2.436
      2     CO           0.0           0.0             0.0
      3     CO2          0.0           0.0             0.0
      4     CH4          0.35344D+04   0.22031D+03    87.861
      5     H2O          0.13803D+03   0.76622D+01     3.056
      6     H2S          0.0           0.0             0.0
      7     C6H14        0.14363D+04   0.16666D+02     6.647
    TOTALS:              0.51210D+04   0.25075D+03

    DEG. F:          223.96
    PSIA :           800.00
    1000 BTU/HR:      -11443.91
    MOL VAP FRCT:      1.000000
    MOL LIQ FRCT:      .0
    AVG MOL WT:      20.423

STREAM SUMMARY: SNG
    STREAM: SNG         FROM: RECYC      TO:

    COMP.   COMP.       LBS/HR          MOLES/HR        MOLE PC
    NO.     ID
      1     H2           0.36949D+02   0.18328D+02     2.436
      2     CO           0.0           0.0             0.0
      3     CO2          0.0           0.0             0.0
      4     CH4          0.10603D+05   0.66092D+03    87.861
      5     H2O          0.41407D+03   0.22985D+02     3.056
      6     H2S          0.0           0.0             0.0
      7     C6H14        0.43088D+04   0.49998D+02     6.647
    TOTALS:              0.15363D+05   0.75223D+03

    DEG. F:          223.96
    PSIA :           800.00
    1000 BTU/HR:      -34331.30
    MOL VAP FRCT:      1.000000
    MOL LIQ FRCT:      .0
    AVG MOL WT:      20.423
```

Figure 6. ASPEN output (excerpt) SNG-process example

Another potential advancement is permitted in the ASPEN system. Tear streams can be designated as desired, so that a user might define blocks or series of blocks and simulate these sets as quasi-linear blocks. The convergence method could utilize this information and solve the material (and energy) balances explicitly. In this way, a simultaneous modular architecture could be utilized. Implementation of these programs will be for later enhancements of ASPEN, not the initial version.

Unit Operations & Streams

A library of generalized models is supplied in ASPEN to allow the user to simulate coal conversion processes as well as chemical and petroleum processes. A listing of ASPEN's unit operations models is given in Table I. Space does not permit descriptions of the models here, however, the ASPEN project reports (2) discuss their capabilities.

All ASPEN models allow solids to be in inlet streams. The user does not need to take special precautions. The solids are normally assumed to be non-distributing in the liquid and vapor phases. That is, the phase equilibria is unaffected by the solid phase. However, the system does allow for the case of solids distributing into other phases. Solids are taken into account in the energy balance around each equipment model.

Although the ASPEN process models can be put together to simulate many types of processes, it may still be necessary to use specialized or proprietary models. Such would be the case, perhaps, for a specific type of a coal gasification reactor. The ASPEN system does facilitate the inclusion of a user's own model either in FORTRAN source code or compiled into an object code. The proprietary model may rely on the entire physical property system to calculate the required properties, however, the user routines must have the correct interface calls. These are to be documented in the ASPEN user manual.

Because ASPEN is to be used with coal conversion processes, its streams can be designated to carry an arbitrary number of solids or solid phases. This is done by specifying any number of substreams. In fact, the conventional vapor/liquid stream is normally assumed as a substream and solids can comprise other substreams. For the conventional vapor/liquid substream, process data is carried on: component molar flows, total molar flow, temperature pressure, specific enthalpy, specific entropy, density, molar vapor fraction, molar liquid fraction, and molecular weight. For solid substreams, which are called "non-conventional substreams," the characterizing data is not as deterministic. The information associated with these substreams is called "attributes". Such attributes may be particle size distribution, ultimate and proximate analyses, or other material specific information. Another type of substream is an "informa-

Table 1

Unit Operations Models in ASPEN

MIXER
- General Mixer

Splitters and Separators
- Flow Splitter
- Substream Splitter
- Component Splitter
- Two Product Separator
- Substream Attribute Separator

Pressure Changers
- Pump/Slurry Pump
- Compressor/Expander

Heaters and Heat Exchangers
- Heat Requirement Heater
- Duty Specified Heater
- Heater for a Vapor Fraction
- Cooler/Condenser
- Direct Contact Heater/Cooler

Flash Models
- Two Phase Flash
- Three Phase Flash
- General Flash

Distillation
- Shortcut, Design
- Shortcut, Rating
- Rigorous Distillation, FRAKB
- Rigorous Absorption, ABSBR
- Rigorous Separation, RADFRAC
- Minimum Reflux, MINREF
- Minimum Stages, MINSTG
- Liquid-Liquid Extraction, EXTRC

Solids Handling
- Cyclone
- Electrostatic Precipitator
- Fabric Filter
- Venturi Scrubber
- Crusher
- Screen/Classifier
- Hydroclone
- Rotary Drum Filter
- Filtering Centrifuge
- Moving Bed Dryer
- Fluid Bed Dryer

Reactors
- Yield-Based Reactor
- Fractional Conversion Reactor
- Combined Specification Model
- Well-Stirred Reactor Model
- Plug Flow Reactor Model
- Two Phase Chemical Equilibrium
- General Phase and Chemical Equilibrium

Stream Manipulators
- Multiplier
- Class Changer
- Duplicator
- Phase Separator
- Phase Mixer

tion substream" which contains no process data. An example is a flow of energy from one process unit to another.

Physical Properties

The physical property monitors of ASPEN provide very complete flexibility in computing physical properties. Quite often a user may need to compute a property in one area of a process with high accuracy, which is expensive in computer time, and then compromise the accuracy in another area, in order to save computer time. In ASPEN, the user can do this by specifying the method or "property route", as it is called. The property route is the detailed specification of how to calculate one of the ten major properties for a given vapor, liquid, or solid phase of a pure component or mixture. Properties that can be calculated are enthalpy, entropy, free energy, molar volume, equilibrium ratio, fugacity coefficient, viscosity, thermal conductivity, diffusion coefficient, and thermal conductivity.

It is believed that ASPEN provides a state-of-the-art capability for thermodynamic properties of conventional components. A number of equation-of-state (EOS) models are supplied to handle virtually any mixture over a wide range of temperatures and pressures. The equation-of-state models are programmed to give any subset of the properties of molar density, residual enthalpy, residual free energy, and the fugacity coefficient vector (and temperature derivatives) for a liquid or vapor mixture. The EOS models (named in tribute to the authors of such work) made available in ASPEN are the following:

(1) Hayden-O'Connell Second Virial Coefficient
(2) Redlich-Kwong
(3) Redlich-Kwong-Soave
(4) Peng-Robinson
(5) Lee-Starling (modified BWR)
(6) Prausnitz Perturbed-Hard-Chain

Correlations for activity coefficients can be selected from the following:

(1) Extended Scatchard Hildebrand
(2) Modified van Laar
(3) Wilson
(4) Modified NRTL
(5) UNIQUAC
(6) Margules (for water-hydrocarbon systems)

Similarly, other property calculations may have a variety of models to choose from.

The physical property system is supported by a data file on both conventional compounds and on typical coal analyses and data. The data bank contains correlation constants for over 400 conventional compounds and extensive data for thirteen characteristic coals. A Data File Management system permits a user to modify the data files, to create his own private data files, and to allow the system to automatically call in these data for a simulation run. Another stand-alone system planned with ASPEN is the Data Regression system. This will allow the user to determine the correlation constants, from any of the system property correlations, to fit experimental data. It would be expected to be widely used for vapor-liquid equilibria fitting.

Cost Estimation and Economic Evaluation

The final measure of merit of a given process is the profitability of the business venture required for its implementation. The purpose of the ASPEN Cost Estimation and Economic Evaluation System is to calculate the profitability of the simulated process. This requires calculation of the total required capital investment and the annual operating expenses.

ASPEN is intended to provide cost estimates of the preliminary study grade with accuracies in the range of +/- 30%. An extensive effort is being made to collect current cost data from the first quarter (June 1976) to 1979. Particular attention has been paid to accurate representation of the effect of different plant locations (land costs, labor productivities, labor costs, construction difficulties, etc.) and date of construction (escalation of the cost of construction material and labor, and installation material and labor, etc.). Also, great care has been given to the proper estimation of the cost of offsites, since they represent a significant fraction of the cost of a new plant, particularly for large coal-conversion plants.

The system is designed so that it can be used independently as a stand-alone system with all equipment sizes and process data supplied as input. Or, it can be run in conjuction with the unit operations simulation portion of ASPEN and can refer to those blocks for equipment sizes and process data.

As with the rest of ASPEN, the cost estimation and economic evaluation system will be modular in design; there will be one program module for each equipment class, and it will be easy for users to add their own costing modules.

A detailed discussion of the cost estimation methodology or system structure is beyond the scope of this paper and can be found in the ASPEN quarterly reports (2).

Advanced Capabilities

In summary, ASPEN has many features, discussed earlier in this paper, which qualify it as a third generation process simulator. A flexible executive system allows the user to have unlimited numbers of dimensions in streams, components, models and stages in equipment models. Solids may be handled in as many phases as desired. Arbitrary properties, called attributes, may be given to these phases and streams to allow handling properties such as particle size distributions. An engineer oriented, easy to use, input language defines a simulation run. The features of FORTRAN insertion, design specifications and convergence options give ASPEN some extremely flexible and powerful simulation capabilities. A library of state-of-the-art unit operations calculations is available. Physical property correlations for conventional components and coals give a wide variety of the best methods for the user. An extensive data bank for physical properties is made automatically available to a user and he may create his own private data banks. A thorough costing and economic evaluation system allows the user to assess the value of a flowsheet in dollars on a consistent basis. Recent cost data are included for estimating equipment and overall plant costs.

Future

A working version of ASPEN will be completed by October, 1979. A future phase of this project is anticipated for the testing of ASPEN and transferring the technology into the public domain. Industrial, government, and academic users will be able to access and use ASPEN in a controlled test environment. The ASPEN staff will maintain the system for testing, train and consult with users, and enhance the system. At the end of the test project a more proven, reliable ASPEN system will then be made available to the public.

Literature Cited

1. Joint MIT-ERDA news release on ASPEN Project. November 10, 1976.

2. Reports MIT-2295T9-1 through 13, Project on Computer Aided Industrial Process Design (U.S. Department of Energy Contract No. E(49-18)-2295, Task Order No. 9. Available through:

   National Technical Information Service
   U.S. Department of Commerce
   5225 Port Royal Road
   Springfield, VA 22161

3. Evans, L. B., Joseph, B., and Seider, W. D., "System Structure for Process Simulation," AIChE Journal 23, (5), pp. 658-666.

RECEIVED November 5, 1979.

17

Counter-Current Equilibrium Stage Separation with Chemical and Ionic Equilibrium and/or Reaction

JOSEPH F. ZEMAITIS, JR.

OLI Systems, Inc., 15 James Street, Morristown, NJ 07960

Various techniques have been developed that allow the rigorous solution of the equations describing the separation of multicomponent mixtures in complex fractionation towers. However, little has been done or published for chemical reactions and/or equilibria occur on each stage of the fractionator. The simulation of fractionators with chemical equilibria occurring in the liquid phase on each stage becomes more important as a result of more stringent effluent water quality regulations focusing attention on stripping processes to remove such volatile materials as ammonia, hydrogen sulfide, sulfur dioxide, and hydrogen cyanide from industrial waste-waters.

Until recently the ability to predict the vapor-liquid equilibrium of electrolyte systems was limited and only empirical or approximate methods using experimental data, such as that by Van Krevelen (7) for the ammonia-hydrogen sulfide-water system, were used to design sour water strippers. Recently several advances in the prediction and correlation of thermodynamic properties of electrolyte systems have been published by Pitzer (5), Meissner (4), and Bromley (1). Edwards, Newman, and Prausnitz (2) established a similar framework for weak electrolyte systems.

Using these methods to describe an aqueous electrolyte system with its associated chemical equilibria involves a unique set of highly nonlinear algebraic equations for each set of interest, even if not incorporated within the framework of a complex fractionation program. To overcome this difficulty, Zemaitis and Rafal (8) developed an automatic system, ECES, for finding accurate solutions to the equilibria of electrolyte systems which combines a unified and thermodynamically consistent treatment of electrolyte solution data and theory with computer software capable of automatic program generation from simple user input.

Several authors have recently reported on the use of Newton-Raphson procedures as applied to the solution of the equation describing a complex fractionator for the separation of multicomponent mixtures. However, it appears that the flexibility to handle various types of electrolyte problems, or to handle

0-8412-0549-3/80/47-124-309$05.00/0

fractionators with chemical reaction is not presently available, partly because of the complexities of handling units where the reaction description or equilibria description is unique to that case. With that in mind the purpose of this work was determined to be the creation of an advanced fractionation program capable of handling different reaction mechanisms that might be investigated in reacting organic systems, and also incorporating the advances made with the development of ECES, so that one could investigate different electrolyte systems with their many competing chemical and ionic equilibria and complex ion-ion interaction. In addition, to enhance the program and reduce the number of case studies necessary, the ability to use the resulting program as a design tool by allowing design to specification calculations to be made was felt to be an important goal.

Modeling a Fractionator with Finite Reaction in the Liquid Phase

The basic equations describing a single stage in a fractionator in which chemical reaction may occur include component material balances, vapor-liquid equilibrium relationships, and energy balance, and restrictions on the liquid vapor phase mol fractions. The model equations for stage j may be expressed as follows:

- Component material balance on stage j:

$$(L_{j+1} - SL_{j+1})\, x_{j+1,i} + (V_{j-1} - SV_{j-1})\, y_{j-1,i} + FL_{j,i}$$

$$+ FV_{j-1,i} - L_j x_{j,i} - V_j y_{j,i} - R_{j,i} = 0 \qquad (1)$$

or in other words:

(Component i in liquid from stage above j)
  + (Component i in vapor from stage below j)
  + (Component i in liquid feed to stage j)
  + (Component i in vapor feed to stage j-1)
  - (Component i in liquid leaving stage j)
  - (Component i in vapor leaving stage j)
  - (Component i disappearing due to reaction on stage j) = 0

- Component Equilibrium Relationship on stage j:

$$y_{j,i} = K_{j,i}\, x_{j,i} \qquad (2)$$

- Energy balance on stage j:

$$(L_{j+1} - SL_{j+1})\, h_{j+1} + (V_{j-1} - SV_{j-1})\, H_{j+1}$$

$$+ QFL_j + QFV_{j-1} - L_j h_j - V_j H_j + Q_j - QR_j = 0 \qquad (3)$$

- Normalization or restriction on fractional concentrations

$$\sum_i x_{j,i} = 1.0 \tag{4}$$

$$\sum_i y_{j,i} = 1.0 \tag{5}$$

The basic model equations are used to solve for the values of:

- T_j the temperature of stage j
- L_j the liquid rate leaving stage j
- V_j the vapor rate leaving stage j
- $x_{j,i}$ the liquid mol fractions of stage j
- $y_{j,i}$ the vapor mol fractions of stage j

When the following are specified:

- P_j the pressure of stage j
- SL_j the liquid side stream withdrawn from stage j
- SV_j the vapor side stream withdrawn from stage j
- $FL_{j,i}$ the liquid feed rate by component to stage j
- $FV_{j,i}$ the vapor feed rate by component to stage j
- QFL_j the heat content of the liquid feed to stage j
- QFV_j the heat contnet of the vapor feed to stage j
- Q_j the heat added or removed with plate heat exchanger on stage j

And the following are functions of the dependent and independent variables of stage j:

$K_{j,i} = f(T_j, P_j, x_{j,i}, y_{j,i})$ = vapor-liquid equilibrium constant

$h_j = (f(T_j, P_j, \sum_i x_{j,i})$ = liquid molal enthalpy

$H_j = f(T_j, P_j, \sum_i y_{j,i})$ = vapor molal enthalpy

$R_{j,i} = f(T_j, P_j, x_{j,i}, L_j, VOL_j)$ = extent of reaction (disappearance)

$QR_i = f(T_j, P_j, x_{j,i}, L_j, VOL_j)$ = heat of reaction

For each equilibrium stage there are NCOMP (where NCOMP is the number of components) component equations (eq 1), NCOMP equilibrium relationships (eq 2), one energy balance (eq 3), one liquid mol fraction normalization equation (eq 4) and on vapor mol fraction equation (eq 5) for a total of 2*NCOMP+3 equations per stage for 2*NCOMP+3 unknowns of $x_{j,i}$, $Y_{j,i}$, T_j , L_j , V_j of each stage. These equations for one stage together with similar equations for other stages result in a total of NSTAGE*(2NCOMP+3) equations. If these equations are to be solved by the Newton-Raphson method, the Jacobian which results from the linearization exhibits the characterisitc block tridiagonal structure:

$$\begin{matrix} B_{NST} & C_{NST} & & & & \\ A_{NST} & B_{NST-1} & C_{NST-1} & & & \\ & A_{NST-2} & B_{NST-2} & C_{NST-2} & & \\ & & \cdot & \cdot & \cdot & \\ & & & \cdot & \cdot & \cdot \\ & & & A_2 & B_2 & C_2 \\ & & & & A_1 & B_1 \end{matrix}$$

Block A_2 for instance consists of the partial derivatives of the model equations 1 through 5 applied to stage 2 with respect to the unknowns of stage 3. Block B_2 consists of the partial derivatives of the model equations applied to stage 2 with respect to the unknowns of stage 2. Block C_2 consists of the partial derivatives of the model equations applied to stage 2 with respect to the unknowns of stage 1.

The Jacobian may be evaluated either analytically or by perturbation of the unknowns to get the partial derivatives of the functional relationships. Since $K_{j,i}$, h_j , H_j , $R_{j,i}$, and QR_j are functions of unknowns, the partials of these functions must also be included in the Jacobian.

The only difference in the Jacobian for a fractionator with reaction is the inclusion of the partial derivatives of the new reaction functions $R_{j,i}$ and QR_j with respect to the unknown variables. In the implementation of this model of a fractionator with reaction occurring in a stage, we require a user written subroutine to supply $R_{j,i}$ and QR_j. The user has the option of also coding $\partial R_{j,i} / \partial T_j$, $\partial R_{j,i} / \partial L_j$, ... or having these partial derivatives evaluated by perturbation. The resulting reaction subroutine is compiled and linked to the main fractionator program at execution time. Figure 1 is an example of a user written subroutine which describes the kinetics of the esterification of ethanol with acetic acid described in the paper by Suzaki, Yagi, Komatsu, and Hirat (6). The fractionator model with reaction is imbedded in the program FRACHEM and reproduces the results obtained in the previously mentioned paper.

Modeling a Fractionator with Chemical and Ionic Equilibria Occurring in the Liquid Phase

For a fractionation tower where the reactions are occurring very rapidly so as to approach equilibrium which occurs in the equilibria of many weak electrolytes and strong electrolytes in solution, equation 1 for the component balance can not be used since the reaction term in not finite. To modify the basic equations of FRACHEM we took advantage of the work we performed in developing ECES which is described in the paper by Zemaitis and Rafal (8). For a system composed of strong and weak electrolytes, ECES using an imbedded thermodynamic framework writes a computer program to calculate the composition of the outlet streams of a single stage flash for a fixed flow of species into the flash unit.

That is, for a given temperature and pressure and inflows of species, the ECES model calculated the flow rates of the species in the liquid phase leaving the flash zone. This information must be in a form suitable for use in the fractionator program and the fractionator program must be able to utilize this information correctly.

To merge the single stage flash capabilities of ECES with the basic fractionator program of FRACHEM, modifications to both ECES and FRACHEM were necessary. A new option was added to ECES, which involved modifying the automatic program writing capabilities of ECES to write two subroutines suitable for interfacing to FRACHEM. These subroutines ACAL and FCAL represent respectively, a subroutine for calculating the activity coefficient and activities of all the species present in solution and a subroutine consisting of the equilibrium expressions and material balances sufficient to define the liquid phase.

The activity coefficients and activities are functions of T_j , P_j , and $x_{j,i}$. The total amount of each species in the liquid phase leaving stage j, $VVEC_{j,i}$ is calculated for each stage by modified Newton-Raphson method using the subroutine FCAL to define the equilibrium expressions and material balances. These component rates, $VVEC_{j,i}$, are functions of T_j , P_j , $\gamma_{j,i}$, and $FIN_{j,k}$, where $FIN_{j,k}$ is the total amount of elemental species k entering the liquid phase of stage j. $FIN_{j,k}$ is a function of $(L_{j+1} - SL_{j+1})$, V_j , V_{j-1} , and all i species containing element k in the streams $FL_{j,i}$, $FV_{j-1,i}$, $x_{j+1,i}$, $Y_{j,i}$, and $y_{j-1,i}$.

```
      SUBROUTINE REACT1(PRESS,TEMP,ISTAGE,NCOMP,NSTAGE,REACT,
     +QREAC,ICTYPE)
C
      DIMENSION REACT(30),FELIQ(40,32),FEVAP(40,32),SSLIQ(40),
     +SSVAP(40),X(40),30),Y(40,30),FLIQ(40),FVAP(40),ICTYPE(30),
     +HOLDM(40),RA(30),RB(30),RC(30),QAD(30),QBD(30),QCD(30),
     +VOL(4)
C
      COMMON /KEYS/X,Y,FLIQ,FVAP
C
      COMMON /FLOWS/FELIQ,FEVAP,SSLIQ,SSVAP
C
      COMMON /KEYIN/ HOLDM,RA,RB,RC,QAD,QBD,QCD,
     + KREAC,KDDX
C
C     VOL(4) ONLY USER INTRODUCED VARIABLE
C
C     ORIGINAL DAT USES DEGREES C SO CONVERT T [M]
C
      TC=(TEMP-32.)/1.8
C
C     HOLDM IS HOLDUP IN THIS CASE IN VOLUME UNITS
C     VOL IS A MOLAR SPECIFIC VOLUME
C
      VOL(1)=0.094*TC+53.12
      VOL(2)=0.0611*TC+57.46
      VOL(3)=0.013*TC+17.4
      VOL(4)=0.2096*TC+89.6
C
C     NOW TO GET RATE CONSTANT
C
      TK=TC+273.
      RATE=60.*(10.**(-2710./TK+3.70))
      VOLM=0.0
      DO 10 1=1,4
   10 VOLM=VOLM+X(ISTAGE,I)*VOL(I)
      REACT(1)=RATE*HOLDM(ISTAGE)*((X(ISTAGE,1)/VOLM)**2)*1.0E=06
      REACT(2)=REACT(1)
      REACT(3)=-REACT(1)
      REACT(4)=-REACT(1)
   20 QREAC=0.0
      RETURN
      END
```

Figure 1. User-written subroutine

To utilize the results of the ECES written subroutines which calculate $VVEC_{j,i}$, equation 1 the component material balance around stage j is replaced by

$$L_j x_{j,i} - VVEC_{j,i} = 0 \tag{6}$$

or the total amount of species i leaving stage j in the liquid phase is equal to $VVEC_{j,i}$ which is calculated by using the ECES written suroutine FCAL. The remaining equations 2 through 5 still are used to describe the model if in performing the energy balance the enthalpy terms are replaced by Heats of Formation. Then the heat of solution or dilution are implicit in this balance and do not need to be computed separately. Thus QR_j equals zero using this basis even if we extend the model to include finite rates of reaction as well as equilibrium.

In forming the Jacobian of equations 2 through 6, the elements of the Jacobian reflecting the partial derivatives of equation 6 with respect to the stage variables (unknowns) are found by using:

$$\frac{\partial\left(L_j x_{j,i} - VVEC_{j,i}\right)}{\partial q} = \frac{\partial(L_j, x_j, i)}{\partial q} - \left[\frac{\partial VVEC_{j,i}}{\partial FIN_{j,i}} \cdot \frac{\partial FIN_{j,i}}{\partial q} + \frac{\partial VVEC_{j,i}}{\partial q}\right] \tag{7}$$

where q is any state variable such that

$$\frac{\partial FIN_{j,i}}{\partial q} \neq 0$$

Most of the partial derivatives required for these elements need to be evaluated by perturbation. The remaining elements of the Jacobian are calculated as for the fractionator with reaction.

Figures 2 and 3 are the subroutines produced by the ECES program generator to describe the activity coefficients and activity of water in SUBROUTINE ACAL and the chemical and ionic equilibria, electroneutrality,and elemental balances in SUBROUTINE FCAL. These subroutines were developed for the ammonia-carbon dioxide scrubber described in Appendix I. The IF conditions are used to eliminate particular functional equations if for some reason particular ionic species are absent from a stage. This occurs for instance on the stages above the stage where NaOH is added for pH control. Since Na^+ is not volatile, any equilibria or elemental balances involving this species are not valid above the caustic feed stage.

The algorithm for FRACHEM-ECES for solving the non linear equations describing the electrolyte system in a fractionator tower can be schematically represented as shown on the next page. After the initialization and setting up of the supporting routines, the program flow is to FRACMD, where the Jacobian for the fractionator is created using equations 2-6 as the basic functional representation. In order to supply the necessary information on the ionic equilibria in the liquid phase, FRACMD

```
      SUBROUTINE ACAL(SE1,SE2,SE3,SE4,SE5,SE6,X,I,NXMAP,AV,AKVC,
     1NKVEC,F,FIN,FINT,CON1)
      DIMENSION AVEC(9),VVEC(13),AKVC(8)
      DIMENSION SE1(30),SE2(30),NXMAP(30),AV(30)
      REAL*8 X(30)
      REAL*4 I
      NAAA=9
      NKKK=8
      NVVV=13
      DO 10 J=1,NVVV
          IF (NXMAP(J),EQ,0) GOTO 10
          VVEC(J)=X(NXMAP(J))
   10 CONTINUE
      SE3=1*VVEC(8)+1*VVEC(10)
      SE4=1*VVEC(9)+1*VVEC(11)+4*VVEC(12)+1*VVEC(13)
      SE5=-CON1*I**0.5/(1+I**0.5)
      SE6=0.01132*VVEC(8 )-0.0388*VVEC(10)+0.009425*VVEC(9 )+0.02616*
      VVEC
     1(11)+0.12528*VVEC(12)+0.04492*VVEC(13)
      AVEC(1 )=EXP(1/(SE3*SE4)*(VVEC(8 )*VVEC(9 )*SE2(1)+VVEC(8 )*
     1*VVEC(1
     11)*SE2(2)+VVEC(8 )*4*VVEC(12)*SE2(3)+VVEC(8 )*VVEC(13)*SE2(4)+
     1VVEC
     1(10)*VVEC(11)*SE2(5))+1/(SE3*SE4)*(VVEC(10)*4*VVEC(12)*SE(6)+
     1VVEC
     1(10)*VVEC(13)*SE2(7)))
      AVEC(2 )=EXP(-.0084*I+2*SE6)
      AVEC(3 )=EXP(0.01300*I+2*SE6)
      AVEC(4 )=10**((1)**2*SE5+(SE1(1)*VVEC(9 )+SE1(2)*VVEC(11)+SE1
     1(3)*V
     1VEC(12)*3**2/4+SE1(4)*VVEC(13)))
      AVEC(5 )=10**((1)**2*SE5+(SE1(1)*VVEC(8)))
      AVEC(6 )=10**((1)**2*SE5+(SE1(5)*VVEC(11)+SE1(6)*VVEC(12)*3**
     12/4+S
     1E1(7)*VVEC(13)))
      AVEC(7 )+10**((1)**2*SE5+(SE1(2)*VVEC(8 )+SE1(5)*VVEC(10)))
      AVEC(8 )=10**((2)**2*SE5+(SE1(3)*VVEC(8 )*3**2/4+SE1(6)*VVEC
     1(10)*3
     1**2/4))
      AVEC(9 )=10**((1)**2*SE5+(SE1(4)*VVEC(8 )+SE1(7)*VVEC(10)))
      DO 310 J=1,NAAA
      AV(J)=AVEC(J)
  310 CONTINUE
      RETURN
      END
```

Figure 2. Subroutine describing activity coefficients and activity of water in ACAL

```
      SUBROUTINE FCAL(SE1,SE2,SE3,SE4,SE5,SE6,X,I,NXMAP,AV,AKVC,
     1    NKVEC,F,
     1FIN,FINT)
      REAL*8 AVEC(9),VVEC(13),AKVC(8)
      DIMENSION SE1(30),SE2(30),NXMAP(30),NKVEC(30)
      REAL*8 F(30),FIN(30),AV(30),X(30),FINT
      COMMON /LIMMY/ NLIM(30)
      REAL*4 I
      NAAA=9
      NKKK=8
      NVVV=13
      DO 10 J=1,NVVV
          IF (NXMAP(J),EQ.0) GOTO 10
          VVEC(J)=X(NXMAP(J))
   10 CONTINUE
      DO 20 J=1,NAAA
          AVEC(J)=AV(J)
   20 CONTINUE
      IF (NLIM(1).EQ.0) F(1)=AKVC(4 )-(1/(AVEC(1 ))*AVEC(4 )*VVEC(8 )
     1   *AV
     1EC(5 )*VVEC(9 )/(AVEC(2 )*VVEC(5 )))
      IF(NLIM(2).EQ.0) F(2)=AKVC(5 )-(1/(AVEC(1 ))*AVEC(6 )*VVEC(10)
     1   *AV
     1EC(7 )*VVEC(11)/(AVEC(3 )*VVEC(7 )))
      IF(NLIM(3).EQ.0 F(3)=AKVC(6 )-(AVEC(8 )*AVEC(6 )/AVEC(7 )*VVEC
     1   ( 12 )*VVEC(10)/VVEC(11))
      IF(NLIM(4).EQ.0) F(4)=AKVC(7 )-(1/(AVEC(1 ))*AVEC(2 )*VVEC (5 )
     1   *AV
     1EC(7 )*VVEC(11)/(AVEC(9 )*VVEC(13)))
      IF(NLIM(5).EQ.0) F(5)= AKVC(8 )-(AVEC(5 )*AVEC(6 )*VVEC(9 )*
     1   VVEC( 10 /AVEC(1 ))
      IF(NLIM(6).EQ.0) F(6)=VVEC(8 )+VVEC(10)-(VVEC(9 )+VVEC(11)+2
     1   *VVEC( 12 )+VVEC(13))
      IF(NLIM(7).EQ.0) F(7)=FINT=(VVEC(1 )/55.5*(4*VVEC(5 )+3*VVEC(1))
     1*VVEC( 7)+5*VVEC(8 )+2*VVEC(9 )+VVEC(10)+5*VVEC(11)+4*VVEC(12)
         +6*V 1VEC(13)
      IF(NLIM(8).EQ.0) F(8)=FIN(1)-(2*VVEC(1 )+VVEC(1 )/55.5*(3*VVEC
     1   (5 )+2*VVEC(13)+4*VVEC(8 )+VVEC(9 )+VVEC(10)+VVEC(11)
      IF(NLIM(9).EQ.0) F(9)=FIN(25)-((VVEC(11)+VVEC(7 )+VVEC(12)+VVEC
     1   (13 ))*VVEC(1 )/55.5)
      IF(NLIM(10).EQ.0) F(10)=FIN(4)-((VVEC(13)+VVEC(5 )+VVEC(8 ))*
     1   VVEC(11 )/55.5)
      RETURN
      END
```

Figure 3. Subroutines describing chemical and ionic equilibria, electroneutrality, and elemental balances in FCAL calls FRACGN to supply the Jacobian elements as in equation 7. FRACGN itself has a Newton-Raphson algorithm embedded in it, much like the overall algorithm and using FCAL and ACAL, FRACGN develops the function values and partial derivatives related to the ionic equilibria required by FRACMD for the overall tower model.

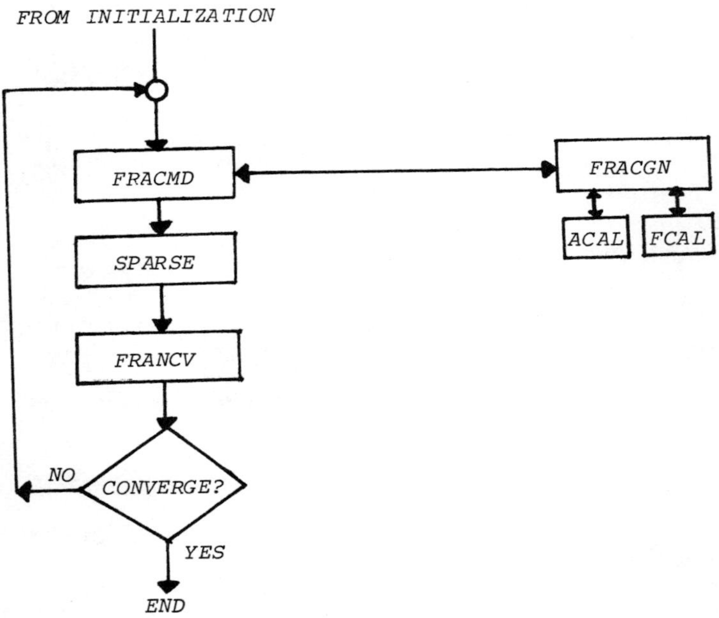

Once the Jacobian for the entire tower is created, the block tridiagonal matrices are solved by routines in SPARSE to give the correction vector for the unknowns. This correction vector is passed to the routine FRANCV where if this were a normal tower, the correction vector would be added directly to the unknown vector and then tested for convergence. However, because of the highly nonlinear behavior of electrolyte systems, the embedding of a Newton-Raphson procedure in FRACGN within another Newton-Raphson procedure originally led to severe convergence problems.

The routine FRANCV has become, in many ways, similar to a traffic control system. In addition to techniques such as bounding of variables, scaling to avoid divergence, etc, FRANCV slows down or speeds up the overall tower convergence to keep pace with the convergence patterns occurring in FRACGN. This is necessary since within FRACGN, the system equations are those describing the ionic and chemical equilibria of the liquid phase of one stage while in FRACHEM proper, all stages are being solved simultaneously and a change vector might easily during the early iterations change a liquid flow rate and a component mol fraction by 50% or more. These changes if allowed could prevent the Newton-Raphson procedure in FRACGN from converging when it used the results of the previous overall tower iteration as its initial guess for solving the ionic equilibria. Thus FRANCV evolved through experiment into a speed controller which slows down the overall tower convergence rate in order to avoid convergence problems within the ionic equilibria model.

The resulting algorithms in FRACHEM/ECES allow for the solution of a wide number of different types of problems involving ionic and chemical equilibria using an established thermodynamic framework that describes with rigor the behavior of the electrolyte system. An example of such an application is an ammonia-carbon dioxide scrubber which is described in Appendix 1. For this case, no attempt was made at improving the initialization and 11 iterations were required with a moderate amount of speed control in the overall model solution.

Design to Specifications - A Useful Option

To extend the applicability of FRACHEM/ECES, the ability to design to specifications was added. This allows a user to investigate rather easily such strategies as pH control in ionic systems to enhance steam stripping. Within the FRAHCEM/ECES program this is done by applying the concepts first presented by Goldstein and Stanfield (3). As used in our algorithm, several specifications can be made at one time with little increase in computer time.

To illustrate, let's assume we wish to fix the liquid rate leaving a particular stage, which is normally an unknown. If an unknown is going to be set, another fixed variable must be allowed to become an unknown, perhaps the duty on an exchanger. We then have a specification equation that must be satisfied:

$$G = \text{Gspec} = \text{SVALUE} - \text{FLIQ}(j) \tag{8}$$

In order to fix this value we need to also develop values of:

$$G_u = \frac{\partial\ (\text{Gspec})}{\partial\ (\text{Unknowns})} = \frac{\partial G}{\partial U}$$

$$G_c = \frac{\partial\ (\text{Gspec})}{\partial\ (\text{Control variables})} = \frac{\partial G}{\partial C}$$

and

$$F_c = \frac{\partial\ (\text{F equations})}{\partial\ (\text{Control variables})} = \frac{\partial F}{\partial C}$$

where our standard algorithm uses the Jacobian

$$F_u = \frac{\partial\ (\text{F equations})}{\partial\ (\text{Unknowns})} = \frac{\partial F}{\partial U}$$

For a standard Newton-Raphson iteration the correction vector is given by

$$\Delta U = -F_u^{-1} \cdot F$$

Applying the method of Goldstein and Stanfield (3) then there are two correction vectors - to the control variables

$$\Delta C = (G_c - G_u F_u^{-1} F_c)^{-1} (-G + G_u F_u^{-1} F)$$

and to the unknowns

$$\Delta U = -F_u^{-1} F - F_u^{-1} F_c \Delta C$$

Since there are at most five or six control variables, the calculations are not increased significantly because the most significant addition to the calculations is the inversion of at most, 6×6 matrix.

Thus this feature is extremely useful with electrolyte systems, reducing the number of computer simulations necessary to find the correct control strategy. With this design to specifications feature, FRACHEM/ECES has been used to determine a proper caustic injection rate to reduce the ammonia and hydrogen sulfide concentrations to meet EPA specifications without inordinate amounts ot stripping steam.

Modeling a Fractionator with Both Finite Reaction and Ionic Equilibria in the Liquid Phase

The ultimate extension of FRACHEM/ECES is to combine the ability to model rate limiting kinetics with the electrolyte equilibria capabilities of ECES. A need for such a tower model arose recently in the development of a new pollution control process. In this process certain organic species in dilute concentrations in waste water streams undergo reactions to form weak electrolytes which are then stripped from the waste water. In order to simulate and optimize the process a suitable tower model was needed.

The same state equations, equations 2 through 6 still apply. However in evaluating the Jacobian, particularly for the elements $VVEC_{j,i}$ in equation 6,

$$L_j x_{j,i} - VVEC_{j,i} = 0 \qquad (6)$$

these component rates calculated by FRACGN from the electrolyte equilibria are now functions of T_j, P_j, $\gamma_{j,i}$, $FIN_{j,k}$, and $R_{j,i}$ the extent of reaction of the species on stage j. The finite rates affect the electrolyte equilibria by changing the elemental amounts of the various species. If we consider reaction occurring in the liquid phase and assume a well mixed stage, then we can represent schematically the process:

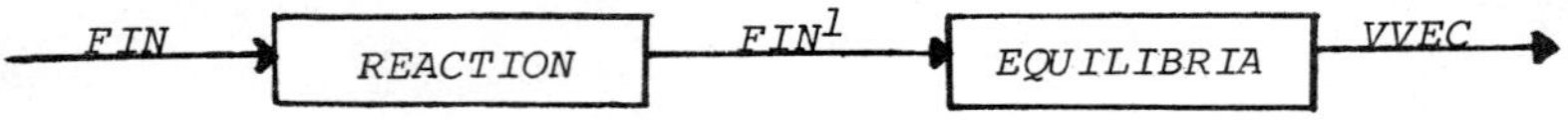

Then:

$$FIN^{1}_{j,k} = FIN_{j,k} - R_{j,k}\ (x_{j,i},\ T_j,\ L_j,\ VOL_j) \tag{8}$$

The elements of the Jacobian reflecting the partial derivative of the NCOMP equation 6 with respect to the state variables are found by using:

$$\frac{\partial\left(L_j x_{j,i} - VVEC_{j,i}\right)}{\partial q} = \frac{\partial(L_j x_{j,i})}{\partial q} - \frac{\partial VVEC_{j,i}}{\partial FIN^{1}_{j,i}}\left[\frac{\partial FIN_{j,i}}{\partial q} - \frac{\partial R_{j,i}}{\partial q}\right] - \frac{\partial VVEC_{j,i}}{\partial q} \tag{9}$$

where q is any state variable such that

$$\frac{\partial FIN^{1}_{j,i}}{\partial q} \neq 0$$

With this change to the Jacobian generator and a user supplied routine to evaluate the extents of reaction and at the user's option the partial derivatives $\partial R/\partial x$, $\partial R/\partial L$, and $\partial R/\partial T$ along with the ECES generated subroutines, this form of process can be successfully modeled. This type of model has been successfully used to simulate proprietary processes developed by our clients.

Conclusions and Significance

A rigorous fractionation tower program has been developed with several features that broaden its applicability far beyond the capabilities of existing commercially available distillation tower simulation programs. In addition to the usual features of most modern multicomponent fractionation programs, the individual component material balances and the enthalpy balance written for each tray have been modified so that the resulting computer program is able to simulate:

(1) Fractionation where chemical reaction occurs in the liquid phase such as in the process involving the production of esters.

(2) Sour water strippers where complex chemical and ionic equilibria occur in the liquid phase affecting the vapor-liquid equilibria.

(3) Strippers of other fractionators where both rate limited reactions and complex chemical and ionic equilibria occur.

(4) Conventional fractionators of various configurations and complexities.

The principal significance of this work is that it allows the user to explore fractionation schemes involving reactions and chemical equilibria with rigorous approach previously not available. Of special significance is that the FRACHEM-ECES option provides a convenient thermodynamically consistent method for the calculation of multicomponent, multistage vapor-liquid equilibria incorporating and extending recent work in order to predict the effects of chemical and ionic equilibria or the phase equilibria and thermodynamics necessary in simulating such a complex fractionation system.

Nomenclature

A block matrix of Jacobian reflects stage j+1's effect on variables of stage j

B block matrix of Jacobian reflects stage j's effect on variables of stage j

γ activity coefficient of liquid species

C block matrix of Jacobian relects stage j-1's effect on variables of stage j

FIN total elemental flow rate into liquid phase of a stage, mols/hr

FL liquid feed flow, mols/hr

FV vapor feed flow, mols/hr

h enthalpy of liquid component, BTU/mol

H enthalpy of vapor component BTU/mol

K vapor-liquid equilibrium constant

L liquid flow, mols/hr

P pressure, psia

Q heat added by plate heat exchanger, BTU/hr

QFL heat content of liquid feed, BTU/hr

QFV heat content of vapor feed, BTU/hr

QR heat removed by reaction on stage, BTU/hr

R disappearance by reaction of species, mol/hr

SL liquid side draw, mol/hr

SV vapor side draw, mol/hr

T temperature of a stage, $^{\circ}F$

V vapor flow, mol/hr

VVEC total flow of species leaving a stage in liquid phase, mol/hr

x mol fraction of liquid component

y mol fraction of vapor component

Subscripts

i component

j stage

k component

NST top stage of tower

References Cited

(1) Bromley, L., "Thermodynamic Properties of Strong Electrolytes in Aqueous Solutions", AICHEJ, vol 19, p 313 (1973)

(2) Edwards, T.J., J. Newman, & J.M. Prausnitz, "Thermodynamics of Aqueous Solutions Containing Volatile Weak Electrolytes", AICHEJ, vol 21, p 248, (1975)

(3) Goldstein, R.P. and R.B. Stanfield, "Flexible Method for the Solution of Distillation Design Problems Using the Newton-Raphson Technique", I&EC Process Design & Development, vol 9, no 1, p 78 (June 1970)

(4) Meissner & Tester, "Activity Coefficients of Strong Electrolytes in Aqueous Solutions", I&EC Process Design and Development, vol 11, p 128, (1972)

(5) Pitzer, K., "Thermodynamics of Electrolytes- I Theroretical Basis and General Equations", J. Physical Chemistry, vol 77, p 268, (1973)

(6) Suzaki, Yagi, Kamatsu, and Hirat, "Calculation of Multicomponent Distillation Accompanied by Chemical Reaction", J. Chem. Eng. (Japan), vol 4, no 1, p 26, (1971)

(7) Van Krevelen, D.W., P.J. Hoftijzer, and F.J. Huntjens, "Composition and Vapor Pressures of Aqueous Solutions of Ammonia, Carbon Dioxide, and Hydrogen Sulfide", Rec. Trav. Chim. Pays-bas, vol 68, p191, (1949)

(8) Zemaitis, J.F., and M. Rafal, "ECES - A Computer System to Predict the Equilibrium Composition of Electrolyte Systems", 68th Annual Meeting, AICHE, Los Angeles, (Nov, 1975)

Appendix I

The tower is a stripper consisting of 5 ideal stages with a liquid feed to the top stage consisting of

22,284	LB/HR	H_2O
540	LB/HR	CO_2
1,794	LB/HR	NH_3

at a temperature of 110°F. FRAHCEM/ECES was used to simulate the performance when the stripper was at 55 PSIA and 8233 LB/HR of steam at 200 PSIA and a temperature of 222° F were introduced below the first or bottom stage.

The input to ECES consisted of the information that follows:

```
FRACHEM
INPUT
H2OIN
NH3IN
CO2IN
SPECIES
H2OVAP
NH3VAP
CO2VAP
NH3AQ
CO2AQ
NH4ION
OHION
HION
HCO3ION
CO3ION
NH2CO2ION
EQUILIBRIUM
H2OVAP=H2O
NH3VAP=NH3AQ
CO2VAP=CO2AQ
NH3AQ+H2O=NH4ION+OHION
CO2AQ+H2O=HION+HCO3ION
HCO3ION=HION+CO3ION
NH2CO2ION+H2O=NH3AQ+HCO3ION
H2O=HION+OHION
```

ECES then using its embedded thermodynamic framework created three computer readable files:

(1) A database containing the parameters needed to compute equilibrium constants, interaction parameters etc.

(2) FCAL - a FORTRAN subroutine which has all the function equations to describe the chemical and ionic equilibria and is developed from the user input.

(3) ACAL - a FORTRAN subroutine which has all the function equations to calculate the activity coefficients of all the ionic and molecular species in the aqueous solution as well as the activity of water.

In running FRACHEM/ECES the user creates an input file in which the necessary information is entered to describe the stripper and its operation. The two FORTRAN files are compiled and linked with FRACHEM to then solve the simulation. For this case the column temperature and flow profiles were found to be:

	STAGE NUMBER	TEMPERATURE (DEGREES F)	VAPOR RATE (LBMOLS/HR)	LIQUID RATE (LBMOLS/HR)
Tower Top	5	239.7	207.08 Off GAS	1492.49
	4	250.5	364.74	1524.29
	3	264.8	387.56	1559.39
	2	276.2	415.46	1582.04
	1	282.2	435.57	1603.93 Bottoms

The liquid bottoms concentration was found to be

COMPONENT	MOL FRACTION
H2O	0.99523E+00
CO2AQ	0.78501E-06
NH3AQ	0.45999E-02
HION	0.60026E-10
OHION	0.11757E-04
NH4ION	0.83742E-04
HCO3ION	0.60582E-04
CO3ION	0.19155E-05
NH2CO2ION	0.75720E-05

and the vapor off had a composition (vapor leaving stage 5) of;

COMPONENT	MOL FRACTION
H2O	.60597
CO2	.04436
NH3	.34966

while the liquid in equilibrium with this vapor had a composition of

COMPONENT	MOL FRACTION
H2O	.91783E+00
CO2AQ	.13093E-04
NH3AQ	.67311E-01
HION	.98140E-10
OHION	.42984E-05
NH4ION	.96984E-02
HCO3ION	.23935E-02
CO3ION	.16617E-03
NH2CO2ION	.70188E-02

RECEIVED November 5, 1979.

18

Leach Process Models

IVAN V. KLUMPAR

Kennecott Copper Corp., Lexington, MA 02173

Chemical processes involving multicomponent multistage leaching were modeled in the following way depending on the process type and complexity:

- Steady state counter-current leaching and washing with specified solid compositions in each phase and a specified material balance of the remaining process units were simulated with simultaneous linear non-differential equations. The model can calculate the material balance including wash and makeup water, losses, number of stages and solute buildup in two recycles. The corresponding ready-to-use computer program can be applied to any process.
- The material balance of steady state leaching and washing units of any configuration integrated with a number of other process units was modeled using iterative simulation. The method is particularly useful if multiple nested and intersecting recycles are involved. The model is applicable if each physical separation is defined by distribution coefficients, and each chemical reaction by stoichiometry and conversion. The corresponding computer program is again general and ready to use.
- Kinetics and mass transfer were simulated for steady and non-steady state cocurrent or cross-current leaching of moving or stationary solids. The model can handle any number of parallel or consecutive reactions. The method is applicable to complex processes that include leaching of desirable and undesirable species, side reactions of a non-leach character, losses, product separation and recovery, lixiviant recycling, etc. While the main-line program is general, specific sub-routines for the individual process steps have to be written.

0-8412-0549-3/80/47-124-327$05.00/0

Linear Equation Model

A typical hydrometallurgical process (1) includes:

- Leaching and washing, represented in Figure 1 by a series of mixers with interstage liquid-solid separation in thickeners.
- Product separation, such as liquid ion exchange (LIX).
- Product recovery, such as electrowinning.
- Waste disposal, such as a tailings pond.

The barren liquors from the product separation or recovery sections, and the waste disposal facility may be recycled as wash liquor to the last stage of the leach-wash system after appropriate treatment. The process may include other streams such as an inlet of lixiviant or additional wash water, and a ficticious exit stream representing losses.

An example of this type of configuration is the Arbiter Process (2) for ammoniacal leaching of copper concentrates. The lixiviant consists of aqueous ammonia flowing counter-currently and oxygen sparged crosscurrently.

Depending on the concentrate nature, the following three and other reactions may occur:

$$CuFeS_2 + 2.25\ O_2 + 4\ NH_3 = Cu(NH_3)_4^{2+} + S_2O_3^{2-} + 0.5\ Fe_2O_3 \quad (1)$$

$$S_2O_3^{2-} + 2\ O_2 + 2\ NH_3 + H_2O = 2\ SO_4^{2-} + 2\ NH_4^+ \quad (2)$$

$$Cu_3AsS_4 + 8.75\ O_2 + 15\ NH_3 + 2.5\ H_2O =$$

$$= 3\ Cu(NH_3)_4^{2+} + NH_4H_2AsO_4 + 4\ SO_4^{2-} + 2\ NH_4^+ \quad (3)$$

Equation 3 is an example of an undesirable reaction because of the solubilization of arsenic. Fortunately, most of it precipitates subsequently as an iron arsenate.

The $(NH_4)_2SO_4$ rich wash liquor passed to a regeneration plant constitutes the exit stream in Figure 1. Regenerated ammonia is the inlet stream. Regeneration is not shown in Figure 1. Neither are other sections such as pregnant solution filtration, tailings flotation and recycle treatment.

In developing the model, the following basic assumptions have been made:

- The solid phase composition in each state is specified.
- The major constituent of the solid phase is an inert solid.

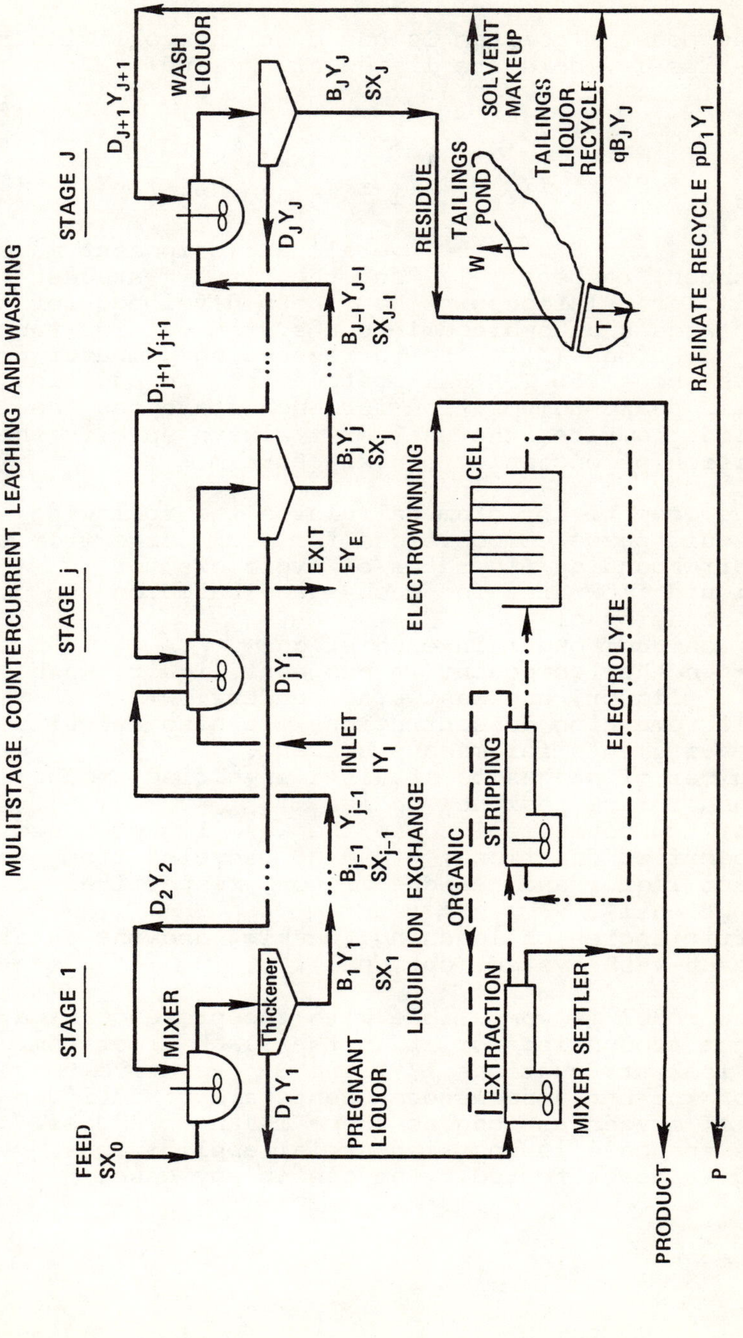

Figure 1. Simple leach process

- Liquid phase compositions of the overflow and underflow liquor are equal.

The model algorithm is based on the following relationships (symbols are listed at the end):

$$D_{j+1} + S + I = D_1 + B_j + E \qquad (4)$$

$$D_{j+1} Y_{j+1} + SX_0 + IY_I = D_1 Y_1 + B_j Y_j + SX_j + EY_E \qquad (5)$$

Equations 4 and 5 are the overall and component material balances, respectively, for the first j stages.

Similar relationships have been developed for the other stages and combined into a system of simultaneous linear equations (3). The corresponding computer program has about 250 FORTRAN statements. In addition to the solid phase composition for each stage and the inert solid flowrate, the user must always specify the composition of one inlet or exit stream.

The computer program calculates the following:

1a. Liquid phase compositions for each stage from the corresponding flowrates of overflows; or

1b. Liquid flowrates for each stage based on the concentration of
   - one component in each stage,
   - another component in pregnant liquor, wash liquor, and last stage underflow;

   All remaining concentrations are also calculated.

2. Requirements for makeup solvent.
3. Number of stages to attain a specified product loss in the residue (optional).
4. Buildup of any solute in the wash liquor based on specified fractions, p and q, recycled from pregnant liquor and residue liquor, respectively (optional).
5. Efficiencies of leaching, washing and the entire leach-wash system (optional).

The model is compatible with the presence of a gas while not accounting for it in the leach algorithm. Let us look at Equations 1, 2 and 3. Relative conversions of the non-gaseous components are fixed independently of oxygen as long as there is an O_2 excess. This is the case in most industrial applications. Washing is not affected by oxygen in any case.

Iterative Simulation Model

A considerably more complex process is shown in Figure 2. It includes:

- Four-stage countercurrent mineral leaching and washing with waste disposal (M,W,D)
- Two separation and recovery circuits for two different products. Each consists of crystallization, intermediate product leaching, final product recovery and extractant regeneration (C,I,P,R).
- Reagent preparation comprising absorption and stripping (A,S).

The process includes a number of recycles that can be interpreted as 18 intersecting and nested loops of a first through fifth order.

It is impossible or impractical to model processes of this type with simultaneous equations. Rather, iterative simulation is used (4). The available computer programs, however, have the following disadvantages:

- They are large and expensive because they were developed for regular process design,
- Becoming well acquainted with the program is time consuming,
- Most cases require writing of specific subroutines.

There is a need for a simple model limited to material balances that could be used for preliminary calculations when there are not enough data or time available for rigorous process design. The proposed model is particularly useful for processes that are controlled by material balance related parameters rather than by the thermal and kinetic nature of the system. Many leach processes fall within this category being sensitive to oversaturation, pH, overall water balance, distribution of impurities, etc.

Any chemical process can be broken down into a number of modules of the following four types (5):

(a) Addition of two or more streams,
(b) Branching of an inlet stream into two outlets of the same composition,
(c) Conversion of a stream by a chemical reaction into another stream of a different composition,
(d) Distribution of an inlet stream into two outlets of different composition.

Each recycle is defined by a module of a fifth type ("e" for encircling to keep the symbols meaningful).

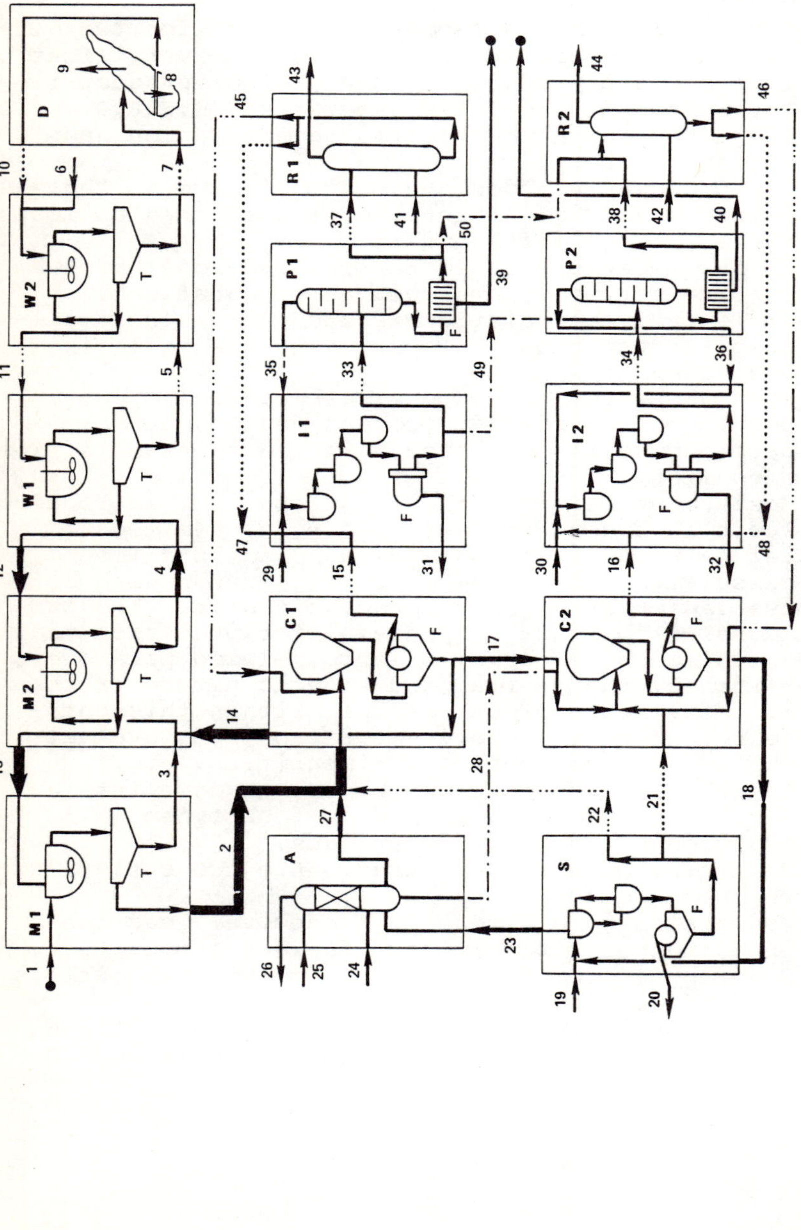

Figure 2. Complex leach process

Figure 3 shows the flowsheet of Figure 2 expressed in terms of the five modules. Lower case letters are used to distinguish modules from plant sections which are denoted by capitals. Some modules are identical with equipment units, e.g. filters and thickeners are distribution modules. In other cases, one piece of equipment is simulated by two or more modules. For example, absorber A is represented by a, c, d and b modules.

The user has to specify a split ratio for each branching module and a distribution coefficient for each component entering a distribution module. The conversion module requires a stoichiometric coefficient for each component involved and the fractional conversion of a selected key component. Each recycle has to be defined by the originating and destination modules, computational sequence, and convergence tolerance of a key component.

After all modules are specified, the computer program calculates flowrates and compositions of all inside and outlet streams based on known inlet streams. Inlet and outlet streams are those which do not originate or end in a module, respectively. The others are inside streams. The program has only 250 FORTRAN statements.

Kinetics and Mass Transfer Model

Reaction kinetics and mass transfer associated with leaching can be expressed using a modified shrinking core model (6). Let us again use the Arbiter Process as an example. The interaction of ammonia and oxygen with a chalcopyrite particle can be visualized as consisting of the following elements (see Figure 4 which is a section of the three phase system):

1. Oxygen dissolution at the liquid interface,
2. Ammonia and O_2 transfer to the reacted particle shell,
3. Aqueous NH_3 and O_2 diffusion to the unreacted particle core,
4. Reaction (Eq. 1) at the core surface,
5. Diffusion of reaction products across the shell,
6. Their transfer to the bulk liquid.

Each element represents a resistance. If one resistance is controlling, a simple rate equation may be derived. For example, if the shell disintegrates and the liquid is vigorously agitated, leaching can be represented by the surface reaction rate equation,

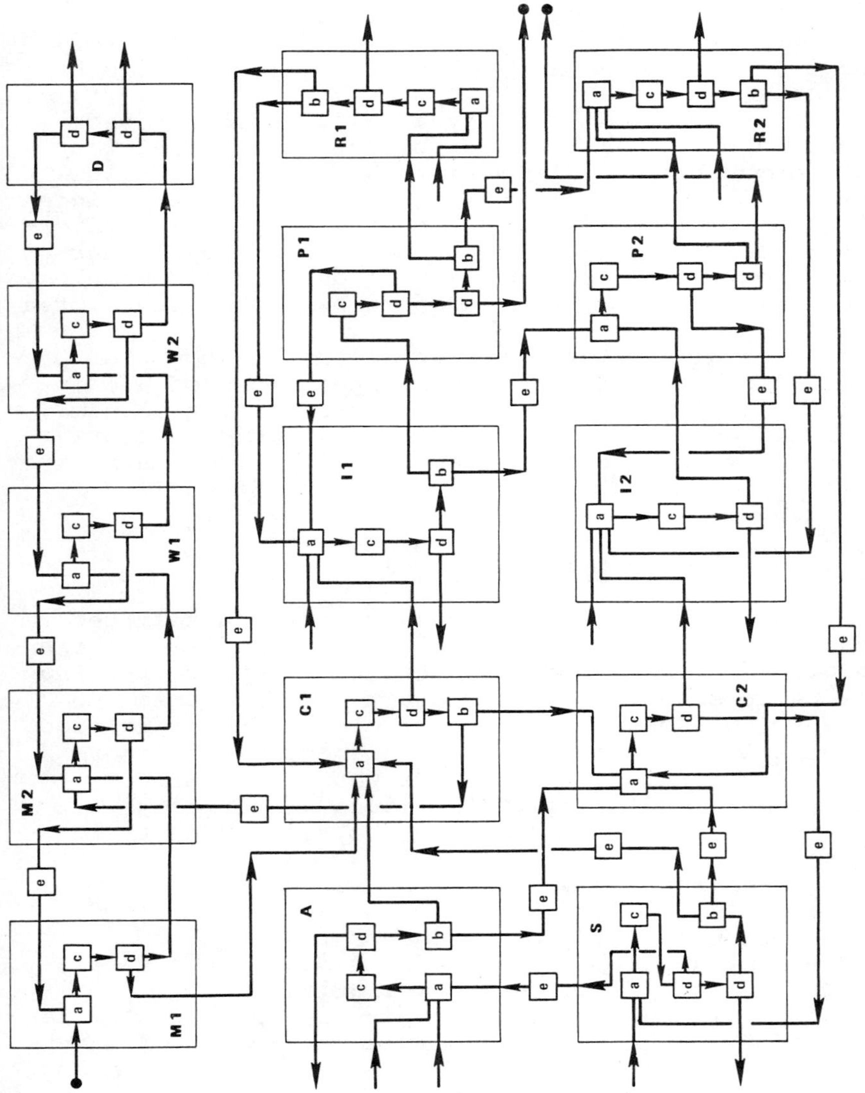

Figure 3. Iterative simulation modules

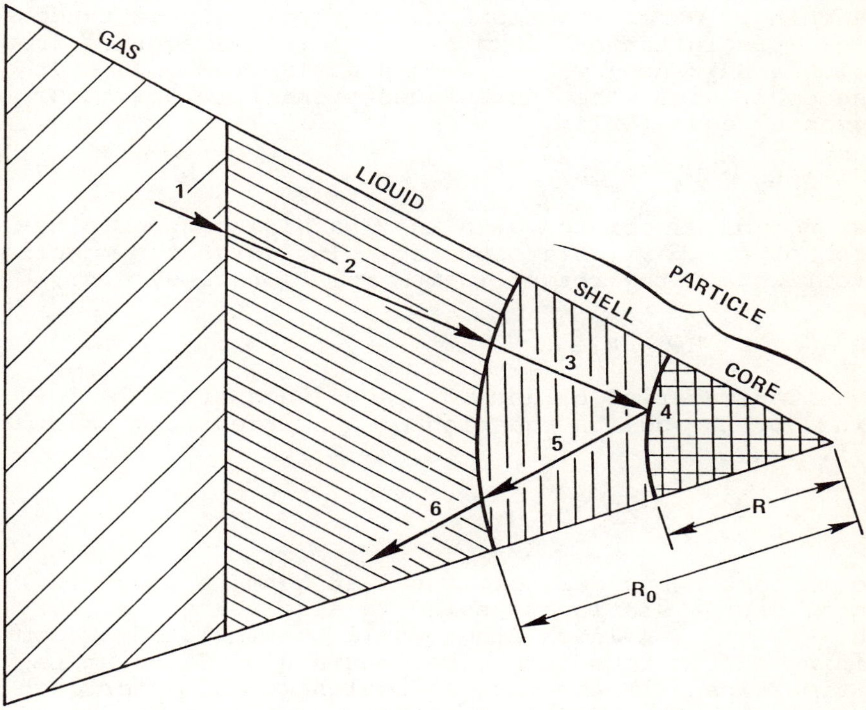

Figure 4. Shrinking-core model

$$r_M = -(dN_M/dt)/V = 4\ \pi\ R^2\ n\ ksC_G/V \tag{6}$$

As the right-hand side expression is based on the mean particle diameter, it has the advantage of eliminating shape factors.

Similar rate equations can be developed for the parallel leaching of other minerals (see Eq. 3). The oxidation of intermediate sulfur compounds to sulfates such as Eq. 2 is a consecutive reaction to mineral leaching. Other reactions are virtually instantaneous, e.g., precipitation of insoluble ferric arsenates from soluble $NH_4H_2AsO_4$ which is consecutive to Eq. 3. On the other hand, the instantaneous shift in the amonia - ammonium equilibrium,

$$NH_3 + H_2O \rightleftharpoons NH_4^+ + OH^- \tag{7}$$

can not be associated with the leaching or precipitation of any particular species. Instantaneous reactions are fixed by their equilibrium relationships, e.g.,

$$C_A = C_U\ C_H\ /\ K_A \tag{8}$$

The material balance of each leach stage is defined by the design equation of a stirred tank reactor (7),

$$F_I = F_E + rV + dN_a/dt \tag{9}$$

For a continuous reactor, the last term in Eq. 9 is zero. For a stationary solid in a semibatch reactor, $F_I = F_E = 0$. A leach stage could be simulated by combining the various rate equations and equilibrium relationships with the material balance, Eq. 9 for each component. The result would be a very complex set of differential equations.

The proposed model takes another approach. It was developed for multistage semibatch reactors with stationary solids and continuous co-current reactors with moving solids. It also allows for a crosscurrent stream such as gas sparged separately into any number of stages. The residence time of each stage is divided into a number of finite time intervals. Within each interval, the individual reactions are treated as successive rather than simultaneous. The model accuracy is controlled by selecting the number of intervals. The higher the number the more closely the model approximates the simultaneous conditions.

A schematic diagram of the entire model is shown in Figure 5. It should be noted that each interval includes also the instantaneous reactions. As the sparged oxygen might have a shorter residence time than the liquid, the number of intervals with respect to the crosscurrent stream might be less than the total number of intervals per stage. Accordingly, the model allows for sequencing of the crosscurrent stream which is simulated as branching. Losses are handled by a fictitious purge step following the last interval.

Model Algorithm and Applications

The algorithm of the kinetics and mass transfer model is a system of algebraic equations that are developed in the following way. The key variable is ΔN, the change in the amount of a particular reactant or product component involved in a reaction step during an interval Δt. The material balance of a step is,

$$N'' = N' + \Delta N \qquad (10)$$

where N' and N'' are the amounts of the component present in the stage before and after the step.

Rate equations such as Eq. 6 are converted to algebraic relationships to give the ΔN of a controlling component by substituting differences, Δ, for differentials, d. For example, in the case of Cu_3AsS_4 reacting according to Eq. 3,

$$\Delta N_M = -4\pi\ R^2\ n\ k\ (8.75)\ C_G\ \Delta t \qquad (11)$$

The other components are related by stoichiometry, e.g.,

$$\Delta N_G = 8.75\ \Delta N_M \qquad (12)$$

$$\Delta N_A = 15.0\ \Delta N_M \qquad (13)$$

where 8.75 and 15.0 are stoichiometric coefficients.

Equilibrium relationships such as Eq. 8 are modified by introducing ΔN as a conversion term. For example, in Eq. 7,

$$\Delta N_A = -\ \Delta N_U = -\ \Delta N_H \qquad (14)$$

Using Eq. 14, Eq. 8 is modified and multiplied by V^2 with the following result, noting that $VC = N$,

$$(N'_A - \Delta N_A)V = (N'_U + \Delta N_A)(N'_H + \Delta N_A)/K_A \qquad (15)$$

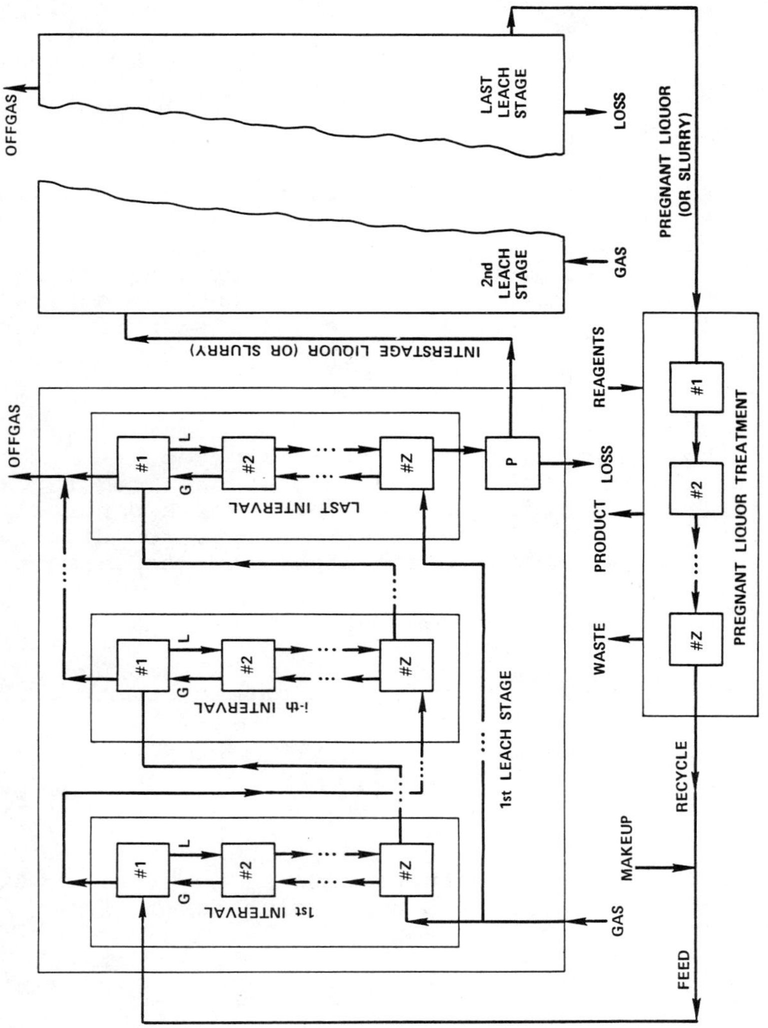

Figure 5. Kinetics and mass-transfer model

Eq. 15 is subsequently solved for ΔN_A. Similar relationships are derived for the pregnant liquor treatment steps.

It would be difficult to establish offhand the number and lengths of intervals for a particular application of the model. Rather, the user specifies the final radius, R_m, of a controlling mineral particle and selects a constant shrinkage factor, f, defined as fractional decrease of the particle surface area per interval. The relationship between radius and shrinkage factor is,

$$R_e = R_b \sqrt{1-f} \tag{16}$$

The shrinkage factor and final radius determine the approximate number of intervals. For example, for f = 10% and R_m = 5% of the initial radius, R_0, the minimum number of intervals is 57 because,

$$R_0(\sqrt{1-0.1})^{56}=0.0523R_0 > R_m=0.05R_0 > R_0(\sqrt{1-0.1})^{57} = = 0.0496R_0 \tag{17}$$

The length of each interval is estimated one at a time in the following way. For a given number of particles per stage, the amount of a mineral is proportional to the cube of an average particle diameter. Selecting a mineral that leaches only in one reaction, e.g., chalcopyrite, Eq. 16 is rewritten as

$$N'' = N' \ (1 - f)^{3/2} \tag{18}$$

Eqs. 10, 18 and 6 for dt = Δt are combined to give

$$\Delta t = N'[1-(1-f)^{3/2}]/9\pi \ R^2 \ n \ k \ C_G \tag{19}$$

As the leach rate is proportional to particle surface area, the intervals established in this way are fairly uniform, though not necessarily equal. They are subsequently used in the other reactions.

The computer program is compact. For a particular application it had only 500 FORTRAN statements including subroutines for seven rate equations and equilibrium relationships, as well as seven pregnant liquor treatment steps. For a series of semibatch reactors, the program calculates the gradual concentration changes of all streams per pass based on initial conditions. For a continuous system, the model can simulate both steady and non-steady state operation. An example of the latter is startup.

The program can also model processes that are not confined to equipment such as dump leaching of low grade sulfide copper ores or tailings. In this process, water is sprinkled on top of the dumps, solubilizing copper as it percolates the porous mass in the presence of air. An acid cupric sulfate solution is collected at the bottom. The leached block of the dumps can be ficticiously divided into a number of sections along the percolation path. Each section is considered a stirred tank reactor stage.

In applying the model, some mineral parameters, such as numbers, n, and mean radii, R_0 of various mineral particles may be estimated by mineralogical techniques. For physical properties such as phase equilibrium constants, K, published ternary and binary data may be used on an approximate basis. Kinetic parameters such as reaction rate constants, k, or mass transfer coefficients can be very roughly estimated based on laboratory experiments. Their values may then be varied in a series of computer runs until the results match pilot plant data. A reasonably good match will, at the same time, confirm the remaining variables, rate equations and other assumptions.

Needless to say, kinetic parameters established in this way are empirical factors dependent on the definition of other variables,e.g., averaging of particle size distribution to determine R. Nevertheless, the model provides a rational approach to the complex physical and chemical phenomena of multistage, multireaction leaching. It is intended to expand the model to include countercurrent operations.

Acknowledgement

The author would like to acknowledge the help of Frank Zybert who did the detailed programing of the kinetics and mass transfer model.

Symbols

A absorption in Figure 2.
B flowrate of underflow solvent, kg/s
C molar concentration, k mol/m^3
  crystallizer in Figure 2
D flowrate of overflow solvent, kg/s
  waste disposal (tailings pond) in Figure 2
E exit solvent flowrate, kg/s
F molar flowrate, k mol/s
  filter in Figure 2
G gas or oxygen

I inlet solvent flowrate, kg/s
intermediate product leach in Figure 2
J number of stages, no dimension
K equilibrium constant, various dimensions
L liquid or slurry
M mineral leach in Figure 2
N amount of substance per stage, k mol
P purge and/or loss or their flowrate, kg/s
product recovery in Figure 2
R particle core radius, m,
extractant regeneration in Figure 2
R_0 mean mineral particle radius, m
S inert solid flowrate, kg/s
stripping in Figure 2
T tailings accumulation rate, kg/s
thickener in Figure 2
V stage volume, m^3
W water evaporation rate, kg/s
washing in Figure 2
X component to inert solid mass ratio, no dimension
Y component to solvent mass ratio, no dimension
#Z last reaction
a addition module
b branching module
c conversion module
d distribution module or differential
e "encircling" or recycle module
f fractional surface shrinkage, no dimension
i interval counter
j stage counter
n number of particles per stage, no dimension
p recycled fraction of pregnant liquor, no dimension
q recycled fraction of the residue liquor, no dimension
r reaction rate, various dimensions
s stoichiometric coefficient, no dimension
t time, s
#1,#2 first and second reaction, respectively

Subscripts
A ammonia
E stage exit
G gas or oxygen
H hydroxyl
I stage inlet
J last stage
M leached mineral
U ammonium
a accumulated
b beginning of interval

e	end of interval
j	stage counter
m	minimum
0	feed stream or initial particle
1,2	stage 1, 2

Streams in Figure 2

1	feed	23	vapor recycle
2	pregnant liquor	24	reagent makeup
3,4,5	underflow	25	absorbent makeup
6	makeup water	26	absorption offgas
7	residue	27,28	reagent
8	tailings	29,30	solvent makeup
9	evaporation	31,32	insolubles
10	water recycle	33,34	product solution
11,12,13	overflow	35,36	solvent recycle
14	lixiviant recycle	37,38	regeneration feed
15,16	crystals	39,40	product
17	semidepleted liquor	41,42	regeneration agent
18	barren liquor	43,44	regeneration offgas
19	desorbent	45,46	regeneration purge
20	strip waste	47,48	adjustment stream
21,22	bottoms recycle		

1st order loop
2nd order loop
3rd order loop
4th order loop
5th order loop
intersecting loop
other streams

Literature Cited

1. Subramanian, K. N.; Jennings, P. H., Can. Metall. Quarterly, 1972, 11 (2) 387.
2. Kuhn, M. C.; Arbiter, N.; Kling, H., CIM Bulletin, 1974, 67 (742) 62.
3. Agarwal, J. C.; Klumpar, I. V., Chem. Engineering 1976, 83 (11) 135.
4. Motard, R. L.; Shacham, M.; Rosen, E. M., AIChE Journal, 1975, 21, 417.
5. Agarwal, J. C.; Klumpar, I. V., Chem. Eng. Progress, 1978, 74,(6) 68.
6. Smith, J. M., "Chemical Engineering Kinetics"; McGraw-Hill, New York, 1970; p.575.
7. Levenspiel, O., "Chemical Reaction Engineering"; J. Wiley, New York, 1962; p.96.

RECEIVED November 5, 1979.

Process Simulation for Mixtures of Hydrocarbons, Petroleum, and Associated Gases in Coexisting Vapor and Liquid Systems

WAYNE C. EDMISTER[1], STERLING H. BOOTH, and ROBERT E. HURNEY

Kaiser Engineers, Inc., Oakland, CA 94623

A computer algorithm has been developed for making multicomponent mixture calculations to predict: (a) thermodynamic properties of liquid and vapor phases; (b) bubble point, dew point, and flash conditions; (c) multiple flashes, condensations, compression, and expansion operations; and (d) separations by distillation and absorption.

Equations and property data are included for 39 hydrocarbons, 10 nonhydrocarbons, and petroleum fractions. Petroleum properties are predicted by equations derived from correlations that are in the Technical Data Book of the American Petroleum Institute (1).

Six alternate methods for predicting the thermodynamic properties are included. These are known by the names of the authors of the methods, which are Chao-Seader (2), Grayson-Streed (3), Lee-Erbar-Edmister (4), Soave-Redlich-Kwong (5), Peng-Robinson (6) and Lee-Kesler-Ploecker (7, 12).

SIMULATION PROGRAMS

Computer simulation programs are useful engineering tools for design, optimization, and control of production and manufacturing processes. A simulator for equilibrium processing of hydrocarbons, petroleum, and associated gases in co-existing vapor and liquid phases will be described in this paper.

To be useful, this type of simulator must calculate the thermodynamic properties of multicomponent mixtures in both liquid and vapor phases while predicting bubble and dew points or partial vaporizations or condensations. Using this basic information, the simulator must then make calculations for other processes, such as gas cooling by expansion, gas compression, multiple flashes or condensations, and separations by absorption

*Current Address: 75 Summit, San Rafael, CA 94901

0-8412-0549-3/80/47-124-343$05.00/0

or distillation. Our simulator has these features and can be expanded to include others, such as reaction equilibrium, flowsheet simulation, etc.

An earlier version of this program, called PROSIM, was described by Edmister and Aguayo at the Summer Computer Simulation Conference in Chicago during July 1977 (8). Additions to and improvements in PROSIM have been made during the past two years. The present version is described and illustrated in this paper.

PROGRAM ORGANIZATION

A computer program of this kind for performing a wide variety of calculations, by alternate methods, is rather large and complicated. The executive, or main, program directs the calculations according to control instructions provided by the user, along with fluid identifications and quantities and the problem conditions. The required physical properties and equation constants are available in data blocks that are integral parts of the program.

There are about 60 subroutines, distributed as follows: 20 on thermodynamic predictions, 12 on petroleum characteristics, 6 on vapor/liquid equilibria, 3 on data, 4 on compression/expansion and multiple flash, 13 on multistage separation, and 3 on output reports.

These subroutines are linked together and to the main for the performance of 16 types of singlestage equilibrium calculations and a like number of multistage calculations. The principal lines of communication between these routines are shown in the schematic diagram, figure 1. It is not possible, in the time and space available, to show all the details of making and converging the calculations.

FLUID COMPONENTS

PROSIM is concerned with vapors and liquids, separately or coexisting in equilibrium, and composed of one or more molecular species in each phase. These mixtures may be made of components from 49 substances consisting of 39 hydrocarbons and 10 nonhydrocarbons. Physical properties are given in block data for these 49 components. It would be a simple matter to expand this list, if needed.

Petroleum may be included in the process fluids of the calculations. Mixtures of petroleum and hydrocarbons, or petro leum fractions alone, with no light hydrocarbons, may constitute the system. For such mixtures the petroleum feed stock must be divided or broken down into small cuts, or "components," and physical properties of these "components" estimated for use in the subsequent calculations. Petroleum breakdowns can be made

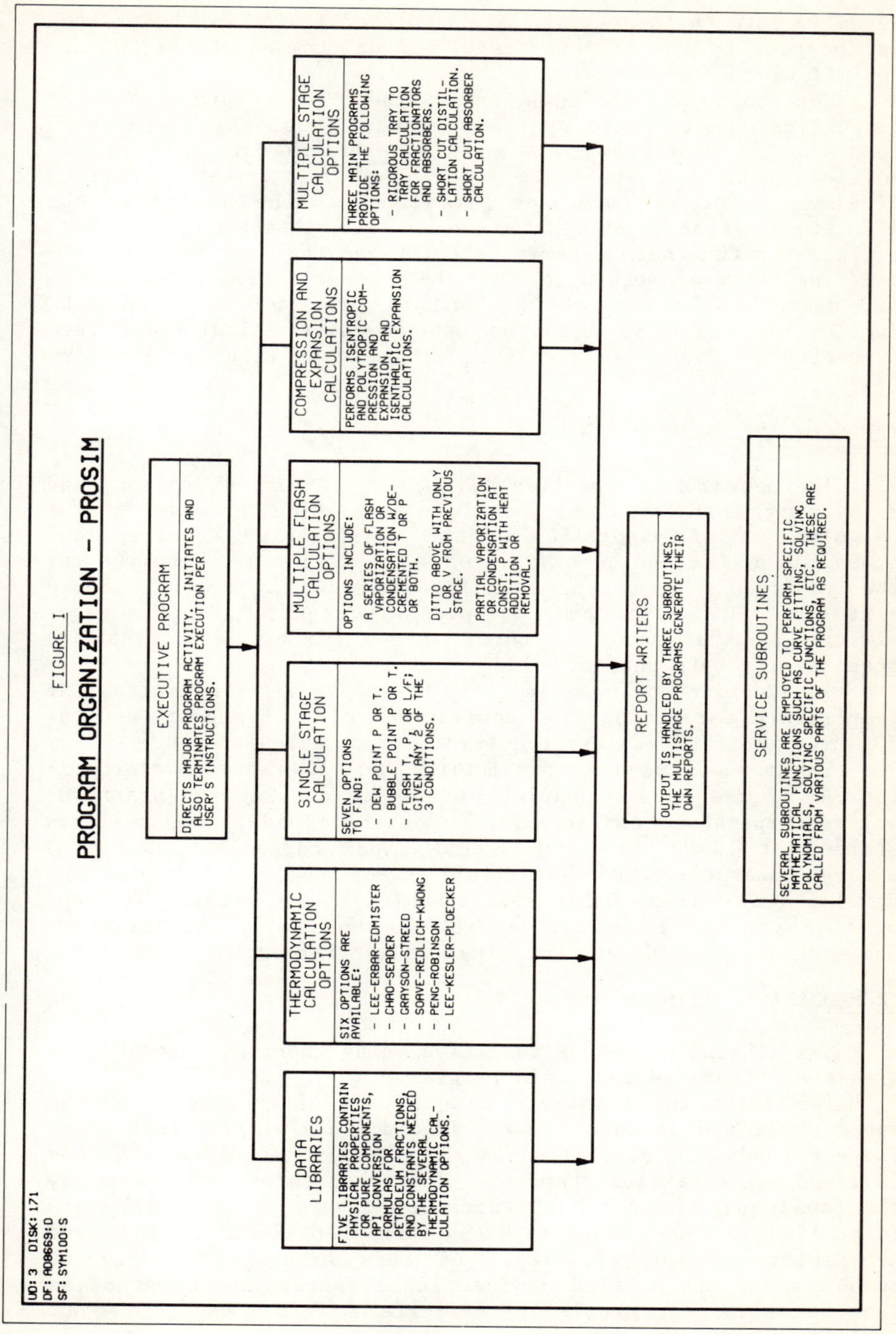

Figure 1. Program organization—PROSIM

manually into "hypothetical" components of finite width, or they may be made by the computer into "pseudo" components of infinitesimal width.

For both hypothetical and pseudo components, physical properties are computed by the same equations, which are based upon the correlations given in the Technical Data Book of the American Petroleum Institute (1). Equivalent molar quantities of these petroleum components are added to the amounts of the discrete (methane, etc.) components, to obtain the complete mixture for the thermodynamic calculations that follow.

The integral method for petroleum vaporization calculations was described by one of the authors previously (9) and the applications of this technique in computer calculations were presented by Taylor and Edmister (10) and by Lion and Edmister (11).

THERMODYNAMIC PROPERTIES

Thermodynamic properties (i.e., fugacities, entropies, and enthalpies) are required by this simulating program in the calculations of vapor/liquid phase equilibrium, compression/expansion paths, and heat balances. Fugacities are required for the individual components of the existing vapor and liquid mixtures. Enthalpies and entropies are required for the vapor mixture or the liquid mixture. Also, mixture densities are required for both phases.

In addition to these ordinary thermodynamic properties, the temperature and composition derivatives of the enthalpy and the fugacity coefficients are required in some calculations.

The methods used in predicting these thermodynamic properties employ: (a) an equation of state, relating the pressure-volume-temperature characteristics of the fluids; (b) ideal gas state heat capacities of the individual components; and (c) binary interaction coefficients between the components. The development of these basic relationships is not within the scope of this paper. Technical literature sources of the thermodynamic equations and data are given in the references.

THERMODYNAMIC METHODS

Six alternate methods for calculating thermodynamic properties are integrated into this program.

The first three methods use one set of equations for the vapor phase and another for the liquid, in a similar technique. These methods are identified as Chao-Seader (2), Grayson-Streed (3), and Lee-Erbar-Edmister (4). The other three methods employ the same equations for both vapor and liquid phases. They are identified as Soave-Redlich-Kwong (5), Peng-Robinson (6), and Lee-Kesler-Ploecker (7, 12). At this writing, the present authors have not settled on one single thermodynamic method as the choice for all problems. A single universal method may not be practical.

The Chao-Seader and the Grayson-Streed methods are very similar in that they both use the same mathematical models for each phase. For the vapor, the Redlich-Kwong equation of state is used. This two-parameter generalized pressure-volume-temperature (P-V-T) expression is very convenient because only the critical constants of the mixture components are required for applications. For the liquid phase, both methods used the regular solution theory of Scatchard and Hildebrand (26) for the activity coefficient plus an empirical relationship for the reference liquid fugacity coefficient. Chao-Seader and Grayson-Streed derived different constants for these two liquid equations, however.

The Grayson-Streed method was developed for hydrogen-petroleum systems at high temperatures and pressures, i.e., hydrofining conditions. Thus the constants, including some critical temperatures and pressures, differ slightly from those recommended in the Chao-Seader method.

The Lee-Erbar-Edmister method is of the same type, but uses different expressions for the fugacity and activity coefficients. The vapor phase equation of state is a three-parameter expression, and binary interaction corrections are included. The liquid phase activity and fugacity coefficient expressions were derived to extend the method to lower temperatures and to improve accuracy. Binary interaction terms were included in the liquid activity coefficient equation.

A multiple-property technique was employed in developing the Lee-Erbar-Edmister equations. In this method, both measured values of the isothermal pressure correction to the enthalpy and the P-V-T data are used.

The Soave modification of the Redlich-Kwong equation is the basis for the fourth thermodynamic properties method. This equation of state is applied to both liquid and vapor phases. Binary interaction coefficients for these applications are from Reid-Prausnitz-Sherwood (13) and the mathematical derivations used here are from Christiansen-Michelson-Fredenslund (14). Temperature and composition derivatives of the thermodynamic functions are included in the later work. These have applications in multistage calculations.

The Peng-Robinson (6) modification of the Soave-Redlich-Kwong equation is a further improvement of the Redlich-Kwong equation.

The Lee-Kesler (7) generalized equation of state, which also applies to both phases, is the basis for the sixth thermodynamic properties method. As originally developed, the Lee-Kesler equation was for predicting bulk properties (densities, enthalpies etc.) for the entire mixture and not for calculating partial properties for the components of mixtures. Phase equilibrium was not one of the uses that the authors had in mind when they developed the equation. Recognizing the other possibilities of the Lee-Kesler equation, Ploecker, Knapp, and

Prausnitz (12) applied the equation to the calculations of vapor/liquid phase equilibria. Component combination rules, interaction coefficients, and pseudo-criticals were all part of the extension of the equation to this new application.

EQUILIBRIUM PREDICTIONS

There are three independent variables in coexisting equilibrium vapor/liquid systems, namely: temperature, pressure, and fraction liquid (or vapor). If two of these are specified in a problem, the third is determined by the phase behavior of the system. There are seven types of vapor/liquid equilibria calculations in our program, as in Figure 1 under "Single Stage Calculation."

There are two "do loops" in each of these equilibrium calculations. When the unknown is either temperature or pressure, the inner "do loop" of 30 iterations converges on the unknown phase compositions and the outer "do loop" of 50 iterations converges on the unknown temperature or pressure. When the unknown of an equilibrium flash calculation is the liquid fraction, the inner "do loop" of 30 iterations converges on the liquid/feed ratio and the outer "do loop" converges on the unknown compositions.

The precedure used in finding an unknown temperature is: (a) Assume a temperature and select a second temperature 1.0% higher to give two values of T, (b) make K value and equilibrium calculations at both temperatures, and (c) the temperatures and summations from these calculations are used in a straight line convergence calculation to estimate a new temperature.

For predicting pressure, the procedure is similar except that a logarithmic convergence method is used in finding the new pressure. In predicting the liquid/feed ratio for flashing at a given temperature and pressure, the Newton-Raphson method is used. This calculation requires one flash at an assumed L/F ratio, 0.5 being assumed in this program. Tolerances for convergence of composition and temperature or pressure may be specified by the user, with default values set at one part in 10,000 for both K-values and temperature or pressure.

Estimating the unknown but required starting values of conditions and compositions is an important and sensitive part of these calculations. The composition of the feed is always known, as is the composition of one of the two phases in bubble and dew point calculations. With the Chao-Seader, Grayson-Streed, and Lee-Erbar-Edmister methods, it is possible to assume that both phases have the composition of the feed for the first trial. This assumption leads to trouble with the Soave-Redlich-Kwong, the Peng-Robinson and the Lee-Kesler-Ploecker methods, however. Accordingly, the Lee-Erbar-Edmister method is used first with the Soave-Redlich-Kwong, the Peng-Robinson and the Lee-Kesler-Ploecker to get starting values of the unknown properties and compositions.

SIMPLE PROCESSES

Subroutines to make simulation calculations for several simple equilibrium processes are included in the PROSIM program. The vapor and liquid properties and the phase equilibrium conditions that are required in these process simulations are found by the thermodynamic and iterative methods described above. The simple processes covered are: (a) multiple-flash vaporizations with incremented temperatures and/or pressures; (b) multiple partial condensations with incremented temperatures and/or pressures on the original vapor or the vapor remaining from the previous condensation calculation; (c) flash vaporization (or condensation) to find the duty of the heater (or cooler) in the process; (d) multistage compression of gas mixtures, with interstage cooling and liquid knockout, if desired; (e) gas expansion through turboexpander for cooling or through throttle valve for cooling or pressure reduction; and (f) flow of compressible fluids through nozzles or transfer lines, where flow is critical at the outlet.

Compression and expansion processes are for light gases only, excluding petroleum fractions, for which the program does not calculate entropy. The flash and condensation processes are for all compositions.

MULTISTAGE PROCESSES

Multiple equilibrium stage processes simulated in this program are distillation, absorption, and stripping. Both simple and reboiled absorbers are included, and multiple feed plus side-stream products are possible from the fractionators. Matrix- and short-cut-type solution methods are provided in separate subroutines.

The matrix method was presented by Taylor and Edmister (15, 10) as a "general" solution for multicomponent multistage separation calculations, being capable of solving various configurations of equilibrium stage processes. Originally, the K and H values were in empirical equation form, with the user providing the constants of the equations. Now that TAYLOR is a subroutine of the larger algorithm, there are two options on thermodynamic properties that the user can choose: (a) polynomial curve fits to values at three points in the column, or (b) direct evaluation of the K and H values every time they are required in the program. The curve fits are made automatically to the results of three equilibrium calculations on the feed, making it unnecessary for the user to do any intermediate data handling.

As it presently operates, TAYLOR uses numerical derivatives in solving the matrix, the original method of the authors. Present improvements include analytical derivatives with both sources of K and H values. Only temperature and liquid rate derivatives are used in this method. Composition derivatives are not used. Instead, the compositions are computed in a separate tridiagonal matrix. This method was used to save time and core in the computer.

With analytical derivatives, the TAYLOR method should handle all problems. If not, composition derivatives will be added to the matrix solution in the manner of Naphtali and Sandholm (16).

In the TAYLOR method, the number of equilibrium stages are fixed by the user, as are the amounts of products, including side-streams. As options, the reflux ratio may be specified or found in the calculations. As output, the program gives the complete temperature, flow rate, and composition gradients.

A short-cut design method for distillation is another subroutine. This method is based upon the minimum reflux of Underwood (17, 18, 19, 20), the minimum stages of Fenske (21) and Winn (22), and the reflux vs stages correlation of Erbar and Maddox (23) and Gray (24). In SHORT, which uses polynomial K and H values, the required number of equilibrium stages may be found for a specified multiple of minimum reflux, or alternatively, the reflux ratio may be found for a given multiple of minimum stages.

Another multistage method included in the program is the absorption and stripping factor method of Edmister (25). ASFPH, as it is called, can simulate simple and reboiled absorbers and also fractionators. The method used does not have very good convergence characteristics; however, it is of value in studying plant performance data.

CONCLUSION

A computer algorithm for simulating equilibrium stage processes while predicting the required thermodynamic properties, has been presented in this paper. With optional methods for properties predictions and for multistage separation calculations and petroleum slicing capability, the program offers the process industries versatile tools for simulation and design.

With all it features, the PROSIM program needs further improvements, especially in the convergence procedures for both single-and multiple-stage calculations. Work is proceeding on this at the present time. Another logical extension of the program is to flowsheet simulation, as well as reduced crude properties prediction.

LITERATURE CITED

1. "Technical Data Book - Petroleum Refining, 2nd Ed."; American Petroleum Institute, Division of Refining: Washington, DC., 1970.

2. Chao, K. C., and Seader, J. D. "A General Correlation of Vapor-Liquid Equilibrium Mixtures," AIChE Journal 1961 7, 598.

3. Grayson, H. C., and Streed, C. W.,. "Vapor-Liquid Equilibria for High Pressure, High Temperature Hydrogen-Hydrocarbon Systems,"; "Proceedings of 6th World Petroleum Congress, Sect III, 1963." Frankfurt, June 1963. p 223.

4. Lee, B. I., Erbar, J. H., and Edmister, W. C. " Prediction of Thermodynamic Properties for Low Temperature Hydrocarbon Process Calculations," AIChE Journal, 1973, 19, 349.

5. Soave, G. "Equilibrium Constants from a Modified Redlich-Kwong Equation of State," Chem Eng Sci, 1972, 27, 1197.

6. Peng, D. Y., and Robinson, D. B., "A New Two-Constant Equation of State," IEC Fundamentals, 1976, 15, 59.

7. Lee, B. I., and Kesler, M. G. "A Generalized Thermodynamic Correlation Based on Three-Parameter Corresponding States," AIChE Journal, 1975, 21, 510.

8. Edmister, W. C., and Aguayo, G. "Computer Simulation of Petroleum Separation Processes,"; "Summer Computer Simulation Conference," Chicago, July 1977.

9. Edmister, W. C., "Improved Integral Method for Petroleum Distillation Calculations," Ind Eng Chem, 1955, 47, 1695.

10. Taylor, D. L., and Edmister, W. C. "Solutions for Distillation Processes Treating Petroleum Fractions," AIChE Journal, 1971, 17, 1324.

11. Lion, A. R., and Edmister, W. C. "Make Equilibrium Calculations by Computer," Hydrocarbon Processing, 1975, August, p 119.

12. Ploecker, U., Knapp, H., and Prausnitz, J. "Calculation of High-Pressure Vapor-Liquid Equilibria from a Corresponding States Correlation with Emphasis on Asymmetric Mixtures," I&EC Process Design and Development, 1978, 17, 324.

13. Reid, R. C., Prausnitz, J. M., and Sherwood, T. K. "The Properties of Gases and Liquids, 3rd Ed." McGraw-Hill Book Company: New York, 1977.

14. Christiansen, L. J., Michelson, M. L., and Fredenslund, A. "Successive Approximation Distillation Calculations Using the Soave-Redlich-Kwong Equation of State," "12th Symposium on Computer Application to Chemical Engineering," Montreaux, Switzerland, 1979.

15. Taylor, D. L. and Edmister, W. C. "General Method for Multicomponent Distillation Calculations," Journal Chem Engr. Sym, Series No. 32," 1969, Inst of Chem Engr. - London.

16. Naphtali, L. M. and Sandholm, D. P. "Multicomponent Separation Calculations by Linearization," AIChE Journal, 1971, 17, 148.

17. Underwood, A. J. V. Journal Inst Petrol, 1944, 30, 225.

18. Underwood, A. J. V. Journal Inst Petrol, 1945, 31, 111.

19. Underwood, A. J. V. Journal Inst Petrol, 1946, 32, 598, 614.

20. Underwood, A. J. V. Chem Eng Progr, 1946, 44, 603

21. Fenske, M. R. "Fractionation of Straight-Run Pennsylvania Gasoline," Ind Eng Chem, 1932, 24, 482.

22. Winn, F. W. "New Relative Volatility Method for Distillation Calculations" Pet Ref, 1958, 37, 216.

23. Erbar, J. H., and Maddox, R. N. "Latest Score: Reflux vs. Trays" Pet Ref, 1961, 40, 183.

24. Gray, J. R. "Reflux-Trays Calculation for Multicomponent Distillation Systems," MS Thesis, Oklahoma State University, May 1968.

25. Edmister, W. C. "Absorption and Stripping Factor Functions for Distillation Calculations by Manual and Digital Computer Methods," AIChE Journal, 1957, 3, 165.

26. Hildebrand, J. H., and Scott, R. L. "Regular Solutions," Prentice Hall; Englewood Cliffs, N. J., 1962.

RECEIVED November 5, 1979.

APPLICATIONS

Utilization of a Computer in On-Line Control and Optimization of a Batch Process for Microbial Conversion of Ethanol to Protein

ARTHUR E. HUMPHREY, DANE W. ZABRISKIE, WILLIAM B. ARMIGER, and WILLIAM M. ZIEGLER

Department of Chemical and Biochemical Engineering, University of Pennsylvania, Philadelphia, PA 19104

The first persons to point out the possibility of computer monitoring of indirectly measured parameters were Yamashito, Hisashi, and Inagaki in 1969 (2). The method was described in a U.S. Patent in 1975 (3). Examples of the application of computer-aided indirectly measured parameters to the control and optimization of batch-fed Baker's Yeast fermentation were described by Jefferis and Humphrey in 1973 (4) and by Wang, Cooney, and Wang in 1977 (5). Background and detailed history of this application can be found in the review by Humphrey (1). The work to be discussed in this report is another example of computer-aided indirectly measured cell biomass and growth rate and the use of this information in the feed back control of the carbon substrate, ethanol, in a process for the production of a yeast single cell protein (SCP).

Production of SCP from Ethanol

Most of the early work on the production of single cell protein (SCP) was based on the conversion of petroleum-derived materials to SCP. In recent years oxygenated compounds such as methanol, ethanol, acetate and cellulosic materials have received increasing attention. Key papers discussing the kinetics of aerobic utilization of ethanol by microorganisms include those of Laskin (6), Mor and Fiechter, (7,8), Prokop, Votruba, Sobotka, and Panos (9), and Ristroph, Watteeuw, Armiger, and Humphrey (10). It was the observation in the Laskin report which specifically led us to undertake this work. Laskin (6) stated that "laboratory experiments have indicated that an SCP fermentation utilizing *A. calcoaceticus* should be operated under ethanol-limiting conditions at high growth rates to maximize the protein content of the product and the amount (yield) of protein and/or biomass produced from ethanol." The problem is two-fold. Firstly, the substrate yield coefficient is strongly dependent on the growth rate. This is so

0-8412-0549-3/80/47-124-355$05.00/0

because the maintenance coefficient for ethanol utilizing microorganisms has been observed to be quite high, i.e. greater than 0.1 g ethanol/g cell biomass/hr. Unless the organism is grown at reasonably high growth rates, i.e. greater than 0.30 hr^{-1}, an appreciable fraction of energy substrate will be diverted to maintenance. For example 17% of the ethanol can be used for maintenance at a growth rate of 0.25 hr^{-1} and 33% at a growth rate of 0.1 hr^{-1}. Secondly, both acetaldehyde and acetic acid, intermediates in alcohol metabolism, are inhibitory to growth. In aerobic ethanol fermentation both compounds can accumulate in the broth when ethanol is in excess. Prokop, Votruba, Sobotka, and Panos (9) report that the growth rate of Candida utilis A-49 is half inhibited at acetaldehyde concentrations of 0.12 g/liter and acetate concentrations of 0.0175 g/liter. Laskin (6) suggested that inhibitory levels for Acinetobacter calcoaceticus are less than 0.1 g/liter for acetaldehyde and 1.0 g/liter for acetate. However, Laskin's data indicated the microorganisms stopped growing when the acetaldehyde concentration was 0.03 g/liter. In general, however, it is acetate not acetaldehyde that is the problem. Ristroph, Watteeuw, Armiger, and Humphrey (10) in their work using NADH-NADPH fluorometric monitoring on the control of batch-fed yeast grown on ethanol found that whenever the ethanol concentration exceeded 4 g/liter, acetate started accumulating and only decreased during periods of ethanol starvation. The data of Prokop, Votruba, Sobotka, and Panos (9) suggests that the saturation constant (point at which the ethanol concentration limits the growth rate to 50% of the maximum) is 0.35 g/liter. This means that at 1 g/liter ethanol concentration the growth rate should be 74% of the maximum and at 2 g/liter it will be 85% of the maximum rate if no inhibition occurs. Obviously it is a thin balance point between striving for maximum growth rate while preventing acetate accumulation which adversely affects the growth rate.

It appears, therefore, that the concept to follow in a fed-batch ethanol SCP fermentation is to determine the substrate demand by indirectly estimating the cell density and growth rate and then to add the substrate in such a way that it satisfies the demand but never exceeds a concentration of 2 g/liter. This strategy should prevent acetate from accumulating towards the end of the batch run affecting the growth rate and ultimately the SCP yield. Normally, one would do this by continuous culture rather than by a fed-batch process. However, there are various process reasons for preferring fed-batch to continuous operation. These reasons may include a desire to finish off the cell culture by letting the cells age without substrate but under full aeration. This is used in Baker's Yeast production to enhance the stability of the cells as well as make them better suited to further processing.

This is the background to the control strategy of this work. A fed-batch process for the conversion of ethanol to a yeast SCP was to be computer controlled by indirectly monitoring the cell density and growth rate and then using these indirect measurements to determine the amount of ethanol to add to the fermentation. The ethanol was to be added at a rate as high as possible but not sufficiently high so that acetate would accumulate allowing it rather than alcohol to ultimately limit growth.

Indirect Measurement of Cell Biomass and Growth Rate

The possibility of indirectly measuring cell biomass and growth rate by computer-aided on-line substrate balancing was first mentioned in a publication by Humphrey (11). The method is discussed in detail in a review by Humphrey (1). The technique operates by noting that the difference in amount of a particular component in and out of the vessel is equal to that used by the organism for growth and maintenance if no appreciable secondary metabolites are formed. Mathematically this can be stated as

$$Q_i X = \frac{1}{Y_{x/i}} \frac{dX}{dt} + m_i X \qquad (1)$$

where

Q_i = specific consumption rate of component i, g_i/g cell-hr
m_i = specific maintenance requirement for component i, g_i/g cell-hr
X = biomass concentration, g cell/liter
t = time, hr
$Y_{x/i}$ = maximum growth yield, g cell/g_i used for growth

The $Q_i X$ is the measured value of the substrate utilization. In the case of oxygen this is simply the difference between the oxygen in and out in the gas streams. The accumulation of oxygen in the broth can be neglected because of its very low solubility. The constants m and $Y_{x/i}$ are obtained from yields vs. growth rate data. Equation (1) can be rearranged to give the growth rate, i.e.

$$\frac{dX}{dt} = Y_{x/i} \, (Q_i X - m_i X) \qquad (2)$$

By initializing the cell concentration, X(0), either from an estimation at some point of the oxygen uptake or carbon dioxide evolution rate or by knowing the inoculum size and then integrating the estimated growth rate over time, both cell biomass and growth rate can be estimated on-line utilizing a computer. The substrate (ethanol) demand, ΔS is then estimated by a similar equation, i.e.

$$V\Delta S = V\left(\frac{\mu X}{Y_{x/s}} + m_s X\right)\Delta t \qquad (3)$$

where

S = substrate concentration, g/liter
V = fermenter volume, liter
μ = specific growth rate = $(1/X)(dX/dt)$, hr^{-1}
$Y_{x/s}$ = maximum growth yield of cells on substrate, g/g
m_s = specific maintenance requirement for substrate, g/g/hr

Materials and Methods

Organism. Candida utilis strain ATCC 26387 was used in this work. The organism was stored on agar slants containing the medium in Table I and 1% dextrose. The culture was initially propagated in 50 ml of media in 250 ml shake flasks on a gyrator shaker at 28°C and 100 rev/min. Four hundred ml of shake flask culture were used to inoculate 7 liters of medium in a 14 liter New Brunswick fermenter. Four liters of this culture were aseptically transferred after 7.5 hours to 40 liters of medium in a 70 liter computer (PDP-11/34 with 3 discs and 96K words of core) - controlled fermentation system. This system has been described in detail elsewhere (12). The details are shown in Figure 1.

Media Composition. All experiments were run with ethanol as the main carbon source. Table I gives a list of components excluding ethanol:

Table I

Media Components in 40 Liters

$(NH_4)_2SO_4$	200 g
KH_2PO_4	300 g
$K_2H\ PO_4$	25 g
$MgSO_4 \cdot 7\ H_2O$	100 g
Yeast Extract	40 g
Sigma Salts (half strength)	1 liter
Antifoam	10 ml

Fermentation Conditions. The control experiments were all carried out in the 70 liter computer-coupled fermentation system. Data were logged and feed-back control commands issued every 2 or 5 minutes. The dissolved oxygen was maintained above 45% saturation by a combined agitation and aeration control. The pH was controlled above 4.3 using NH_4OH. The temperature was maintained initially at 28.0°C. However, in the latter stages of the run cooling water and refrigeration systems could not keep up with the heat evolution rate from the fermentation. This caused the tem-

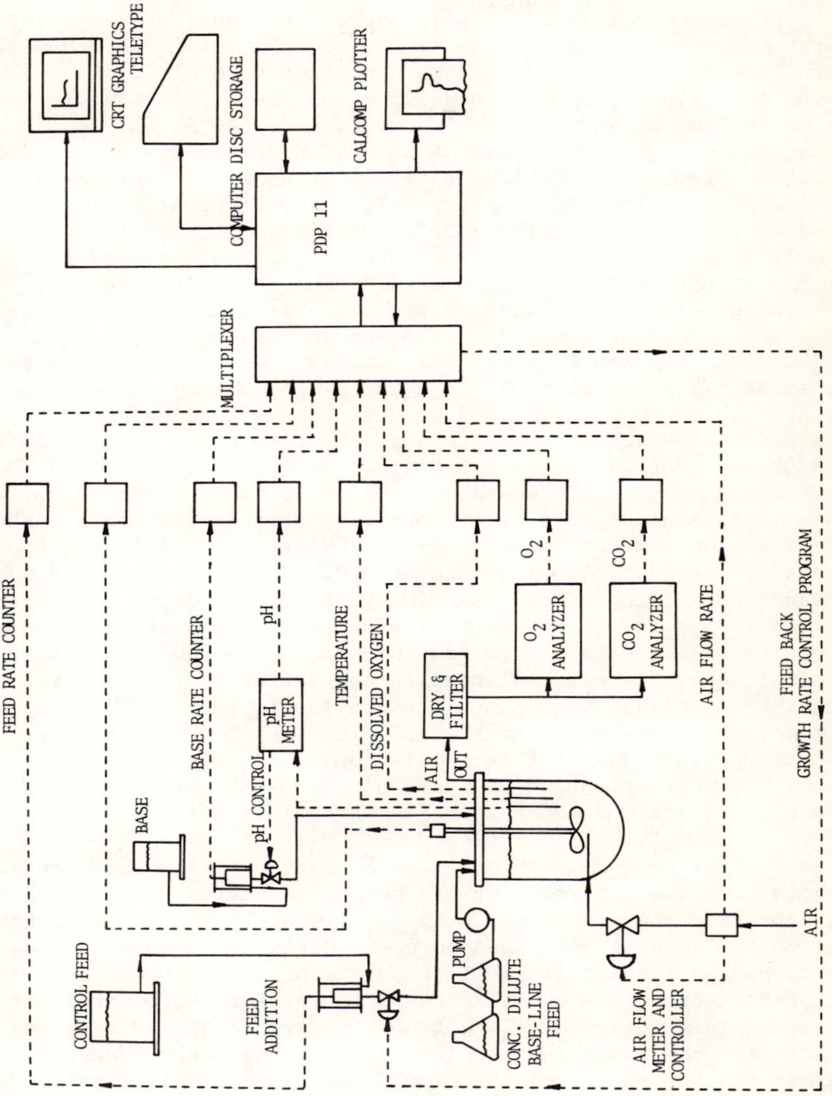

Figure 1. Schematic of 70-liter computer-coupled fermentation system

perature to increase to 32°C and higher. It was then no longer possible to maintain the growth rate within the desired range of 80-95% of the maximum.

Ethanol Feed. The ethanol was fed by a dual system. The first was a two flask feed system, used to maintain an ethanol base line feed rate. Each flask contained 4000 ml of ethanol solution. Flask one contained 125 ml of 190 proof ethanol (93.7 g) in 3875 ml of H_2O. Flask two contained 3500 ml of 190 proof ethanol (2623 g) plus 500 ml of water. When the system is feeding solution from flask one to the fermenter at a constant rate, r liter/hr, a linear increase in alcohol concentration in the feed will occur. See Figure 2 for details. This will result in a cell mass increase in the fermenter that is second order in time. This base line feed rate will always be less than the desired growth rate as long as the correct feed rate is selected.

The second or control system was comprised of a constant volume feeder. See Figure 3 for details. This system will add 20 ml of 171 proof alcohol (13 g shots) on demand from the computer. This could be every 2 or 5 minutes depending upon the frequency at which the computer was set to monitor the system and update the control. When the cell concentration approached 20 g/liter, the system was demanding nearly 400 g of ethanol/hour. This is equivalent to a cell productivity of 5 g/liter-hr. When the shots were being added every five minutes, some triple shot doses were required to meet the demand. This meant that 39 g of ethanol were being added essentially instantaneously. This could result in a brief period where the ethanol concentration reached 1 g/liter. Since this bordered on being undesirable, 2 minute control intervals were selected in the latter runs. This meant that there would never be more than a double shot, i.e. 26 g alcohol/40 liters added to the vessel. These conditions were expected to avoid acetate accumulation. This was borne out in the hourly gas-chromatograph analysis of the broth. At the end of the run the acetate was found to be less than 0.1 g/liter.

Control Program. This system is part of a fermentation data acquisition and control software package entitled DATAN developed by Zabriskie and described in his Ph.D. thesis (13). The particular control scheme described here and detailed in Table II was developed by Ziegler (14). It consists of 129 instructions. The first 9 instructions read in the variables for the control integration. Instructions 10-16 set up the permanent disc file and initialize the variables for integration. Next the necessary parameters are read from the disc file. Then instructions 37 to 49 determine the proper density of ethanol being used, i.e. whether it is 190 or 200 proof. Next the program initializes the variables for the integration of the cumulative base line ethanol feed and then calculates the amount of ethanol feed since the last time interval. Instructions 69 and 70 call for actual

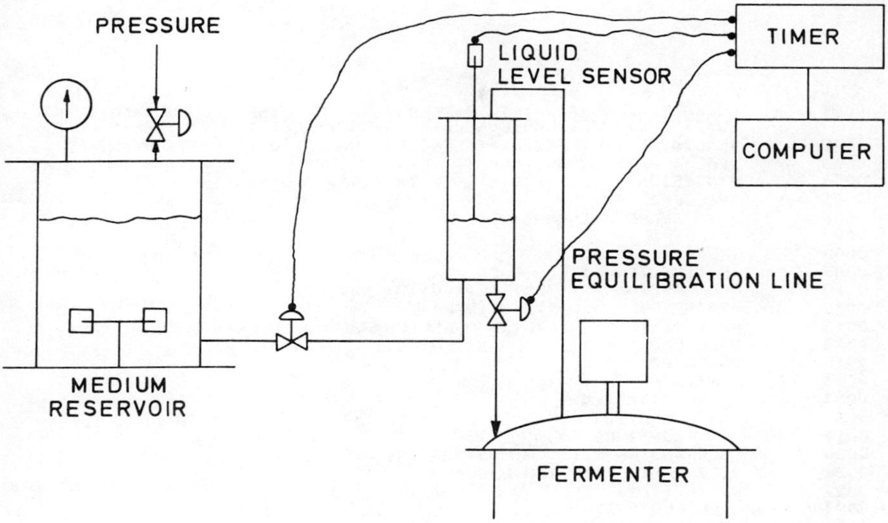

Figure 2. Constant-volume-feed system control

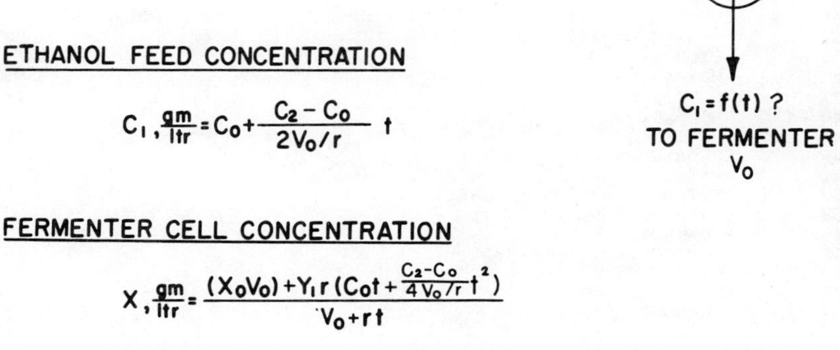

Figure 3. Mathematical representation of two-flask-feed system

TABLE II

```
      SUBROUTINE CONTRL (ETHRS,IFLAG7,IFLAG8,LINFED,CUMSUB,MU,MUACTL,
     2NUM7,NUM8,CUM7,CUM8,CUMFED,GRO2,XO2,IREC,FILOUT,NSIZE,VOLFER)
      REAL*8 DELT,LINFED,LINFDL,NB,NBO,NBL,ET,DNBDT,VBT,CB,DLINFD
      REAL*4 MAIN,MU,MUO,MUACTL
      DIMENSION I(10),P(38),D(57),INT(14),ZERO(20),IMSG(13)
      INTEGER*2 SWITCH
      DATA DNAME/6RDKHAND/
      IFLAG7=0
      IFLAG8=0
      IF (IREC.EQ.1) GO TO 50
      CALL ASSIGN (1,'CONTROL.000',0,'RDO',0,1)
      DEFINE FILE 1(2,128,U,IVAR)
      READ (1'1) ETHRSL,LINDFL,NBL,CUMSBL
      CALL CLOSE (1)
      GO TO 60
   50 CONTINUE
      DO 40, IJ=1,20
           ZERO(IJ)=0.0
   40      CONTINUE
      CALL ASSIGN (1,'CONTROL.000',0,'NEW',0,1)
      DEFINE FILE 1(2,128,U,IVAR)
      WRITE (1'1) ZERO
      CALL CLOSE (1)
      CUMSBL=0.0
      ETHRSL=0.0
      LINFDL=0.0
   60 CONTINUE
      CALL ASSIGN (1,'PARAMS.000',0,'OLD',0,1)
      CALL FDBSET (1,'OLD')
      DEFINE FILE 1(2,256,U,IVAR)
      READ (1'1) (I(N),N=1,10),NUM7,NUM8,SWITCH
      READ (1'2) FILNAM,(P(N),N=1,38),SIZE7,SIZE8,ETH7,ETH8,VOL7,VOL8,
     1VOLTOT,VOLETA,VOLETB,FLOWRT,GROMAX,YXS,MAIN,CODE9
      CALL CLOSE(1)
      VOLO=P(3)
      XINO=P(33)
      IPROOF=I(10)
      RHO190=0.7493
      RHO200=0.7893
      IF (IPROOF.EQ.200) GO TO 100
      IF (IPROOF.EQ.190) GO TO 110
      TYPE 120
  120 FORMAT (' THIS SYSTEM IS ONLY CAPABLE OF OPERATING WITH',
     1/,' 190 OR 200 PROOF ETHANOL.')
      RETURN
  100 RHO=RHO200
      GO TO 111
  110 RHO=RHO190
  111 CONTINUE
      CONCA=RHO*VOLETA/VOLTOT
      NB=NBL
      DELT=0.001*(ETHRS-ETHRSL)
      LINFED=LINFDL
      NBO=RHO*VOLETB
      IF (IREC.EQ.1) NB=NBO
      VBO=VOLTOT
      ET=ETHRSL
      Q=FLOWRT
```

TABLE II Continued

```
      DO 2000 J=1,1000
           ET=ET+DELT
           DNBDT=(Q*CONCA/2.0)-(Q*NB)/(VBO-Q*ET/2.0)
           NB=NB+DNBDT*DELT
           VBT=VBO-Q*ET/2.0
           CB=NB/VBT
           DLINFD=CB*Q*DELT
           LINFED=LINFED+DLINFD
 2000      CONTINUE
      MUACTL=GRO2/XO2
      X=XO2
      MU0=0.9*GROMAX
      MU=MU0
      IF (ABS(MU-MUACTL).LE.0.1*GROMAX) GO TO 800
      MU=MUACTL
      IF (MU.LT.0.80*GROMAX) MU=0.8*GROMAX
      IF (MU.GT.GROMAX) MU=GROMAX
  800 CONTINUE
      IF (IREC.LE.SWITCH) MU=MU0
      IF (IREC.LE.SWITCH) X=XINO*EXP(MU*ETHRS)
      MU=1.01*MU
      DELSUB=VOLFER*(MU*X/YXS+MAIN*X)*(ETHRS-ETHRSL)
      CUMSUB=CUMSBL+DELSUB
      GO TO 900
  900 CONTINUE
      WTSH7=SIZE7*RHO*ETH7/VOL7
      WTSH8=SIZE8*RHO*ETH8/VOL8
  950 CONTINUE
      ETHXS=WTSH7*NUM7+WTSH8*NUM8+LINFED-CUMSUB
      IF(ETHXS.GT.0.0) GO TO 500
      IF(0.66*WTSH8+ETHXS.LT.0.0) GO TO 300
      GO TO 500
  300 CONTINUE
      IMSG(1)=2
      IMSG(2)=1920
      CALL SEND (DNAME,IMSG,IDSW)
      TYPE 1008
 1008 FORMAT (' TANK 8 ADDITION IN PROGRESS')
      CALL WAIT (13900,1,IXX)
      IMSG(2)=28672
      CALL SEND (DNAME,IMSG,IDSW)
      NUM8=NUM8+1
      IFLAG8=1
      GO TO 950
  500 CONTINUE
      CALL ASSIGN (1,'PARAMS.000',0,'OLD',0,1)
      CALL FDBSET (1,'OLD')
      DEFINE FILE 1(2,256,U,IVAR)
      WRITE (1'1) (I(N),N=1,10),NUM7,NUM8
      WRITE (1'2) FILNAM,(P(N),N=1,38),SIZE7,SIZE8,ETH7,ETH8,VOL7,VOL8,
     1VOLTOT,VOLETA,VOLETB,FLOWRT,GROMAX,YXS,MAIN,CODE9
      CALL CLOSE (1)
      CALL ASSIGN (1,'CONTROL.000',0,'OLD',0,1)
      CALL FDBSET (1,'OLD')
      DEFINE FILE 1(2,128,U,IVAR)
      WRITE(1'1) ETHRS,LINFED,NB,CUMSUB
      CALL CLOSE(1)
      GO TO 600
  600 CONTINUE
      CUM7=WTSH7*FLOAT(NUM7)
      CUM8=WTSH8*FLOAT(NUM8)
      CUMFED=CUM7+CUM8+LINFED
      RETURN
      END
```

growth rate and cell mass determined by the oxygen balance subroutine CMO2 in the DATAN. When the growth rate is within ±5% of 90% of the maximum growth rate, then the control scheme proceeds with this growth rate in estimating the required substrate feed. If the growth rate deviates from this range, then the actual growth rate is used unless it falls outside the range of 80 to 100% of the maximum growth rate. An exception is the first four hours of the run where the oxygen balance data is not sufficiently accurate for control. During this period it is assumed that the growth rate is 90% of the maximum. After the first four hours the growth rate is set at 1% greater than the actual growth rate if it is less than the desired 90% of the maximum growth rate. This increase of the growth rate by 1% is designed to prevent the positive feedback control from driving the system to the lower growth, i.e. 80% of the maximum growth rate.

Once the desired amount of substrate to be added is estimated, step 86, the system then determines the number of shots to be added. If the shortage of ethanol in the fermenter is greater than 2/3 of a shot, it is added; if not then the shot is not added, steps 100-111. Any excess added is calculated and the program continues by updating the permanent disc files with the newly calculated values of the appropriate variables. When this is done, the program returns at the next time intervals, either in 2 or 5 minutes, to the beginning of the control subroutine.

Results

In the course of this work, 10 fed batch runs were made under computer control. Figure 4 presents 10 minute interval data for a typical run in which the maximum growth rate was set at 0.3 hr^{-1}. In the beginning, the growth rate exceeded 0.3 hr^{-1}, however, after the initial 5 or 6 hours the system mostly stayed at 80% (i.e. 0.24 hr^{-1}) of the set maximum growth rate. It appears that the actual maximum is greater than 0.3 hr^{-1}. In fact it may be closer to 0.4 hr^{-1}. It appears that the cell yield must be near the 0.7 g cell/g ethanol assumed in the latter runs. Prokop, Votruba, Sobotka, and Panos (9) found that their species of *Candida utilis* had a $Y_{x/c}$ = 0.5 to 0.6 g biomass/g ethanol depending upon the feeding conditions. Mor and Fiechter (7) found that with *Saccharomyces cerevisiae* grown on ethanol the $Y_{x/c}$ = 0.4 to 0.5 g cells/g ethanol. Laskin (6) reported a $Y_{x/c}$ = 0.7 g cell/g ethanol for *Acinetobaetic calcoaceticum*. These data tend to agree with the $Y_{x/c}$ = 0.7 found in these experiments. For the run in Figure 4, 2452 g of ethanol gave 1692 g of cells. The 20 hour integrated maintenance requirement of the cells in this run, assuming m = 0.02 g/g-hr, is equivalent to approximately 134 g of ethanol. This means that the maximum growth yield is 1692/(2452 - 134) = 0.73 g/g which is consistent with the literature.

In conclusion we believe that our control system outlined and tested here will prove useful in optimizing fed-batch fermen-

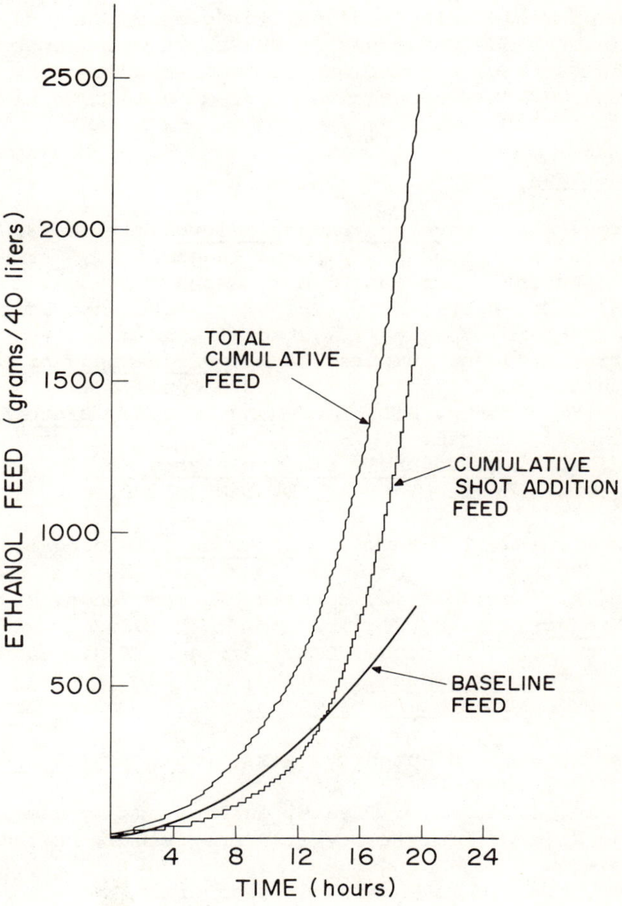

Figure 4. Typical results of controlled batch-fed fermentation

tation to produce yeast SCP on ethanol. It needs to be further refined, i.e. more accurate yield and maximum growth rate parameters determined, before it is finally optimal.

Acknowledgment

This work was part of a M.S. thesis by one of the authors in the College of Engineering and Applied Science, University of Pennsylvania, Philadelphia, PA, in 1979. It was supported in part by a National Science Foundation grant NSF-ENG 76-17181 in connection with the US/USSR cooperative program in Industrial Microbiology.

Literature Cited

1. Humphrey, A.E., Develop. Indust. Microbiol., 1977, 18, 58.
2. Yamashita, S., Hishashi, H., and Inagaki, T., Ferment. Adv. (Proc. 4th Intl. Ferment. Symp., Japan), 1969, 441.
3. Wilson, J.D., Nyiri, L.K., Humphrey, A.E., and Harmes, C.S., U.S. Patent No. 3, 926, 737, Dec. 16, 1975.
4. Jefferis, R.P. and Humphrey, A.E., Proceedings of GIAM-IV, Sao Paulo, Brazil, July 23-28, 1979, pp. 767-774.
5. Wang, H.Y., Cooney, C.L., and Wang, D.I.C., Biotechnol. Bioengr., 1977, 19, 55.
6. Laskin, A.I., Biotechnol. Bioengr. Symp. No. 7, 1977, 91.
7. Mor, J.R., and Fiechter, A., Biotechnol. Bioengr., 1968, 10, 159.
8. Mor, J.R., and Fiechter, A., Biotechnol. Bioengr., 1968, 10, 787.
9. Prokop, A., Votruba, J., Sobotka, M. and Panos, J., Biotechnol. Bioengr., 1978, 20, 1523.
10. Ristroph, D.L., Watteeuw, C.M., Armiger, W.B., and Humphrey, A.E., J. Ferment. Technol., 1977, 55, 599.
11. Humphrey, A.E., Proc. Labex Symp., Earls Court, London, 1971, 1.
12. Armiger, W.B., and Humphrey, A.E., Proc. 1st Intersect. Congress IAMS, 1974, 5, 99.
13. Zabriskie, D.W., Ph.D. Thesis, College of Engineering and Applied Science, University of Pennsylvania, Philadelphia, PA, 1976.
14. Ziegler, W.M., M.S. Thesis, College of Engineering and Applied Science, University of Pennsylvania, Philadelphia, PA, 1979.

RECEIVED November 5, 1979.

21

Modeling and Control of the Activated Sludge Process

GUSTAFF OLSSON

Department of Automatic Control, Lund Institute of Technology, Lund, Sweden

During recent years there has been an increasing demand for better operation of wastewater treatment plants in order to guarantee satisfactory effluent quality at minimal cost. The renewed interest in instrumentation and control comes after a period of huge investments in sewer networks and treatment plants. Several factors have contributed to the potential for better operation and control, such as cheap computing power, improving sensors and better knowledge of process dynamics and control.

The activated sludge process is recognized as the most common and major unit process for the reduction of municipal and industrial waste. Although its dynamics is complex the potential for control is far from exhausted. Already with commercially available instrumentation today its operation can be significantly improved.

In the process heterotrophic microorganisms react with organic pollutants in the wastewater and with oxygen dissolved in the water to produce more cell mass, carbon dioxide and water. Nitrifying organisms oxidize ammonia-nitrogen in the water to form nitrite, nitrate and more cell mass. The reactor effluent flows to a sedimentation basin, where the solids are separated from the liquid. A portion of the concentrated sludge is recycled in order to maintain enough mass of viable organisms in the system and a reasonable food-to-mass ratio.

Some typical disturbance patterns and control difficulties are summarized here. A detailed discussion is made in (1). Hydraulic disturbances are significant in amplitude. Diurnal variations as well as shock loads from rain storms or melting snow may cause major upsets. Significant disturbances also appear from internal sources like primary pumps, back-washing of deep bed filters or return sludge flow rate changes. The amplitudes are such, that quasi-stationary of linear control methods are seldom adequate.

Concentration and composition changes of the influent wastewater create tremendous control problems. The number of components in the sewage water is so large and the concentrations generally so low, that the measurement problems seem prohibitive. Microbial composition and concentrations in the reactor can be significantly

0-8412-0549-3/80/47-124-367$05.00/0

changed, sometimes very fast, sometimes gradually. As a result the pollutant elimination will be affected and the process may turn into undesirable operational conditions. These properties of the process imply, that disturbance detection and parameter and state estimation are the major control problems.

The process dynamics is complex. Typical response times vary from a few seconds up to weeks and months. Nonlinear phenomena are common due to large disturbance amplitudes. Measurement and transport time delays cannot be neglected. Sensor location is important due to spatial distributions. As living organisms are part of the system, not only parameters but the structure of the system dynamics can be changed during the operation.

Dissolved oxygen (DO) concentration is a key variable in the operation. It affects the economy and is also closely related to the biological activity of the organisms in the sludge. The main emphasis in this paper is to show how the DO concentration distribution in a non-homogeneous aerator can be used for the estimation of the activity of both heterotrophic and nitrifying organisms.

The paper is organized as follows. A brief review of the activated sludge dynamics and DO control is made. Then the properties of DO profiles in organic removal are discussed. In the last section it is shown, that the DO profile concept can be favourably used in reactors with combined organic and nitrogen removal, and the implications for control are discussed.

Activated sludge dynamics

Large efforts have been spent to develop highly structured models of activated sludge dynamics. The results in this paper are based on models derived elsewhere (1-4). The models consider the flow regime in the aerator, oxygen dynamics, cell growth and basal metabolism. It also includes degradation of pollutants - both organic carbon and nitrogen - as well as the solubilization of particulate degradable organic carbon.

Organic waste removal. It is assumed that the organic waste is degraded into the following steps:

(i) substrate penetration of the cell membrane

(ii) metabolism of the mass stored in the floc phase, the "growth" phase

(iii) decay of the organisms to inert mass

Typical response times are of the order 15-30 minutes (i), days (ii), and several days (iii). The net microorganism growth rate is a key design parameter and its reciprocal is termed sludge age or mean cell residence time. An adequate control has to take into consideration, that the growth rate is time-varying (5).

Nitrification. In biological nitrification ammonium nitrogen is oxidized into nitrite by means of Nitrosomonas species, and nitrite is oxidized to nitrate by the Nitrobacter species.

The first reaction is slower than the second one, so the resulting effluent nitrite concentration will be small. Nitrate is the final oxidation state if no denitrification takes place in the aerator. In the oxidation of ammonium nitrogen alcalinity is lost. About half of it could be restored if the system is designed for denitrification as well.

The growth rates of Nitrosomonas and Nitrobacter species are assumed to follow the so called Monod kinetics with respect to ammonia and nitrite concentrations respectively as well to oxygen. The growth rates are much more sensitive to pH, toxic and temperature changes than the heterotrophic organisms. Moreover, the growth rates are significantly smaller than for heterotrophic organisms. Therefore a fully nitrifying plant requires much longer mean cell residence times than plants designed for only carbonaceous removal.

Dissolved oxygen. The structure of the DO material balance in a reactor can be divided into three parts, the hydraulic transport of DO, the production related to the air supply and the consumption (uptake rate) due to organism growth and decay. The hydraulic transport of DO is negligible in comparison with the oxygen transfer (production).

The oxygen uptake rate terms are considered in more detail. Here they are represented by four terms, oxygen uptake rate due to heterotrophic growth (r_1), oxygen consumption rate by organism decay (r_2), oxygen uptake rate due to Nitrosomonas (r_3) and Nitrobacter (r_4) respectively. The specific carbonaceous oxygen uptake rate (SCOUR) is defined by (6)

$$\text{SCOUR} = (r_1 + r_2)/\text{cell mass}$$

and the specific nitrogeneous oxygen uptake rate (SNOUR) in analogous way by

$$\text{SNOUR} = (r_3 + r_4)/\text{cell mass}$$

In a fully nitrifying reactor SNOUR is of the same order of magnitude as SCOUR. Consequently, full nitrification will almost double the air demand.

The DO dynamics include phenomena in widely different time scales, from fractions of hours to weeks. This means, that a control system designed for the control of DO as a physical variable has to gradually change its parameters.

Process identification and parameter estimation has been applied in water quality and wastewater treatment systems (7-9). The overall oxygen transfer coefficient can be determined on-line. The hydraulic dispersion has been identified by manipulation of the influent flow rate or the return sludge flow rate (9).

Reactor model. The structured model of the activated sludge process includes the following concentration variables: soluble organic substrate, stored mass, ammonia nitrogen, nitrite and nitrate nitrogen. Organisms are represented by viable Heterotrophs, Nitrosomonas and Nitrobacter. Inert (non-viable) nitrifying bacteria are lumped together with the non-viable heterotrophic organisms. Dissolved oxygen is included, but alcalinity, pH and gas changes are neglected, as non-covered aerators are used.

The concentration dynamics is described by (5),

$$\frac{\partial c}{\partial t} = E \frac{\partial^2 c}{\partial z^2} - v \cdot \frac{\partial c}{\partial z} + f_1(c) - f_2(c)$$

where c = c(z,t) = a space and time variable concentration
E = dispersion coefficient
v = stream velocity
f_1= production rate
f_2= consumption rate

Dissolved oxygen control

DO dynamics and control is of vital importance for the operation of an activated sludge system. The relation between DO and biological activity will be summarized here.

The choice of an adequate DO concentration. The desired DO concentration is a compromize between economy demands and hydraulic and biological demands. Moreover, in a non-homogeneous reactor a proper spatial distribution has to be determined.

Aerobic organism growth demands a certain amount of DO concentration. In the aerator the growth of specific types of organisms should be favoured. At low DO concentrations heterotrophic organisms may be inhibited while filamentous organisms are favoured, resulting in bulking sludge. The DO demand of nitrifying organisms is higher than that of heterotrophic organisms. Consequently in a nitrifying reactor bulking sludge seldom appears.

To find the proper degree of mixing is complex. It is related to both DO concentration, floc size, floc formation as well as hydraulic dispersion and bulking sludge formation. Experiments seem to indicate that the sludge volume index (SVI) will increase sharply for an increasing air supply. The reason for this is unclear (10). Either it is due to the addition of oxygen as such or by the increasing mixing. The air supply thus has to be optimized between the growth limitation and the SVI increase.

Nitrates can be used as complements or alternatives to oxygen, both for energy saving and for the treatment of industrial wastes. This is the background for the Kraus modification of the activated sludge process. Similarly an anoxic zone in the inlet area of the aerator can be applied (11).

DO control. Most of available commercial DO control systems today are based on an analog PI controller and one DO probe. Quite often they give unsatisfactory result due to difficult process and measurement noise, time delays in the plant and insufficient measurement information. Unsuitable actuators are common, but actuator limitations can be overcome easier with digital control.

A typical result from direct digital control of the Käppala plant outside Stockholm is shown in fig 1. The DO concentration target is 3 mg/l. The controlled concentration (curve B) has an average value of 3.0 mg/l with a standard deviation of 0.28 mg/l. Encouraging and remarkable results of DO control have been achieved recently (12), where two aeration basins were compared in parallel, one manually and the other automatically controlled. With automatic control energy was saved and significant improvements of the water quality were obtained. Further practical DO control experiences are reported in (13 - 15).

The biodegradable load to the reactor can be indicated by the air flow rate if a DO control system is active. In fig 2 dry mass of COD from municipal sewage is compared to such an air flow signal in Käppala.

Dissolved oxygen concentration profiles with organic removal

The spatial variation of concentrations in a non-homogeneous reactor will increase the control complexity. However, the concentration profiles provide extra information. The immediate problem of locating the DO probes suitably must be considered, which was remarked already in 1964 (16): "there is no one position in the tank which is representative of the whole tank all the time. This means that a one-point control system applied to a piston-flow plant is likely to be, at best, a compromise...".

DO profile characterization. In an aerator with a uniform air supply distribution the DO concentration has a typical profile, fig 3. The oxygen content is close to zero at the inlet but is often in excess close to the outlet. As the influent wastewater is mixed with recycled sludge the soluble substrate is captured quickly by the floc. Within 30 minutes it is significantly reduced. No oxygen is needed for this process. Stored mass is consumed as a result of cell growth, and the DO uptake rate is proportional to the growth rate. As the stored mass concentration decreases towards the outlet, the growth rate will decrease. Consequently the DO consumption at the outlet is mainly due to cell decay.

As most organism are satisfied with 2-3 mg/l of DO the natural question arises, how to define a suitable DO concentration target. Normally the air flow distribution cannot be controlled but only the total air flow. An analysis of DO profiles and systematic ways to define desired setpoints were made in (5, 17).

Five features of the DO profile are important, fig 3:
(i) the position of the maximum slope, which is related to the organic or hydraulic load. It is pushed towards the outlet by an increasing load, which can be compensated by an increasing air supply.

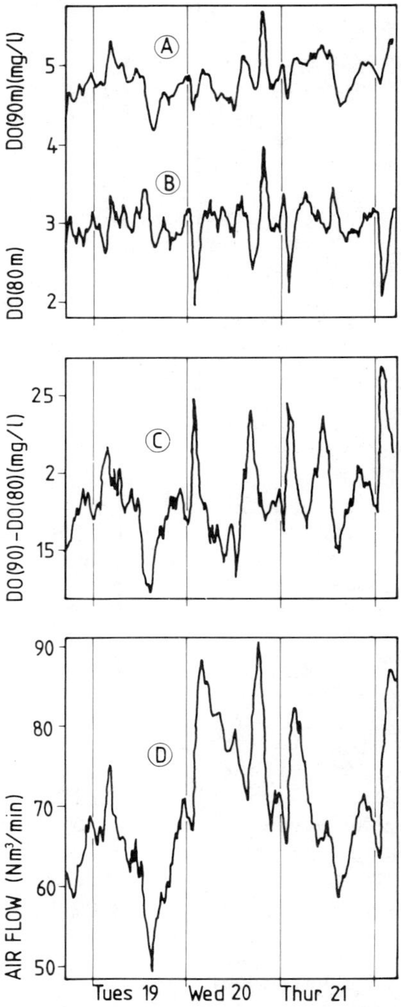

Figure 1. DO control at Käppala. The DO at 80 m downstream is controlled (B) and DO at 90 m is only recorded (A). The slope (C = A − B) is correlated with the air flow rate (D).

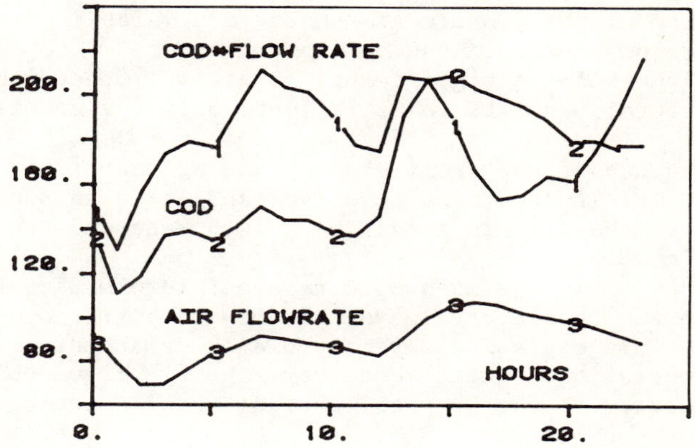

Figure 2. DO control at Käppala with hourly data of COD concentration, dry-mass flow of COD and air flow rate

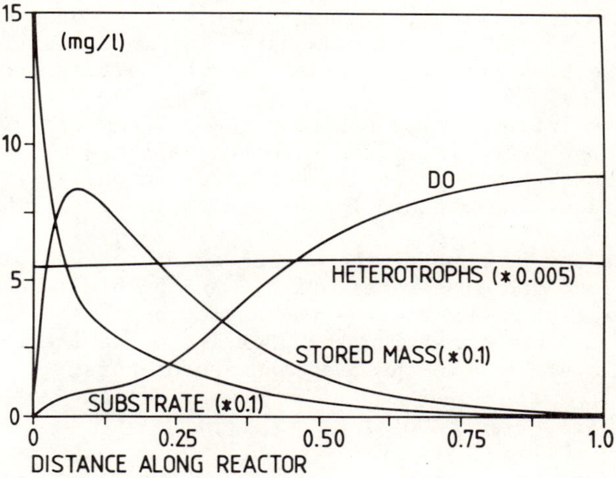

Figure 3. Typical concentration profiles in a plug flow aerator with organic removal

(ii) the value of the maximum slope, which is approximately proportional to the value of the maximum specific growth rate.
(iii) the DO concentration at the outlet, which is relatively high if the growth is close to completion.
(iv) the slope of the profile at the outlet, which is directly related to the completion of the cell growth. For incomplete reactions the slope will be significantly positive. Otherwise it is almost horizontal, and its value reflects only the endogeneous respiration.
(v) the curvature of the profile down-stream, quantified by a second derivative. It should be negative, and its sign and value are related to the organic load, the growth rate and the degree of reaction completion.

The slope (iv) is a much more relevant information for DO control than the absolute value of the DO concentration at the reactor end. This explains, why many plants demand high excess DO concentrations at the outlet. Moreover, the air flow demand is primarily related to the DO profile slope than to the concentration value at the outlet.

The qualitative features of the DO profile are the same, even if the dispersion coefficient E is increasing. All gradients are decreasing until they approach zero for the extreme case, a CSTR.

DO profile dynamics. The dynamical characteristics of the DO profile are illustrated by simulation of a plant, consisting of an aerator with four CSTR in series, each one having organic removal but no nitrification. The return sludge concentration is assumed constant, which is a reasonable assumption as long as the return sludge flow rate is constant and the settler sludge buffer is positive.

Both the influent flow rate and the biochemical oxygen demand (BOD) have been varied 25 % around their average values in 24 hour periods. As they are assumed to be in phase with each other, the resulting organic load varies a factor of three during the period.

Without any DO control the DO concentrations decrease for an increasing load and vice versa, figs 4 and 5. Fig 5 shows, how the DO profile is pushed towards the outlet as a result of increasing load. To parametrize the profile, we define two variables, the "slope" between the two last subreactors,

$$d_{43} = DO(4) - DO(3)$$

and an estimate of the second spatial derivative of the profile of the three last subreactors,

$$d_{42} = DO(4) - 2 \cdot DO(3) + DO(2)$$

where DO(i) is the DO concentration in subreactor i. The value of

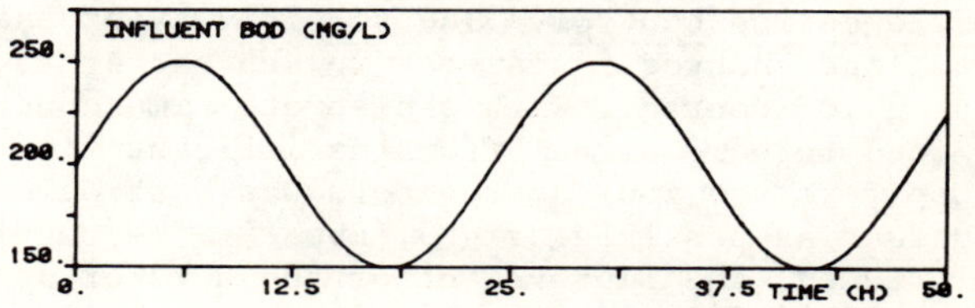

Figure 4. Influent BOD disturbance (in phase with the flow rate disturbance)

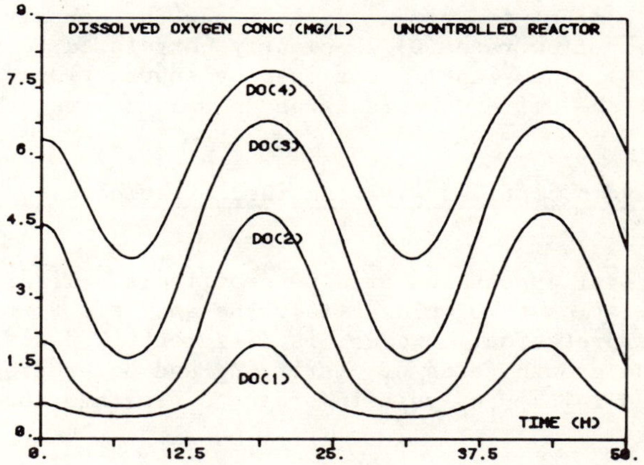

Figure 5. The DO concentrations of the uncontrolled reactor, disturbed by the diurnal load changes

d_{42} should be negative, but tends to positive values for high loads, fig 6, and is very sensitive to the biodegradable load. The slope d_{43} is also correlated to the load change (figs 4 and 6).

As SCOUR is an alternative measure of biodegradable load, it has been illustrated in fig 7 for the same simulation. The SCOUR profile (fig 8) is closely related to the DO profile (fig 6). The latter, however, is directly measurable, and the oxygen transfer coefficient does not have to be estimated.

DO profile during control. The sensitivity of the DO concentration to load changes is largest in subreactor 3, so DO(3) has been chosen for control. With a discrete time controller supplied with a feed-forward signal from the influent flow rate the total air supply to the four subreactors was controlled to keep DO(3) as constant as possible, fig 9. Even if the regulator is not optimally tuned, the improvement of the aerator performance as seen by the DO profile is significant.

The load variations are detected by the air flow rate (fig 9) as well as by d_{43} and d_{42}, fig 10. The slope is now less positive and d_{42} is satisfactory (negative) all the time. As the profile parameters are more directly related to the oxygen uptake rates, they are better indicators of the biodegradable load than the air supply, particularly when the control signal is limited.

The DO control experiment, fig 1, can further verify the results. The air flow rate (D) is clearly correlated to the DO profile slope (C). The slope is obtained by subtraction of two noisy signals, but the difference is large enough to give satisfactory accuracy.

Dissolved oxygen profiles with combined organic removal and nitrification

The general appearance of a DO profile is similar as before when nitrification takes place (16). The analysis of the profile and its interpretation, however, is different. The different time scales of the growth rates of nitrifying and heterotrophic organisms is used in the analysis in order to separate the different phenomena.

Characterization of a satisfactory DO profile. A typical steady-state simulation of DO profiles in a plug flow reactor will illustrate the ideas. Ammonia-nitrogen from the influent flow is consumed by both nitrifying and heterotrophic organisms, fig 11. In a completed reaction the NH4-N approaches zero before the outlet. The Nitrosomonas oxygen demand is shown in fig 12. The characteristic break of the oxygen demand is caused by the disappearance of NH4-N. As the nitrite is oxidized to nitrate a sharp break appears in both the nitrite concentration and the oxygen demand due to Nitrobacter growth, figs 11 and 12.

With the long mean cell residence time, necessary to complete

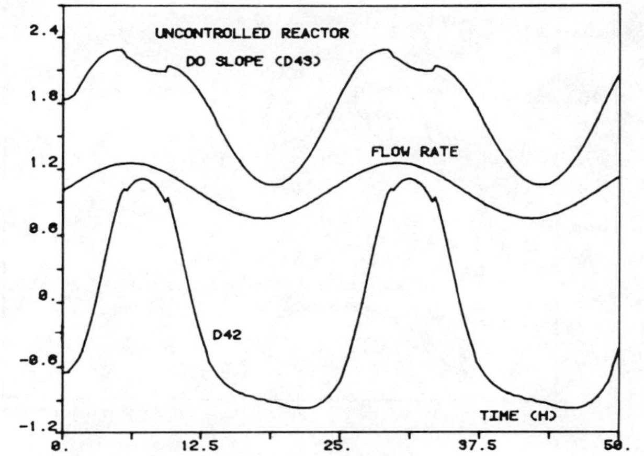

Figure 6. The DO profile parameters d_{43} and d_{42} and the influent flow rate of the uncontrolled reactor

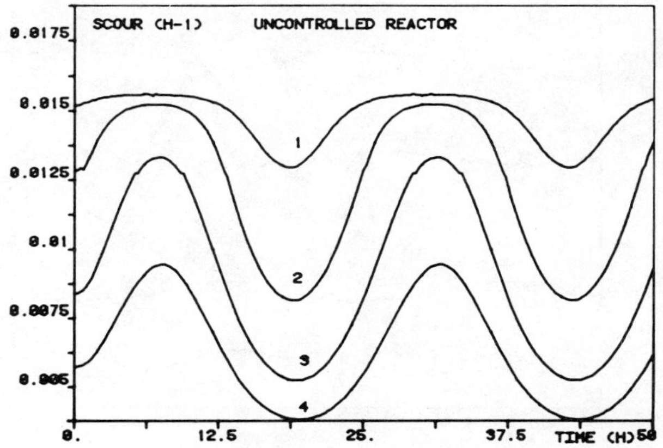

Figure 7. SCOUR of the four subreactors in the uncontrolled reactor

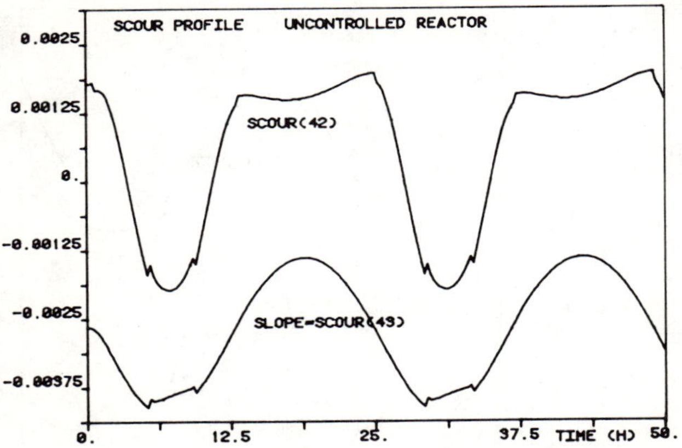

Figure 8. The SCOUR profile parameters of the uncontrolled reactor. SCOUR(43) = SCOUR(4) − SCOUR(3); SCOUR(42) = SCOUR(4) − 2 · SCOUR(3) + SCOUR(2)

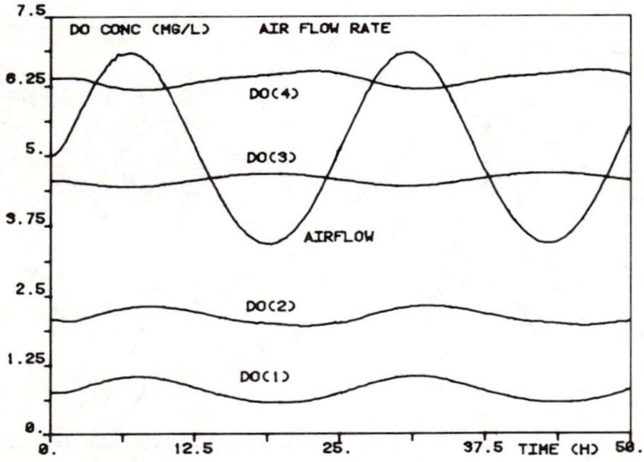

Figure 9. The aerator with DO control of DO(3), showing the total air-flow rate and the DO concentrations

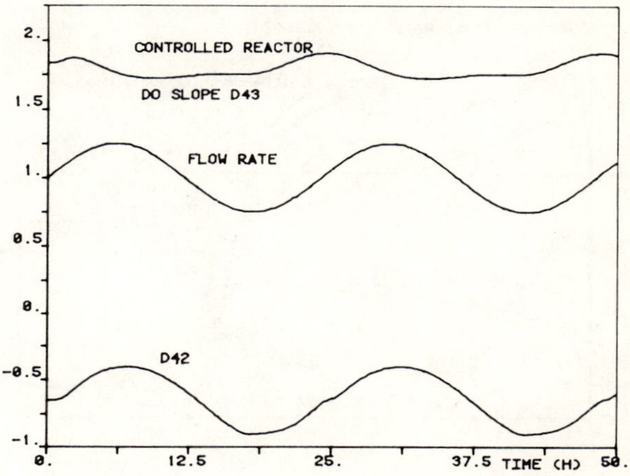

Figure 10. Same simulation as Figure 9. DO profile parameters and the influent flow rate

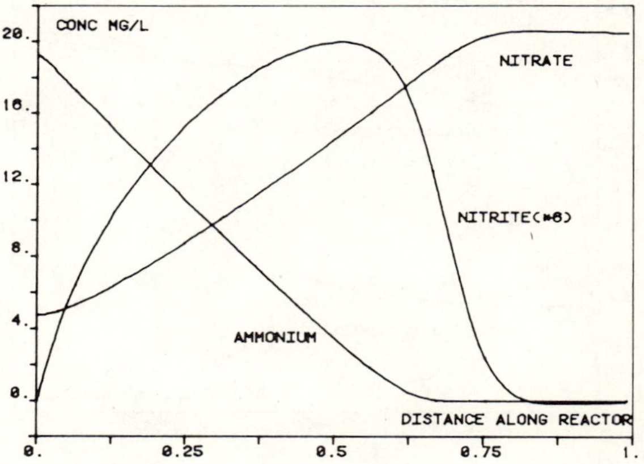

Figure 11. Concentration profiles of ammonium, nitrite, and nitrate in a plug flow reactor

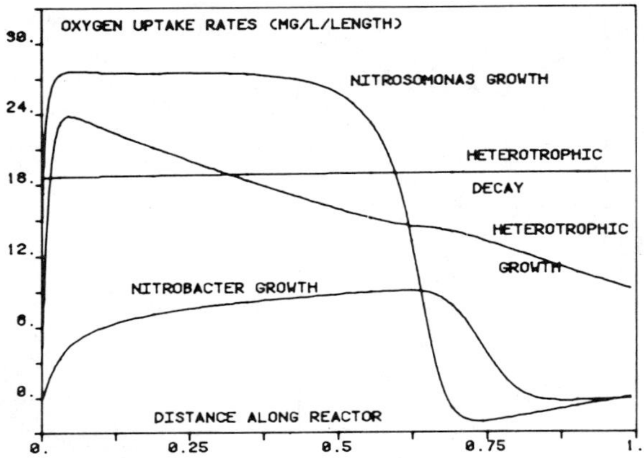

Figure 12. Steady-state spatial distribution of different oxygen uptake rates in a plug flow reactor

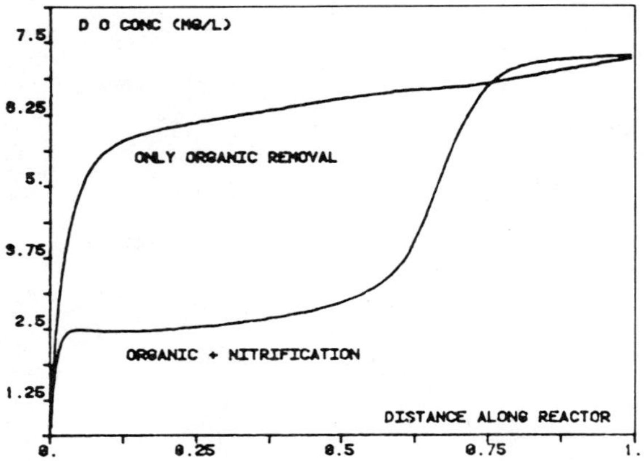

Figure 13. DO profiles in a plug flow reactor with both full nitrification and with no nitrification

the nitrification, the oxygen demand due to heterotrophic growth has its maximum early. If no nitrification would occur, the corresponding DO profile would have its characteristic break close to the inlet, fig 13.The amplitudes of SCOUR and SNOUR are similar in the first part of the reactor, but the SNOUR has a clear break, which makes it possible to separate carbonaceous and nitrogeneous oxygen demands.

Disturbances of the DO profile. The result of different types of disturbances in operational conditions or influent flow characteristics have been studied extensively by simulation. The DO profile behaves qualitatively as for carbonaceous removal (5), but the resulting control action may be different with nitrification. A load increase in terms of higher influent flow rate or a larger BOD or ammonia concentration will push the break of the profile towards the outlet. A BOD increase will favour more heterotrophic growth, thus increasing SCOUR. This limits the oxygen available for nitrification. An incomplete reaction is noted as increasing ammonium and nitrite concentrations at the outlet as well as by an increasing DO slope at the outlet. A decrease of the sludge age or insufficient air flow rate or decrease in temperature will make the nitrification chain of reactions incomplete. This will be noted in the profile in a similar way as a load change.

Implications for control. If the DO concentration has a non-homogeneous profile at least three DO probes are needed to monitor the profile changes. They give clear indications of variations in the biological activity. The control of the air flow is crucial, not only for the economy but for the effluent water quality. For on-line control the DO probes are sufficient indicators of external disturbances, but the measurements can be regularly cross-checked with alcalinity, nitrite or nitrate measurements of the effluent. By controlling the alcalinity by nitrification the cost for chemical dosage in post-precipitation plants can be significantly reduced.

Acknowledgment

Part of the research has been supported by the Swedish Board for Technical Development. Many of the results presented here have grown out of cooperative research with professor John F Andrews, Univ of Houston, Texas.

Literature cited

1. Olsson, G. "State of the art in sewage treatment plant control"; AIChE Symp. Series, 1977, 72, 52-76.

2. Stenstrom, M.K. "A dynamic model and computer compatible control strategies for wastewater treatment plants"; PhD thesis, Department of Environmental Systems Engineering, Clemson University, Clemson, S.C., 1975.
3. Sharma, B.; Ahlert, R.C. "Nitrification and nitrogen removal"; Water Res., 1977, 11, 897-925.
4. EPA "Process design manual for nitrogen control"; U.S. Environmental Protection Agency, Oct. 1975.
5. Olsson, G.; Andrews, J.F. "The dissolved oxygen profile - a valuable tool for control of the activated sludge process"; Water Res., 1978, 12, 985-1004.
6. Andrews, J.F.; Stenstrom, M.K.; Buhr, H.O. "Control systems for the reduction of effluent variability from the activated sludge process"; Prog. Water Tech., 1976, 8, 41-68.
7. Beck, M.B. "Identification and parameter estimation of biological process models"; In "System Simulation in Water Resources", Vansteenkiste, G.C., Ed.; North Holland, 1975, 19-44.
8. Olsson, G.; Hansson, O. "Modeling and identification of an activated sludge process"; Proc. IFAC Symp. on Identification and System Parameter Estimation, Tbilisi, USSR, Sep. 1976, 134-146.
9. Olsson, G.; Hansson, O. "Stochastic modeling and computer control of a full scale wastewater treatment plant"; Proc. Symp. on Systems and Models in Air and Water Pollution, The Institute of Measurement and Control, London, Sep. 1976.
10. Bosman, D.J.; Kalos, J.M. "Relation between aeration rates and settling characteristics of activated sludge"; Water Poll. Control, (S. Africa), 1978, 101-103.
11. Tomlinson, E.J.; Chambers, B. "The use of anoxic mixing zones to control the settleability of activated sludge"; Paper 23, Int. Environmental Colloquium, Liége, Belgium, May 1978.
12. Wells, C.H. "Computer control of fully nitrifying activated sludge processes"; Instrumentation Tech., April 1979, 32-36.
13. Flanagan, M.J.; Bracken, B.D. "Design procedures for dissolved oxygen control of activated sludge processes"; USEPA, 600/2-77-032, June 1977.
14. Gillblad, T.; Olsson, G. "Computer control of a medium sized activated sludge plant"; Prog. Water Tech., 1977, 9, 427-433.
15. Gillblad, T.; Olsson, G. "Implementation problems for activated sludge controllers"; Report TFRT-7137, Department of Automatic Control, Lund Institute of Technology, Sweden, 1978.
16. Jones, K.; Briggs, R.; Carr, J.G.; Potten, A.H. "Automatic control of aeration in a fully nitrifying activated-sludge plant"; Inst. Publ. Health Engr. J., 1969, LXIII, (4), 271-295.
17. Olsson, G.; Andrews, J.F. "Estimation and control of biological activity in the activated sludge process using dissolved oxygen measurements"; Proc. IFAC Symp. Environmental Systems Planning, Design and Control, Kyoto, Japan, 1977.

RECEIVED November 5, 1979.

22

A Systems Approach to Optimum Pyrolysis Reactor Design

K. H. PANG, W. R. SHU, and M. I. KOENIG

C. F. Braun and Co., Alhambra, CA 91802

Traditionally, olefins in the United States have been produced from light hydrocarbon pyrolysis. Earlier publications on computer simulation and control of pyrolysis reactors were addressed primarily to pyrolysis furnaces using ethane, propane and butane as feedstocks (1,2,3). Recent changes in world energy supply, however, had accelerated a shift to petroleum-based feedstocks (4), such as naphthas, gas oils, and even whole crudes (5), resulting in the need for a high degree of feedstock flexibility in reactor design and operation. In addition, soaring costs of new construction have spurred interests in the revamping and optimization of existing olefins plants (6-10).

These studies can best be carried out in a generalized computational framework applicable over a wide range of feedstocks and capable of accurate predictions of furnace performance.

A computer simulation program for predicting furnace performances was developed by Usami (11). However, the program could only be run in a simulation mode and also lacked flexibility in handling feedstocks other than gases. Goossens et al (9) described a commercial program which was developed to cover to a wide range of feedstocks. Reactor design and process optimization was carried out using a case-by-case approach.

Figure 1 shows a computational framework, representing many years of Braun's research and development efforts in pyrolysis technology. Input to the system is a data base including pilot, commercial and literature sources. The data form the basis of a pyrolysis reactor model consistent with both theoretical and practical considerations. Modern computational techniques are used in the identification of model parameters. The model is then incorporated into a computer system capable of handling a wide range of industrial problems. Some of the applications are reactor design, economic and flexibility studies and process optimization and control.

0-8412-0549-3/80/47-124-383$05.00/0

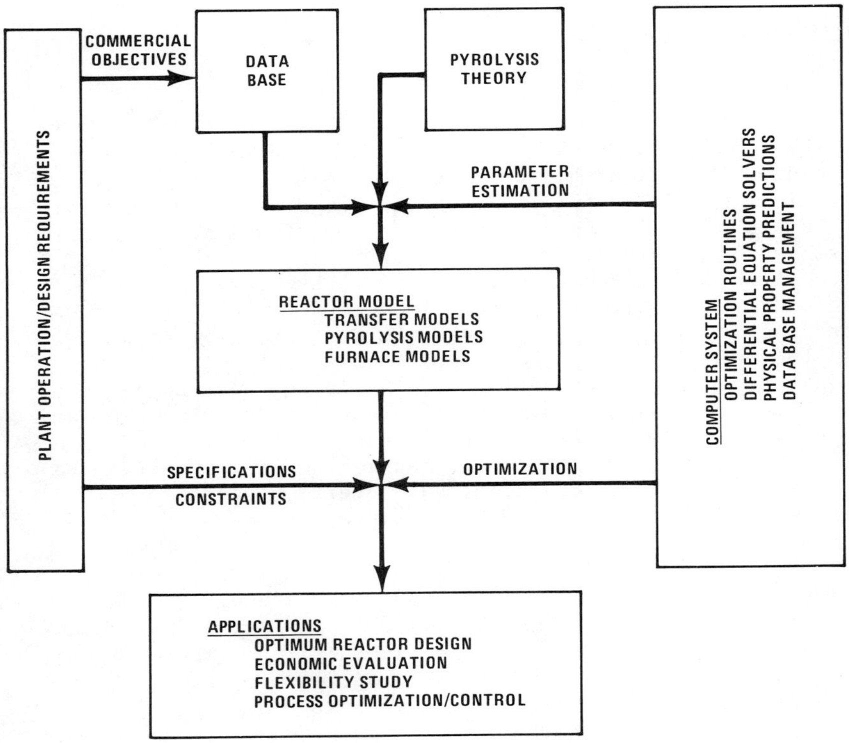

Figure 1. Pyrolysis technology development system

In a previous publication (12,13) the development of pyrolysis models for gas and liquid feeds was discussed. In a more recent paper, Shu and Ross (14) described a generalized model for predicting the rate of thermal decomposition of naphthas. These models become key components of Braun's computational system.

The focus of this paper is to provide a description of the computational system and solution techniques employed. Its applications in commercial design and simulation are also demonstrated.

Pyrolysis Furnace

Modern olefins plants now under design and construction employ indirectly fired tubular pyrolysis reactors. A typical furnace housing the reactor is illustrated schematically in Figure 2.

The furnace is divided into a convection section and a radiant section. The tubular reactor, generally 3 to 5 inches in diameter and over 150 feet in length, is situated in the radiant section. Feed enters the convection section where it is preheated and mixed with dilution steam.

The reactant mixture then enters the tubular reactor or the radiant coil at the cross-over temperature generally above 1000° F. It is rapidly heated to the cracking temperature by radiant heat supplied by burners in the combustion chamber. The gas leaving the reactor enters the transfer line exchanger where it is rapidly quenched to avoid decomposition of valuable products.

The following discussion is confined to the radiant coil where pyrolysis reactions occur.

Reactor Model

The simulation of tubular pyrolysis reactors is well established (15). It is basically composed of differential equations governing energy, mass, and momentum balances. These equations are summarized as follows. The energy balance equation may be written as

$$C_p G \frac{dT}{dz} = - \sum_i H_i r_i + Q \qquad (1)$$

where r_i and H_i are the rate and heat of reaction i, respectively, in a reaction network, and Q is the heat input to the reactor governed by radiant and corrective heat transfer principles.

The momentum balance may be expressed as

$$\frac{dP}{dz} = - \frac{fG^2}{2g_c \rho D} + \frac{G^2}{\alpha g_c \rho 2} \frac{d\rho}{dz} \qquad (2)$$

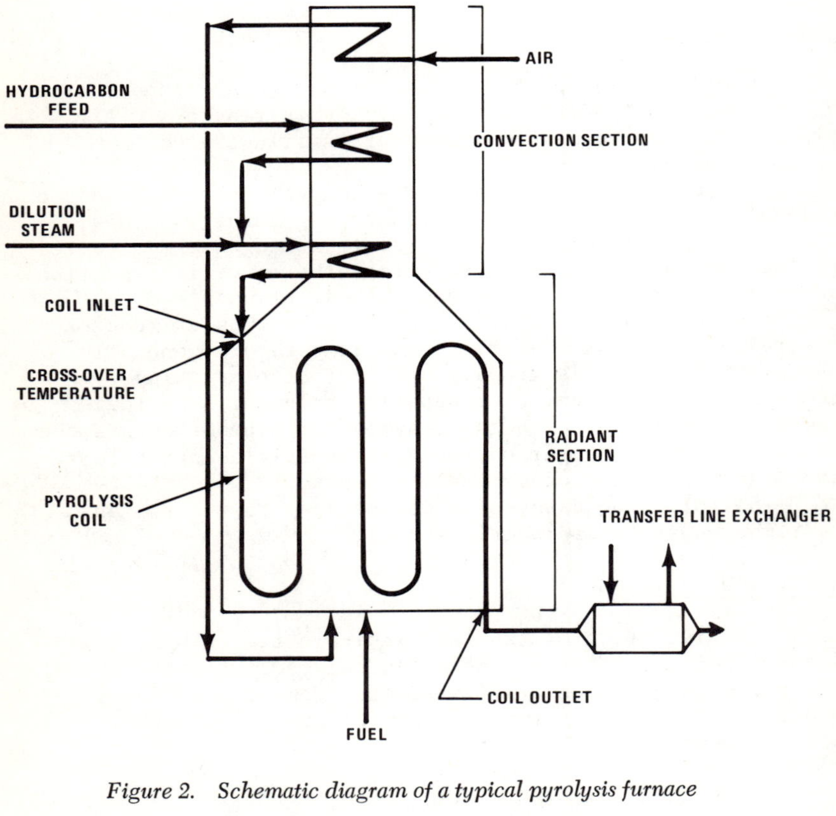

Figure 2. Schematic diagram of a typical pyrolysis furnace

where the first term on the right hand side represents the pressure drop due to friction loss while the second term represents the kinetic energy loss. Friction coefficients for elbows, U-bends and fittings are incorporated into the friction factor f.

Assuming ideal gas behavior, the density variation may be expressed as follows.

$$\frac{1}{\rho}\frac{d\rho}{dz} = \frac{1}{P}\frac{dP}{dz} - \frac{1}{T}\frac{dT}{dz} - \frac{1}{n_T}\frac{dn_T}{dz} \qquad (3)$$

where

$$\frac{dn_T}{dz} = \sum_i \sum_j S_{ij} \frac{d\xi_i}{dz} \qquad (4)$$

and S_{ij} is the stoichiometric coefficient for component j in reaction i.

The reaction extent ξ_i is governed by the mass balance equation

$$G \frac{d\xi_i}{dz} = r_i \; (C_1, C_2, \ldots, C_m, P, T) \qquad (5)$$

The concentration C_j is related to ξ_i as follows.

$$C_j = C_{jo} + \sum_i S_{ij} \; \xi_i \qquad (6)$$

The present system is sufficiently general to incorporate any type of reaction network, from simple empirical models to detailed mechanistic ones. However, in the following discussion, pyrolysis models reported in our previous publications (12,13,14,16) will be employed.

Solution Method

Equations 1 through 6 can be solved numerically for temperature, pressure and concentration profiles along the length of the reactor, provided that the appropriate initial or boundary conditions are given. Solution methods are now discussed for both the initial and boundary value problems. However, emphasis will be placed on the latter since most industrial applications fall into this category.

Initial Value Problem. If the conditions at the inlet of the reactor are known, the problem is classified as an initial value problem. One example is the simulation of an operating reactor (11). In this case, the temperature, the pressure, the feed composition at the reactor inlet, and the heat flux to each section of the reactor are known. The yield structure and fluid temperature at the end of the reactor can be calculated. The maximum tubeskin temperature, often the bottleneck of the operation, can also be predicted. The solution of this type of problem involves only a once-through integration.

Boundary Value Problem. If the conditions at the reactor outlet, such as the pressure and feed conversion are specified, it becomes a boundary value problem. This situation often arises when a reactor is to be designed for a specified feed conversion and for an outlet pressure dictated by downstream equipment. The solution of this problem involves a trial-and-error procedure and several passes of integration.

Most industrial applications are boundary value problems. These include (1) the determination of reactor diameter and length while meeting specifications in residence time and maximum tubeskin temperature, (2) the determination of maximum feed rates subjected to a constraint in the tubeskin temperature and (3) the estimation of model parameters, such as reaction rate constants from yield data. Examples 1 and 2 will be discussed in detail later.

A boundary value problem can be reconstructed in the following manner. For any set of initial conditions, Equations 1, 2 and 5, can be integrated to the outlet of the reactor to give the following relationships.

$$g_i = f_i(\underline{x}) - \bar{f}_i = 0 \tag{7}$$

where $\underline{x}$ is a vector of n unknown initial conditions or other independent variables, and $\bar{f}_i$ could be the known boundary conditions or design specifications.

An example is the design of a pyrolysis reactor. In this case, the elements of the vector $\underline{x}$ consist of the diameter, the length of the reactor, the pressure at the inlet and the heat flux into the reactor. The known boundary conditions and design specifications are the pressure and the feed conversion at the outlet, the residence time and maximum tubeskin temperature.

To solve the set of non-linear transcendental equations the Newton Raphson's iterative method is used. First, Equation 7 is linearized about an initial point $\underline{x}_o$ using the Taylor series expansion.

Hence

$$g_i(\underline{x}) = g_i(\underline{x}_o) + \sum_{j=1}^{m} \frac{\partial g_i(\underline{x}_o)}{\partial x_j} \Delta x_j = 0 \tag{8}$$

$$i = 1, 2, \ldots, m$$

where

$$\Delta \underline{x} = \underline{x}_1 - \underline{x}_o \tag{9}$$

Equation 8 is linear and can be solved for Δx_j by standard methods such as Gaussian elimination (17). An improved set of independent variables $\underline{x}_1$ can then be calculated using Equation 9. Equation 7 is again linearized about the improved point $\underline{x}_1$ and the process of calculation repeated until convergence on $\underline{x}$ is obtained.

For practical reasons, all independent variables must have physical limits as represented by the following equation.

$$\underline{x}_L \leq \underline{x} \leq \underline{x}_U \tag{10}$$

where $\underline{x}_L$ and $\underline{x}_U$ are the lower and upper bounds of the vector $\underline{x}$.

As a result, a feasible solution of Equation 7 may not exist. When this situation arises, one or more equations in the set may be replaced by an objective function

$$\Phi = \Phi\ (\underline{x}) \tag{11}$$

Then the problem is transformed into one of optimizing a function with respect to the independent variables subjected to some constraints governed by their physical limits. Equations 8, 10 and 11 constitute a typical linear programming problem which can be readily solved by the simplex method (18). An example is the design problem where the residence time is minimized if its specification cannot be met.

Program Structure

Based on the concepts described in the previous section, a computer system PF60 has been developed. A modular approach was used for the construction of the system to facilitate maintenance and modification. The input and output features are simple and user oriented, as illustrated in Figure 3. The inputs are reactor geometry, feed rate and composition, and inlet and outlet conditions, whether they be specified or estimated. The outputs are a description of the geometry, operating conditions, heat requirement, various profiles along the reactor and finally the yield structure. The program is accessible from time-sharing terminals and provides sufficient man-machine interactive capability so that a simulation or design problem can be solved with minimum elapsed time.

The overall structure of the program is illustrated in Figure 4. The program starts off with the calculation of

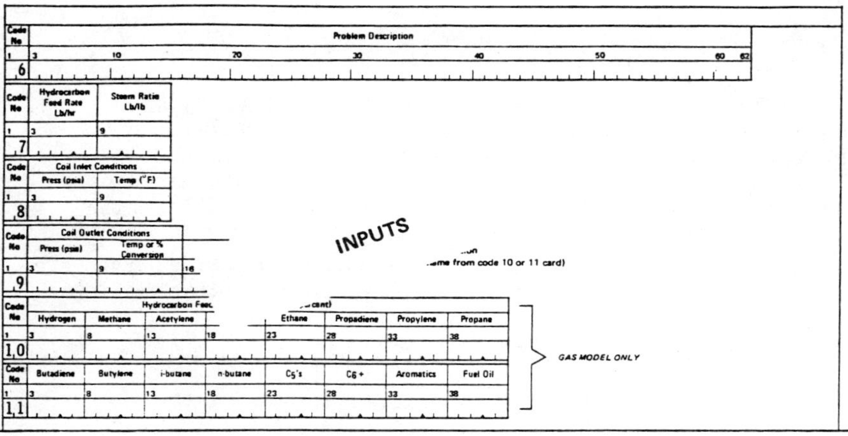

PF60 REV 20
KHP2527
GAS MODEL

ETHYLENE PLANT CRACKING COIL

PAGE 0
3700100
06/18/79

COIL GEOMETRY

SECTION NO.	NO. OF SECTIONS	PARALLEL TUBES	ID INCHES	LENGTH FEET	WALL INCHES	FITTINGS
1	1	4	4.750	38.00	0.3750	1/2 180 U-BEND 14.00 INCHES C TO C
2	1	4	4.750	38.90	0.3750	ONE 180 U-BEND 14.00 INCHES C TO C
3	1	4	4.750	38.90	0.3750	ONE 180 U-BEND 14.00 INCHES C TO C
4	1	4	4.750	38.90	0.3750	ONE 180 U-BEND 14.00 INCHES C TO C

PF60 REV 20
KHP2527
GAS MODEL

ETHYLENE PLANT CRACKING COIL

PAGE 1
3700100
06/18/79

HYDROCARBON FEED, LB/HR-COIL	7500.0
STEAM, LB/HR-COIL	2250.0
STEAM/HC RATIO, LB/LB	0.300
TIME ABOVE 1200F, SEC.	0.81438
EQ. TIME AT 1553 F, SEC.	0.31162
TOTAL HEAT ABSORB, MBTU/H	42287.62
TOTAL HEAT LOSS, MBTU/H	7.79
CIRCUM. HEAT FLUX FACTOR	1.4[illegible]
TOTAL FIRED AREA, FT2 OD	[illegible]
TUBE ROUGHNESS FACTOR	

STREAM PHYSICAL PROPERTIES	COIL INLET	COIL OUTLET
HYDROCARBON MOLECULAR WE[illegible]	32.160	19.815
VISCOSITY, CP	0.0265	0.0308
THERMAL CONDU[illegible]	0.0618	0.0995
SPECIFIC [illegible]	0.8366	0.9032
REY[illegible]	489898.	190509.
	0.867	0.676
	31.79	17.25

OUTPUTS

SEC NO	LENGTH FEET	TEMPERATURES F STREAM	SKIN	FIR BO.				./SEC	..EAM DENSITY LB/FT3	PROPANE PERCENT CONV.	PRESS OUT PSIA	EXPAND FACTOR M/M	RES-TIME SEC.
0	0.0	1200.0	1619.5	2017			..0	295	0.0746	0.00	48.81	1.000	0.000
1	38.9	1338.0	1709.1	2041			365.9	331	0.0666	1.92	47.38	1.007	0.127
2	79.6	1414.3	1756.3	2050		..0.3	2361.9	369	0.0597	12.46	45.64	1.054	0.246
3	120.4	1436.8	1762.8	2047		4830.9	4051.6	409	0.0538	28.45	43.72	1.132	0.353
4	161.1	1451.6	1767.0	2047	20167	4756.1	4240.3	455	0.0484	43.53	41.57	1.215	0.449
5	201.8	1467.5	1771.5	2045	19849	4693.9	4139.1	509	0.0432	56.82	39.16	1.296	0.535
6	242.6	1485.6	1782.1	2049	19743	4644.1	4006.7	576	0.0382	68.26	36.41	1.374	0.612

PF60 REV 20
KHP2527
GAS MODEL

ETHYLENE PLANT CRACKING COIL

PAGE 2
3700100
06/18/79

REACTOR PROFILE, MOLES/100 MOLES OF FEED

SEC.	H2	CH4	C2H2	C2H4	C2H6	C3H4	C3H6	C3H8	C4H6	C4H8	N-C4H10	I-C4H10	GASOLINE C5'S	GASOLINE BTX	GASOLINE OTHER	FUEL OIL
0	0.00	0.96	0.00	0.34	82.74	0.08	0.37	15.46	0.00	0.05	0.00	0.00	0.00	0.00	0.00	[illegible]
1	0.55	1.18	0.00	0.95	82.26	0.08	0.49	15.16	0.00	0.04	0.02	0.00	0.00	[illegible]		
2	4.17	2.35	0.00	4.93	79.06	0.08	1.09	13.53	0.01	0.03	0.08	0.[illegible]				
3	10.43	4.25	0.01	11.85	73.41	0.08	1.86	11.06	0.04	0.06	[illegible]					
4	17.04	6.25	0.04	19.15	67.30	0.09	2.42	8.73	0.[illegible]							
5	23.58	8.29	0.07	26.28	61.11	0.09	2.77									
6	30.02	10.39	0.12	33.12	54.[illegible]											
7	36.43	12.59	0.1[illegible]	[illegible]												
8	[illegible]															

Figure 3. Examples of inputs and outputs, Braun's pyrolysis program FP60

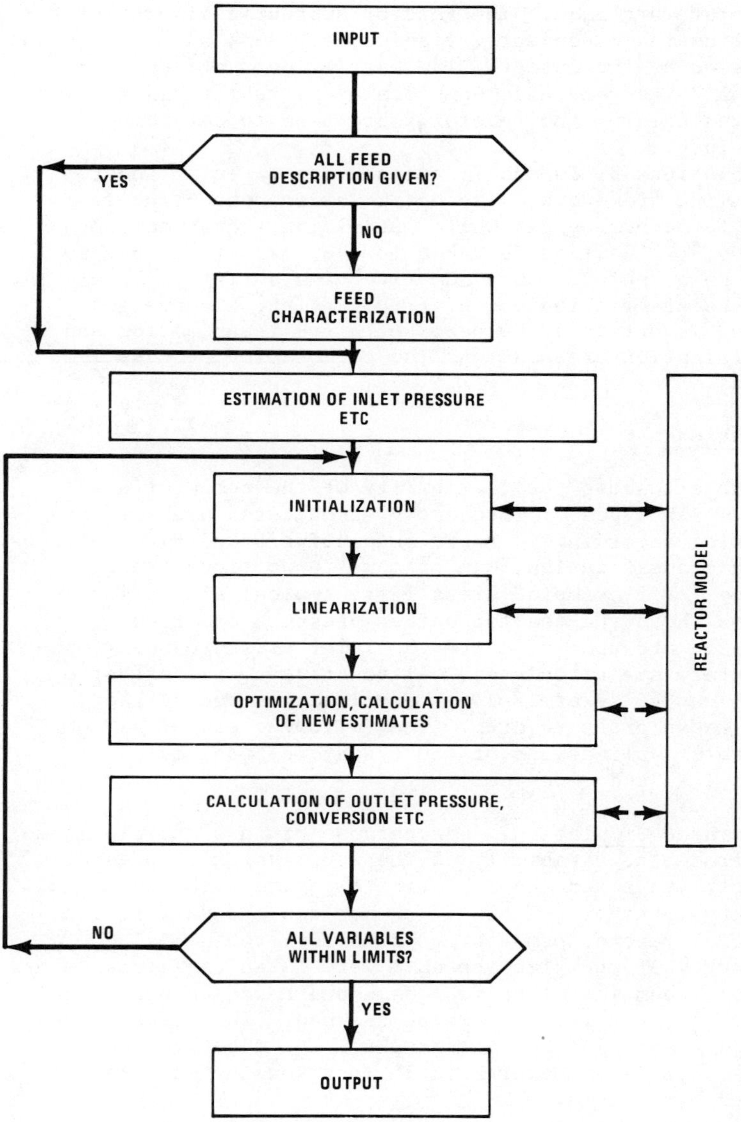

Figure 4. Program structure

the feed characteristics, if not given, from known properties. Initial estimates of the independent variables, x_i, such as inlet pressure, heat flux, or radiant temperature are made. The reactor subroutine is called to initialize the dependent variables, f_i, such as pressure and conversion at the outlet. The partial derivatives in Equation 8 are generated numerically by taking small steps in x_i and calling the reactor subroutine to determine the change in g_i.

Equations 8, 10 and 11 are now set up in a linear programming framework. The optimization subroutine is called to determine Δx_j while optimizing the objective function Φ. Caution is taken to keep Δx_j within pre-set limits so as not to cause numerical instability. If all variables are within bounds at this point, a converged solution is obtained. Otherwise, a re-linearization and re-optimization are made and the calculation process is repeated.

Applications

Table I shows the flexibility of the computational system. Six types of frequently encountered problems are classified according to their respective boundary conditions. In each classification, one or more run options can be selected. For example, Class 1 are typical simulation problems where the reactor outlet pressure and feed conversion are specified and the inlet pressure and radiant temperature are calculated. Alternatively, the effect of fouling can be determined by calculating a coking factor from a known pressure drop. The following examples illustrate applications of the system in problems under Classes 1, 5 and 6 respectively.

Furnace Simulation. The purpose of this example is to demonstrate the capability of the PF60 system to predict yield structure and the tubeskin temperature in commercial operating furnaces. Table II summarizes the data from a commercial reactor processing primarily propane as feedstock. At the time the data were taken, the furnace had been on stream less than four days and hence an unfouled radiant coil condition could be assumed. The yields were recorded by on-line chromatographs. The tubeskin temperatures were measured in 15 locations by calibrated infrared pyrometers.

In the simulation, feedrate, steam ratio, and inlet temperature were maintained constant. The pressure and feed conversion at the coil outlet were specified. The inlet pressure and firebox temperature profile were adjusted to meet the two specifications. A uniform temperature profile

TABLE I

PROBLEM CLASSIFICATIONS
BRAUN PYROLYSIS PROGRAM PF60

CLASS	SPECIFICATION					VARIABLES						APPLICATIONS
	TUBESKIN TEMP	OUTLET PRESS	CONVERSION	OUTLET TEMP	RESIDENCE TIME	COKING FACTOR/ INLET PRESS	HEAT FLUX/ TEMP PROFILE/ FIREBOX TEMP	FEED RATE	MODEL PARAMETERS	DIAMETER	LENGTH	
1		X	X			X	X					Simulation
2		X		X		X	X					Simulation
3		X	X			X			X			Modeling
4	X	X	X			X	X	X				Flexibility Studies
5	X	X		X		X	X	X				Flexibility Studies
6	X	X	X		X	X	X			X	X	Design

in the firebox was assumed. Table II compares predicted and observed yields. Tubeskin and fluid temperature profiles are shown in Figure 5. It can be seen that the agreement between the predicted and observed results was good in all comparisons.

Furnace Flexibility. In this example, the flexibility of an operating propane furnace in handling alternate feedstocks was examined. Specifically, the maximum loading of the furnace subjected to various process and physical constraints is to be determined. The alternate feedstocks are ethane, n-butane and a naphtha. The following physical and process limits are observed.

1 The furnace radiant duty, due to design limitation, does not exceed the design base by more than 4%.

2 The maximum clean tubeskin temperature is limited at 1900° F by the metallurgy of the reactor.

3 The crossover temperature must satisfy an overall heat balance around the furnace system, including transfer line heat exchangers and feed preheaters, and must not exceed certain limits to avoid incipient cracking.

4 Specifications on feed conversion and outlet pressure must also be met.

Results of the flexibility study are shown in Table III. The furnace loading is limited by the crossover temperature and radiant duty in the case of both ethane and n-butane feeds. In the case of naphtha feed, the constraints were the radiant duty and the tubeskin temperature. The maximum loadings for the ethane, n-butane and naphtha feeds were 4,450, 5,000, 4,625 lb/hr/coil, respectively compared to 4750 lb/hr/coil for the propane feed.

The PF60 system has been proven most useful in process analyses such as this one where interactions among process variables are complicated by the existence of many constraints.

Design of Pyrolysis Reactors. A pyrolysis reactor is to be designed for a fixed feed rate and composition such that the outlet conditions must meet the specifications, $\bar{P}_{out}$. $\bar{C}_{out}$, and $\bar{\theta}$. Furthermore, the maximum tubeskin temperature must not exceed a limit $\bar{T}_P$ dictated by the metallurgy of the reactor. Specifications, process information, and key calculated variables are listed in Table IV.

TABLE II

OPERATING DATA AND YIELDS
FURNACE SIMULATION

Hydrocarbon Feed, lb/hr/coil	4000
Steam Ratio, lb/lb	.375
Crossover Temperature, °F	1150
Outlet Pressure, psia	28.5
Feed Purity, mole % Propane	95.0
Propane Conversion, weight %	92.0

Yields, weight %

	Observed	Predicted
Hydrogen	1.64	1.56
Methane	25.67	24.26
Acetylene	.87	.74
Ethylene	35.56	35.75
Ethane	4.12	4.09
Propylene	12.93	13.43
Propane	7.62	7.63
Butadiene	2.85	2.97
C4+	8.78	9.57

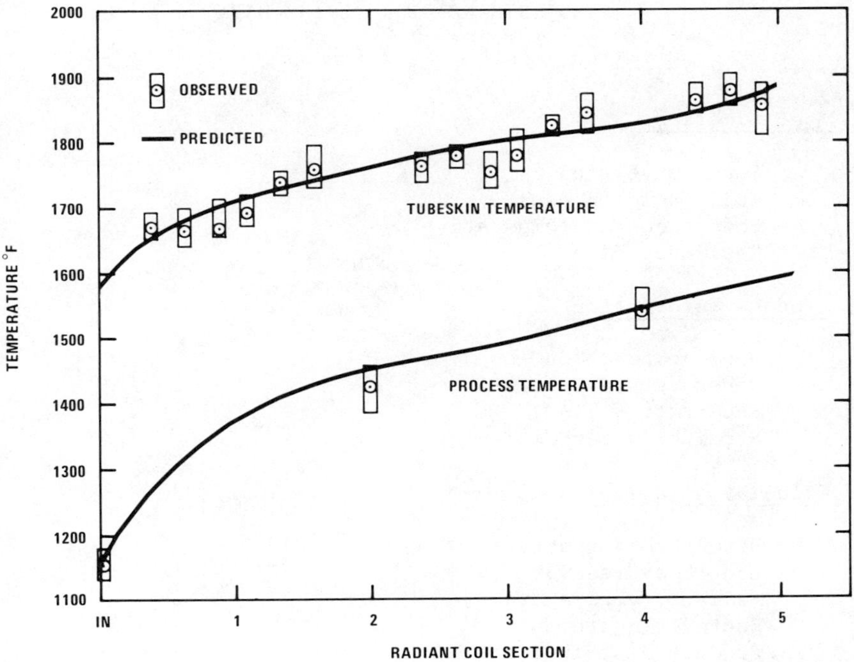

Figure 5. Temperature-profiles furnace simulation

TABLE III

RESULTS OF FLEXIBILITY STUDY

FEEDSTOCK	PROPANE (BASE)	ETHANE	N-BUTANE	NAPHTHA
Crossover Temp, °F	1200	1242*	1150*	1084
Radiant Duty, mm Btu/hr/coil	6.00	6.24	6.05*	6.08*
Feed Rate, lb/hr/coil	4750	4450	5000	4625
Maximum Clean Tubeskin Temp, °F	1900	1885	1845	1900*

*Limiting Constraints

TABLE IV

PYROLYSIS REACTOR DESIGN DATA

Specifications

- Feed conversion
- Coil outlet pressure
- Peak tubeskin temperature
- Residence time

Process Information

- Feed composition
- Crossover temperature
- Feed rate
- Steam/hydrocarbon ratio

Calculated Variables

- Diameter of reactor
- Length of reactor
- Radiant temperature
- Energy requirement
- Product distribution
- Temperature, pressure, gas velocity, and concentration profiles
- Hydrocarbon partial pressure at the outlet

The generalized Equation 7 can be rewritten as follows.

$$g_1 = P_{out} (P_{in}, q, D, L) - \bar{P}_{out} = 0 \quad (12)$$

$$g_2 = X_{out} (P_{in}, q, D, L) - \bar{X}_{out} = 0 \quad (13)$$

$$g_3 = \theta (P_{in}, q, D, L) - \bar{\theta} = 0 \quad (14)$$

$$g_4 = T_P (P_{in}, q, D, L) - \bar{T}_P = 0 \quad (15)$$

where D and L are the diameter and length of the reactor, respectively.

If the residence time $\bar{\theta}$ cannot be met due to physical constraints, the shortest possible value would be chosen. Then Equation 14 will be replaced by the following objective function.

$$\Phi = \theta (P_{in}, q, D, L) \quad (16)$$

The problem now becomes one of minimizing the residence time θ subjected to the equality constraints represented by Equations 12, 13 and 15 and inequality constraints governed by the physical limits of the variables P_{in}, q, D, and L.

Figure 6 shows the variations of the independent variables in successive iterations and their effects on the dependent variables. Initial estimates of the diameter, length, radiant temperature and inlet pressure have resulted in a reactor with residence time, outlet pressure, conversion, and tubeskin temperature all higher than specified values.

In the first iteration, a reduction in the length and diameter successfully corrected most discrepancies except those of the outlet pressure and the residence time. To meet the latter specification, the diameter was further reduced. Simultaneously, the inlet pressure had to be increased to account for the resulting increase in pressure drop. Furthermore, as the heat transfer area decreased with diameter, the radiant temperature must also increase to supply a constant total heat input.

As shown in Figure 6, the system successfully selects the diameter and length of the reactor which meets all specifications in five iterations. The method is stable and converges rapidly despite the non-linear relationship of the variables.

To show the uniqueness of the optimum design, a comparison was made with an alternate reactor of identical volume but shorter length and larger diameter. Results are shown in Table V. The alternate reactor does not meet both the residence time and tubeskin temperature specifications. As expected, because of its larger diameter it shows a lower

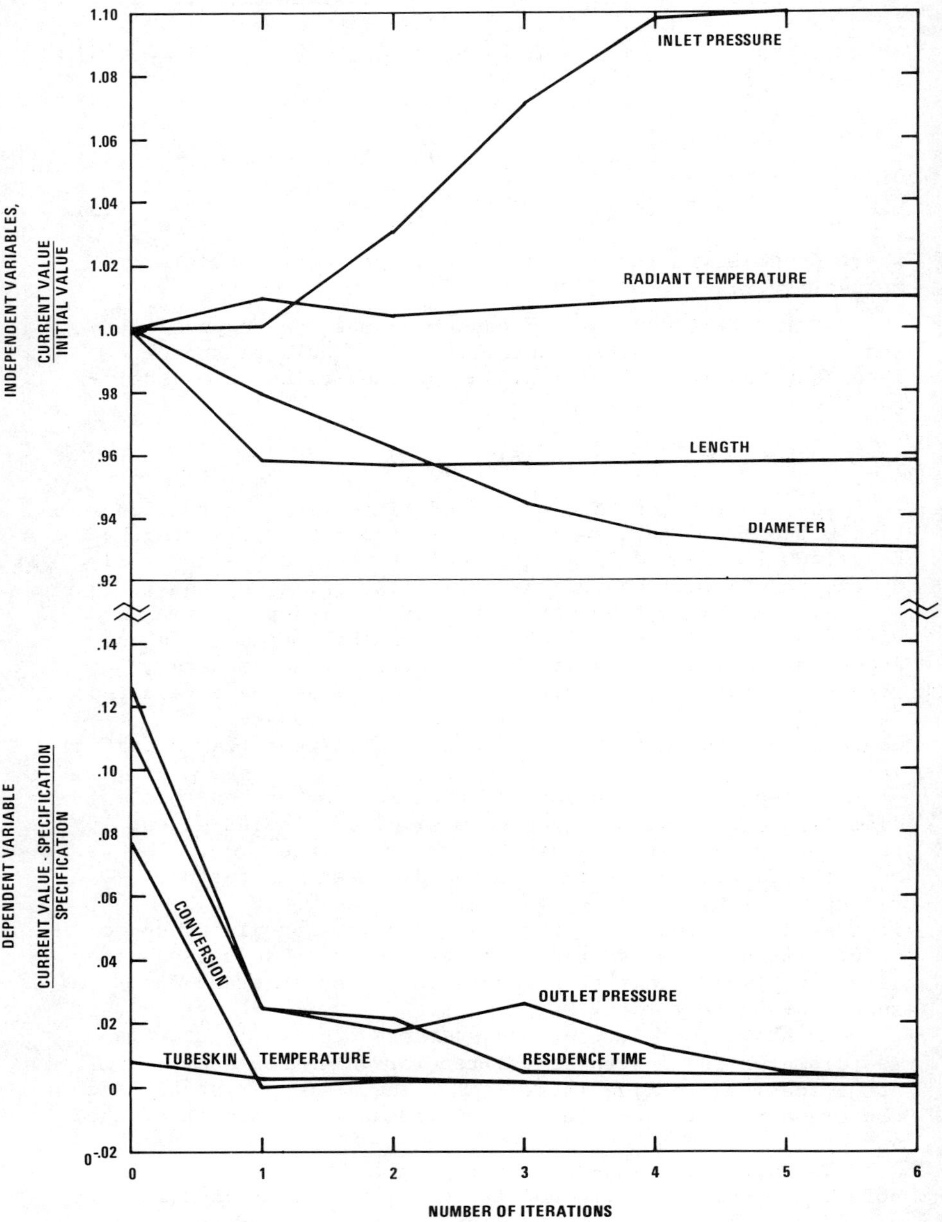

Figure 6. Variations of dependent and independent variables, pyrolysis reactor design

TABLE V

COMPARISON OF ALTERNATE AND OPTIMUM REACTORS

	ALTERNATE	OPTIMUM
Diameter Ratio	1.057	1.0
Length Ratio	0.895	1.0
Pressure Drop, psia	27.3	35.4
Inlet Pressure, psia	50.6	58.5
$\theta/\bar{\theta}$	0.91	1.0
$T_p-\bar{T}_p$, °F	43	0

pressure drop giving rise to a high volumetric flow rate and a lower residence time.

For a constant volume, it can be shown that the surface area is proportional to $\sqrt{L}$. With a shorter reactor, and hence a smaller area, the radiant temperature must increase in order to provide a constant heat input causing the tubeskin temperature to rise beyond the design limit.

Using dimensions of the alternate reactor as initial estimates, the system again converges to the unique optimum design in five iterations.

The present method can be extended easily to other applications. For instance, a cost function involving operating and capital expenses, and product distribution may be minimized with respect to operating and design variables.

Conclusions

Braun's computer system PF60 is a useful tool for designing and simulating pyrolysis reactors. It provides sufficient flexibility to accommodate a wide range of problems. The solution method employed is numerically accurate, efficient and stable. Its capability has been fully demonstrated by the three industrial applications discussed in this paper.

Nomenclature

C_j Concentration of component j

C_{jo} Initial concentration of component j

C_p Average heat capacity, Btu/lb-°F

D Reactor diameter, ft

f Friction factor, dimensionless

f_i	i-th dependent variable
G	Mass velocity, lb/ft^2-hr
g_c	Gravitational constant
g_i	i-th equality constraint, Equation 7
H_i	Heat of i-th reaction, Btu/lb-mole
L	Length of reactor, ft
m	Number of chemical species
n	Number of unknown initial conditions
n_T	Total molal rate, lb-moles/lb-hr
P	Pressure, lb/in^2
Q	Heat input per unit volume, Btu/ft^3-hr
q	Heat flux factor
r_i	Rate of reaction i, lb-moles/ft^3-hr
S_{ij}	Stoichiometric coefficient for Component j in Reaction i
T	Process temperature, °R
T_P	Maximum tubeskin temperature, °R
X	Feed conversion, fraction
x_i	i-th independent variable
$\underline{x}$	Independent variable vector
$\underline{x}_L$	Lower bound of independent variable vector
$\underline{x}_U$	Upper bound of independent variable vector
$\underline{x}_o$	Initial value of independent variable vector
$\underline{x}_1$	Improved value of independent variable vector
z	Axial distance, ft
α	Kinetic energy correction factor, dimensionless

ρ Density, lb/ft^3

ξ Extent of reaction, lb-moles/lb

θ Residence time, sec

Φ Objective function, Equation 11

Δ Difference operator

Subscripts and Superscripts

in Inlet condition

out Outlet condition

– (overbar) Specifications

– (underbar) Vector quantity

Literature Cited

1 Lichtenstein, I., Chem. Eng. Prog. 1964, 60, (12), 64.

2 Shah, M. J., Ind. Eng. Chem. 1967, 59 (5), 70.

3 Freedman, B. G., Paper No 61C, "Nonlinear Distributed Parameter Model of a Gaseous Feed Ethylene Furnace", the 68th AIChE National Meeting, Houston, 1971.

4 Minet, R. G., Tsai, F. W., The Oil and Gas J., March 21, 1977, p.135.

5 Ishikawa, T., Keister, R. G., Hydrocarbon Processing, December 1978, p.109.

6 Langeback, H. P., Deubler, U., The Oil and Gas J., April 2, 1979, p.74.

7 Mol, A., The Oil and Gas J., April 2, 1979, p.82.

8 Hammett, J. L., Jr., Lindsay, L. A., Chem. Eng., November 8, 1976, p.115.

9 Goosens, A. G., Dente M., Ranzi E., Hydrocarbon Processing, September 1978, p.227

10 Stewart, W. S., Funk, G. L., Huang, W. H., The Oil and Gas J., April 2, 1979, p.98

11 Usami H., Hydrocarbon Processing, January 1972, p.103.

12 Ross, L. L., and Shu, W. R., The Oil and Gas J., October 17, 1977, p.58

13 Shu, W. R., Ross, L. L., Pang, K. H., "A Naphtha Pyrolysis Model for Reactor Design". Paper No 27d, the 85th National Meeting of AIChE, Philadelphia, June 1978.

14 Shu, W. R., Ross, L. L., "A Feed Decomposition Model for Naphtha Pyrolysis". Paper No 62c, the 71st Annual Meeting of AIChE, Miami Beach, November 1978.

15 Aris, R., "Elementary Chemical Reactor Analysis". Prentice Hall, Englewood Cliffs, N J, 1969.

16 Ross, L. L., Shu, W. R., Computer Modeling of Hydrocarbon Pyrolysis for Olefins Production, in "Thermal Hydrocarbon Chemistry", Adv. Chem. Series, 1979, In Press.

17 Lapidus, L. "Digital Computation for Chemical Engineers". McGraw-Hill Book Co, New York, 1962.

18 Gass, S. L. "Linear Programming, Methods and Applications". McGraw-Hill Book Co., New York, 1964.

RECEIVED December 26, 1979.

23

Computer Simulation of Heterogeneous Nitration of Toluene to Dinitrotoluene

A. K. S. MURTHY

Allied Chemical Corporation, Morristown, NJ 07960

Nitration of aromatic compounds, particularly that of benzene and toluene, has been extensively studied partly because of its industrial importance in the manufacture of explosives, solvents, pharmaceuticals and organic intermediates and partly because of its theoretical significance in studying electrophilic substitution (1,2). Most of the reported studies of toluene nitration are under conditions where only mononitration reactions take place. Studies on dinitration reactions have been carried out (3,4,5), but starting with mononitrotoluenes. In commercially important nitration of toluene for the manufacture of tolylene diisocyanate (TDI) and trinitrotoluene (TNT), dinitrotoluene (DNT) is the desired product and mononitration is simply an intermediate reaction. Optimum use of a DNT production facility can be achieved by integrating the mononitration and the dinitration steps. A computer simulation model, to be useful in such optimization studies, must be capable of representing the various physical and chemical phenomena over a wide range of operating variables where mono- and dinitration reactions take place. This paper describes a successful effort to synthesize such a model using information on the various phenomena available in the open literature, reported by different authors. A systems engineering approach was used to decompose the total phenomenon into subphenomena which have been separately studied. The original intention of developing the model was to form a basis for planning and designing a process development study in a pilot plant for optimizing an existing DNT production facility. However, when a computer simulation program for the plant based on the model was developed, it was found to be already adequate for

0-8412-0549-3/80/47-124-403$05.25/0

trouble shooting and preliminary optimization of plant operation.

The nitrating agent used in most industrial nitration processes is a mixture of aqueous nitric and sulfuric acids, commonly known as the mixed acid. The organic species are only sparingly soluble in the acid and hence industrial nitration is heterogenous involving two liquid phases. The term "macrokinetics" has been used (6) to describe overall kinetics of nitration under heterogenous conditions. In microscopic scale, two phenomena, viz., chemical reaction or "microkinetics" and interphase mass transfer, simultaneously occur. Early workers (7,8,9) assumed that mass transfer effects can be eliminated by strong agitation. Hanson and Albright (6,10) have pointed out that this assumption is erroneous. More recently Hanson and co-workers (11) and Cox and Strachan (12,13) have developed models for heterogeneous nitration of toluene by interpretting the overall rate as a combination of chemical kinetics with simultaneous mass transfer. These models are applicable for the initial phases of toluene nitration where the toluene content of the organic phase is high. In the industrial production of DNT, the reactors are run such that the toluene concentration is very small, even in the first nitrator.

To be useful in the optimization of DNT production, the mathematical model which forms the basis of the simulation program must be applicable over a wide range of operating conditions and should provide detailed accounting of the following effects.

a. Chemical kinetics is a strong function of sulfuric acid strength in addition to being dependent on the concentration of the reactants in the acid phase and the temperature of the system.
b. Isomer distribution is generally important; certainly in the case of intermediates production.
c. The solubility of the organic species in the acid is a function of the temperature and the composition of the acid phase.
d. The solubility of nitric acid in the organic phase is quite significant and the distribution constant is a function of the acid and organic phase compositions as well as the temperature.
e. Interphase mass transfer rates depend both on agitation rate and phase compositions which determine the drop size distribution and phase ratio.

Prior to the present study no author is known to have studied all of these effects nor proposed a model which simultaneously included all of these effects. Information on many of the above effects are available through studies reported by various authors (e.g., 3,4,5,8,12,15,17,19,20,25,26,27)

Chemistry of Toluene Nitration

The mechanism of aromatic nitration is shown in Figure 1. It may be seen that the sulfuric acid serves as the source of hydrogen ion (proton) which protonates nitric acid to form nitronium ion and water. The NO_2^+ ion is the active electrophile that causes the nitration reactions.

Chemical species that are present in the system are:

1. Toluene (T)
2. p-Nitrotoluene (p-MNT)
3. o-Nitrotoluene (o-MNT)
4. m-Nitrotoluene (m-MNT)
5. 2,4-Dinitrotoluene (24-DNT)
6. 2,6-Dinitrotoluene (26-DNT)
7. Ortho Dinitrotoluenes (mixture of 2,3-DNT & 3,4-DNT) (O-DNT)
8. Water
9. Nitric Acid
10. Sulfuric Acid

The following are the reactions that take place to any significant extent.

$$T + HNO_3 = p\text{-}MNT + H_2O \quad (1)$$
$$T + HNO_3 = o\text{-}MNT + H_2O \quad (2)$$
$$T + HNO_3 = m\text{-}MNT + H_2O \quad (3)$$
$$p\text{-}MNT + NHO_3 = 24\text{-}DNT + H_2O \quad (4)$$
$$o\text{-}MNT + HNO_3 = 24\text{-}DNT + H_2O \quad (5)$$
$$o\text{-}MNT + HNO_3 = 26\text{-}DNT + H_2O \quad (6)$$
$$m\text{-}MNT + HNO_3 = O\text{-}DNT + H_2O \quad (7)$$

Typical distribution of isomers when toluene is dinitrated is shown in Figure 2.

The reactions are essentially irreversible and take place only in the acid phase. Thus transport of mass between the phases and phase equilibrium which defines the driving force for mass transfer are also important. Heterogeneous nitration is illustrated in Figure 3.

$$H_2SO_4 = H^+ + HSO_4^-$$

$$HNO_3 + H^+ = NO_2^+ + H_2O$$

$$ArH + NO_2^+ = ArNO_2 + H^+$$

Figure 1. Chemistry of nitration

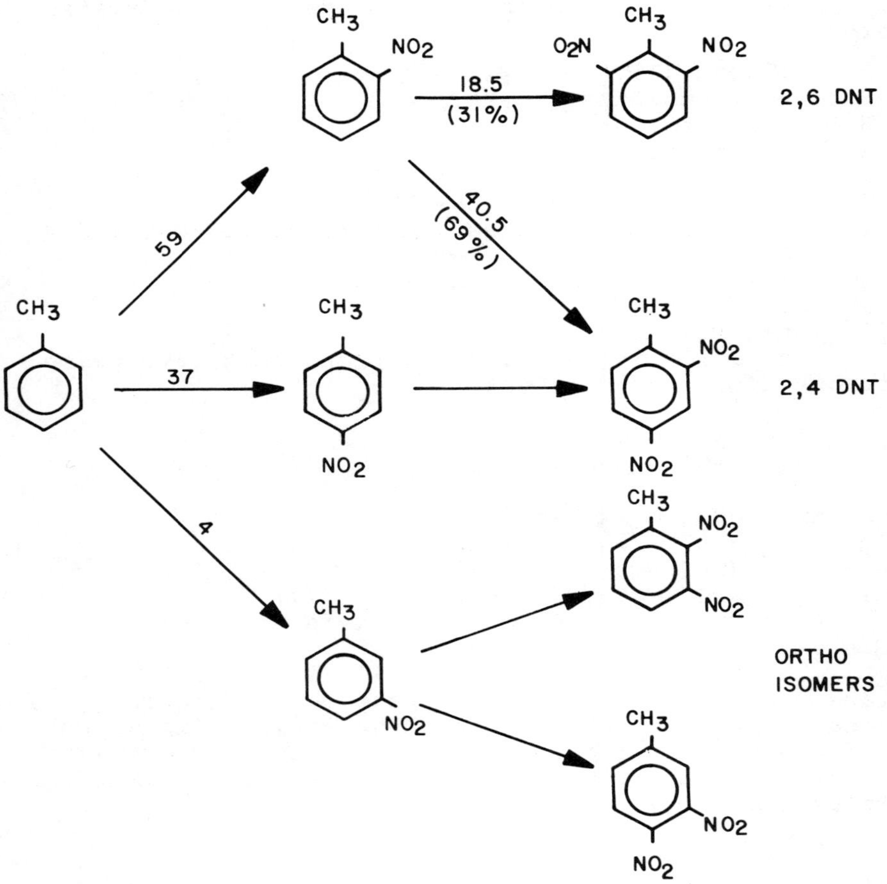

Figure 2. Typical distribution of isomers

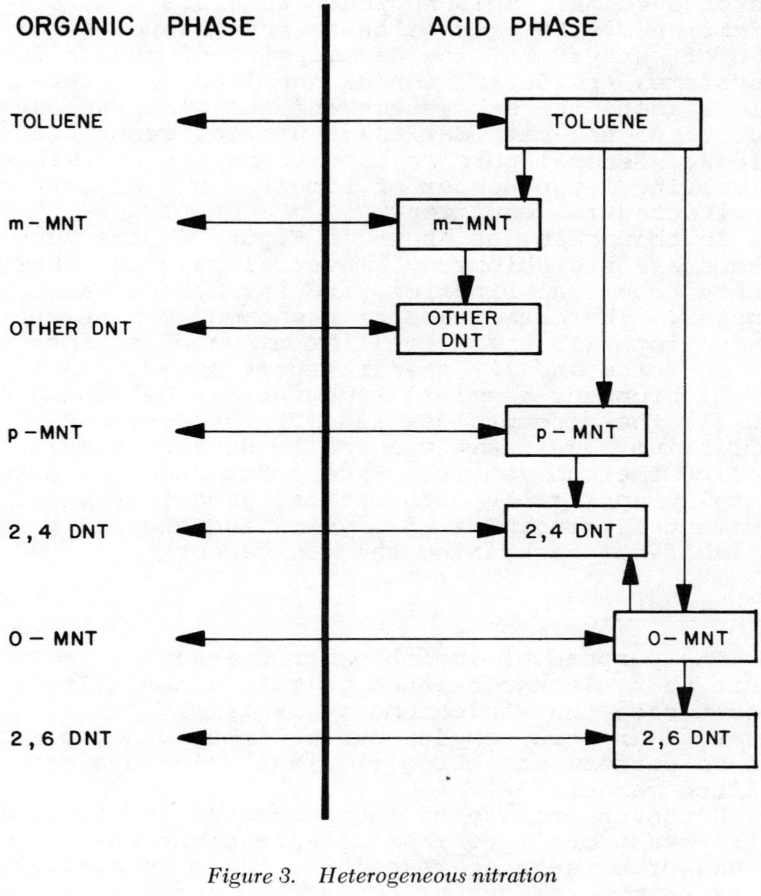

Figure 3. Heterogeneous nitration

Systems Engineering Approach

Analysis of multiphase reaction systems using fundamental laws of physics and chemistry is complex since several phenomena such as kinetics, phase equilibria, and interphase transport are simultaneously significant. The use of systems engineering approach greatly simplifies the modeling of complex reactor systems. This approach consists of the definition of the process as an ensemble or assemblage of subsystems, the development of models for subsystems, the definition of coupling or interactions among the subsystems and the simultaneous solution of the mathematical equations describing the systems. General purpose algorithms are available for solving large number of simultaneous algebraic and differential equations.

In this case, as shown in Figure 4, the subsystems are stoichiometry, material balance, energy balance, chemical kinetics, and interphase mass transfer. The mass transfer phenomena can be subdivided into (1) phase equilibrium which defines the driving force and (2) the transport model. In a general problem, chemical kinetics may be subdivided into (1) the rate process and (2) the chemical equilibrium. The next step is to develop models to describe the subsystems. Except for chemical kinetics, generally applicable mathematical equations based on fundamental principles of physics and chemistry are available for describing the subsystems.

Stoichiometry

The purpose of stoichiometric analysis is to insure that element balance is maintained. In the present case the stoichiometry is fairly straightforward. In more complex cases linear algebra can be used to perform stoichiometric analysis in a generalized manner (14).

A convenient way of incorporating the stoichiometry of the reactions into mass balance equations is by the use of extents of reaction. The composition of the organic phase can be defined by specifying extents e_1 through e_7 which are simply the moles undergoing reactions (1) through (7) respectively, per mole of total organics, in order to produce that composition starting from pure toluene. Acid free organic mole fractions, y_i, can be calculated from the extents using the following equations.

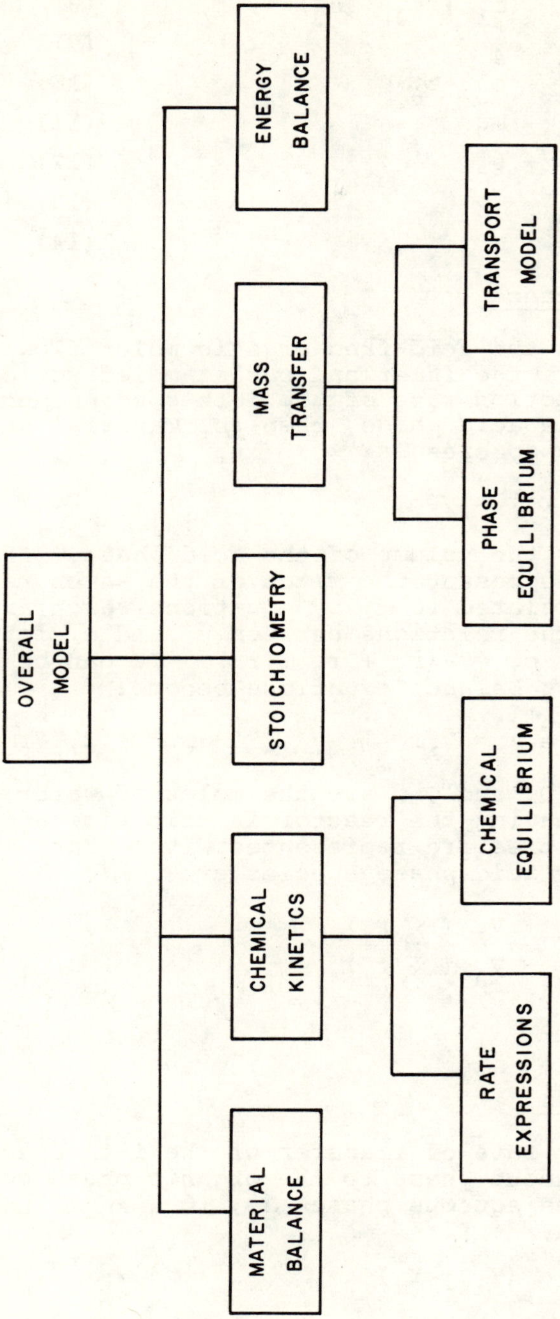

Figure 4. Systems approach to kinetic modeling

$$Y_1 = 1 - (e_1 + e_2 + e_3) \quad (8)$$
$$Y_2 = e_1 - e_4 \quad (9)$$
$$Y_3 = e_2 - e_5 - e_6 \quad (10)$$
$$Y_4 = e_3 - e_7 \quad (11)$$
$$Y_5 = e_4 + e_5 \quad (12)$$
$$Y_6 = e_6 \quad (13)$$
$$Y_7 = e_7 \quad (14)$$

Material Balance

If G is the acid-free organic molar flow rate, $\bar{Y}_i$ amd Y_i are the inlet and outlet molefractions and p_i the production rate of the i-th species per unit volume of the acid phase, then the material balance for the i-th species is:

$$G(Y_i - \bar{Y}_i) = V_A p_i \qquad i = 1,2..7. \quad (15)$$

where, V_A is the volume of the acid phase. If r_1 through r_7 represent the rates of the seven reactions, the p_i are related to r_i by equations which are similar to the relations between y_i and e_i for example, for toluene, $p_1 = -(r_1 + r_2 + r_3)$. It can be shown that the mass balance equations become:

$$G(e_i - \bar{e}_i) = V_A r_i \qquad i = 1,2..7. \quad (16)$$

If Q_8, Q_9 and Q_{10} are the moles of water, HNO_3 and H_2SO_4 leaving the reactor in unit time and the inlet quantities are represented with a bar, mass balances for acid phase species are:

$$Q_8 = \bar{Q}_8 + V_A \left(\sum r_i\right) \quad (17)$$
$$Q_9 = \bar{Q}_9 - V_A \left(\sum r_i\right) \quad (18)$$

and

$$Q_{10} = \bar{Q}_{10} \quad (19)$$

Mass Transfer

The net rate of transfer of the i-th species from the aqueous phase to the organic phase per unit volume of the aqueous phase, J_i, is assumed to follow the equation:

$$J_i = k_o a \,(Y_i - Y_i^*) \qquad i = 1,2..7. \quad (20)$$

where, k_o is a transport coefficient for the organic species, a is the interfacial area per unit volume of reactor and Y_i^* are the equilibrium molefractions of the organic species in the aqueous phase. Nitric acid is assumed to be in phase equilibrium at all times.

The model presented in this paper assumes that mass transfer and chemical kinetics are consecutive steps. Primary resistance for the transfer of toluene from the organic phase into the acid phase lies in the organic phase film. This mass transfer step is followed by homogeneous reaction uniformly over the entire acid phase. If the nitrator is run such that the toluene content of the organic phase is high, diffusional resistance as well as the chemical reaction will be confined to an acid film and the model proposed by Hanson and co-workers (11) will be more applicable.

If the nitrator is assumed to be an ideal backmix reactor, the aqueous and the organic feeds should instantaneously reach the compositions of the aqueous and the organic phases, respectively, upon entering the reactor. Then at steady state,

$$p_i = J_i \qquad (21)$$

Phase Equilibrium

It may be seen that components of the model discussed thus far are generally applicable to any two phase ideal back-mixed reactor. Chemical kinetics and phase equilibrium are the two components which make the model unique.

For all practical purposes, water and sulfuric acid are totally in the acid phase.

Data on the distribution of nitric acid between the organic and acid phases are available at low sulfuric acid concentrations (15). The thermodynamic procedure used to extrapolate these data to regions of interest in the reactors is given below. The activity of HNO_3 in the two phases must be equal at equilibrium.

$$\gamma_{(acid)}\, X_9 = \gamma_{(organic)}\, Y_9 \qquad (22)$$

Using the Redlich-Kister expansion (16) an expression for $\gamma_{(acid)}$ was developed at high temperatures using vapor-liquid equilibrium data for the mixed acid (17). As illustrated by Figure 5, heat of mixing data and the Gibbs-Helmholtz equation were used to extrapolate the activity coefficients to lower temperature. $\gamma_{(acid)}$ was then used in equation (22) to calculate $\gamma_{(organic)}$. Activity coefficient of nitric acid in the organic phase was then correlated as a function of organic composition using the NRTL equation (18).

Solubility of toluene and p-mononitrotoluene in sulfuric acid are available (19,20). These data were correlated using the NRTL equation. In the preliminary model presented here, all mononitrotoluenes are assumed to have the same solubility and nitric acid is assumed to have negilgible effect on their solubility.

Chemical Kinetics

Nitration of an aromatic compound, ArH, takes place by electrophilic attack by NO_2^+ to form $ArHNO_2^+$, followed by the decomposition of this activated complex to give $ArNO_2$ and H^+. The decomposition of the activated complex is the rate determining step(1,2). It can be shown that the nitration reaction is a second order reaction, proportional to the concentrations of the organic species and the nitronium ion in the acid phase. In the model presented here molefractions are used instead of concentrations. Since the solubility of the organic species in the acid phase are small, concentrations of any species is directly proportional to its molefraction.

Nitric acid forms nitronium ions in sulfuric acid according to the reaction

$$HNO_3 + H^+ = NO_2^+ + H_2O \qquad (23)$$

It can be seen from Equation (23) that the fraction of nitric acid which is converted into nitronium ions depends on the activities of the hydrogen ion and water. The activity of the hydrogen ion increases sharply with the increase in sulfuric acid content and this accounts for the sharp increase in the rate of nitration with sulfuric content. A characteristic of nitration using mixed acids, which has not been fully understood, is that the rate of reaction reaches a maximum at about 90% sulfuric acid. At least two different theories have been proposed to explain this behavior (3,21). Fortunately

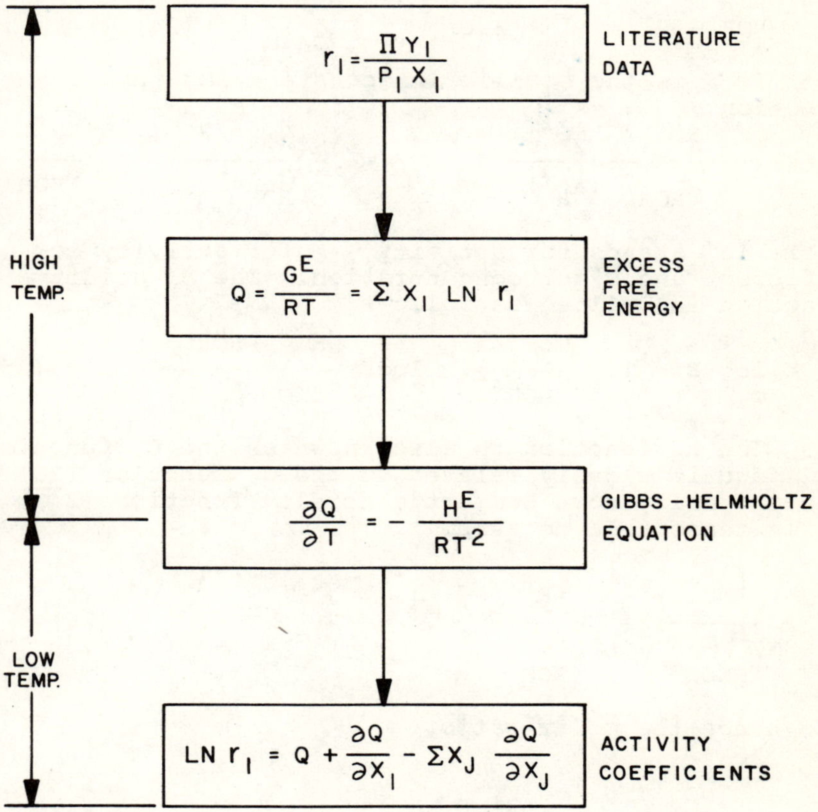

Figure 5. Activity coefficient of nitric acid in mixed acid

the sulfuric acid content of the nitrator media is well below 90%, generally in the range of 70-80%, and hence the kinetic model need not be consistent with this characteristic of aromatic nitration. It is reasonable to assume that the behavior of nitric acid in sulfuric acid should be similar to that of a base ROH which is protonated according to the following equation.

$$ROH + H^+ = R^+ + H_2O \qquad (24)$$

If K is the equilibrium constant for this reaction,

$$K = \frac{(a_{R^+})\ (a_{H_2O})}{(a_{ROH})\ (a_{H^+})} = \frac{\gamma_{R^+}\ a_{H_2O}}{\gamma_{ROH}\ a_{H^+}}\ \frac{C_{R^+}}{C_{ROH}}$$

where 'a' stands for activity, γ for activity coefficient and C for concentration. The H_R acidity function is defined as follows:

$$H_R = \log K - \log \frac{C_{R^+}}{C_{ROH}} = -\log \frac{a_{H^+}\ \gamma_{ROH}}{a_{H_2O}\ \gamma_{R^+}} \qquad (25)$$

The H_R function is also known as the C_o function (22) and is closely related to the J_o function (23) and the well known Hammett's acidity function (24). It is reasonable to assume

$$\frac{\gamma_{NO_2^+}}{\gamma_{HNO_3}} = \frac{\gamma_{R^+}}{\gamma_{ROH}} \qquad (26)$$

and hence the molar ratio,

$$\frac{(NO_2^{\ +})}{(HNO_3)} = 10^{K'-H_R} \qquad (27)$$

Where, K' is the logarithm of the equilibrium constant for the reaction represented by Equation (23). In the present model the equilibrium constant is "lumped" into the velocity constant, thus resulting in the following kinetic expression for nitration.

$$r = k.10^{-H_R} \cdot X_{HNO_3} \cdot X_{ArH} \quad (28)$$

where, r is the rate of reqction per unit volume and the X's are the molefractions. The constant k is a function only of the temperature of reaction and is assumed to be given by the Arrhenius equation.

The H_R acidity scale for sulfuric acid has been measured by a number of research workers, particularly by Arnett and Bushick (25) who determined this function at different temperatures in the range of 0 to 45°C. Their data were curve-fitted over the range of 68-92% sulfuric acid using the following equation.

$$H_R = (A_o + B_o/T) + (A_1 + B_1/T) \cdot W + (A_2 + B_2/T)\, W^2 + A_3 \cdot W^3 \quad (29)$$

where, W = weight percent sulfuric on nitric free basis.

Figures 6 and 7 show the dependence of $r/(X_{HNO_3} \cdot X_{ArH})$ on H_R. It can be seen that Equation (28) is a very reasonable expression for the rate of nitration and that H_R scale can be used to correlate and extrapolate the rate constant as a function of sulfuric acid strength.

The relative reactivities of the ortho, meta and para carbon atoms should be proportional to the electron availability at these sites and the relative probability of collision with a given NO_2^+ ion. It should be possible to calculate the steric effects on the collision frequencies and apply molecular orbital theory to determine the relative reactivities of the various sites. However, a more practical approach would be to assume that all the five carbon atoms encounter nitronium ions at equal frequencies and attribute the differences in the reactivity of the different positions to differences in the activation energies associated with those sites. According to this model, temperature is the only variable that affects the isomer ratio. Our plant experience indicates that this model is adequate for explaining and controlling the variation of the isomer ratio.

If E_o, E_p and E_m are respectively the activation energies for ortho, para and meta substitution during mononitration, the corresponding velocity constants k_o, k_p and k_m are given by:

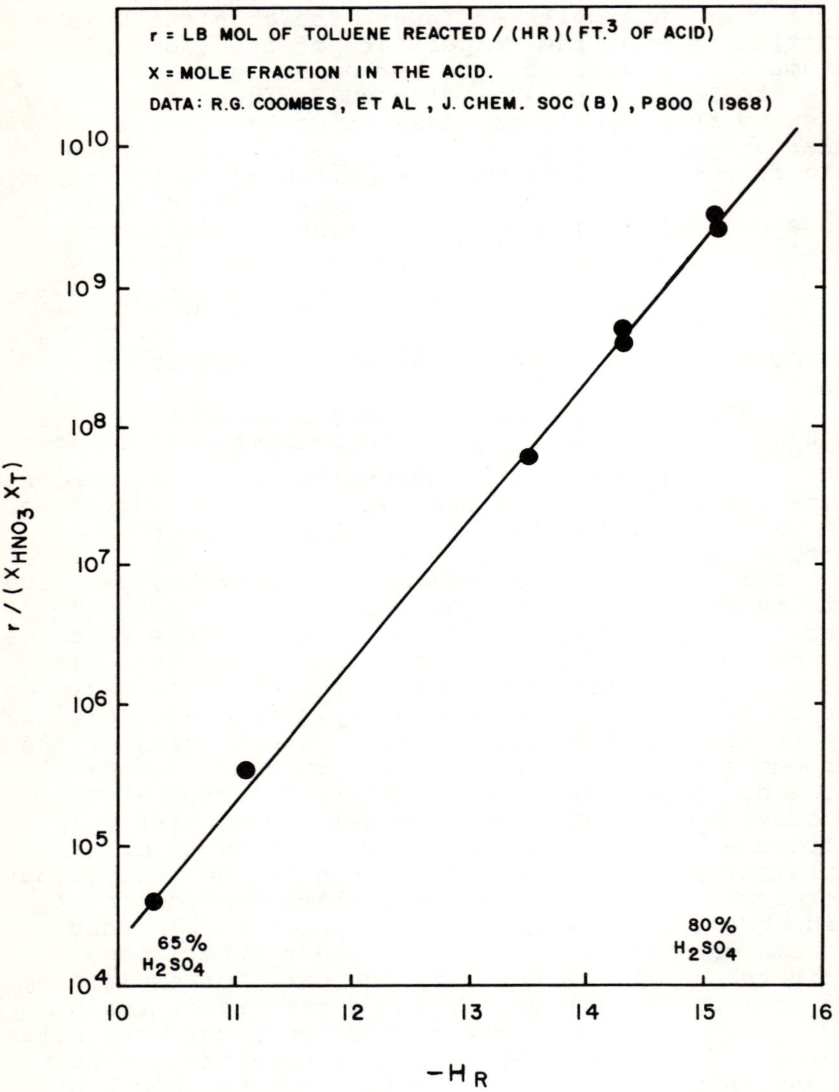

Figure 6. Dependence of $r/(X_{HNO_3} \cdot X_{ArH})$ on H_R. r = lb mol of toluene reacted/(HR) (ft^3 of acid); X = mole fraction in the acid (27).

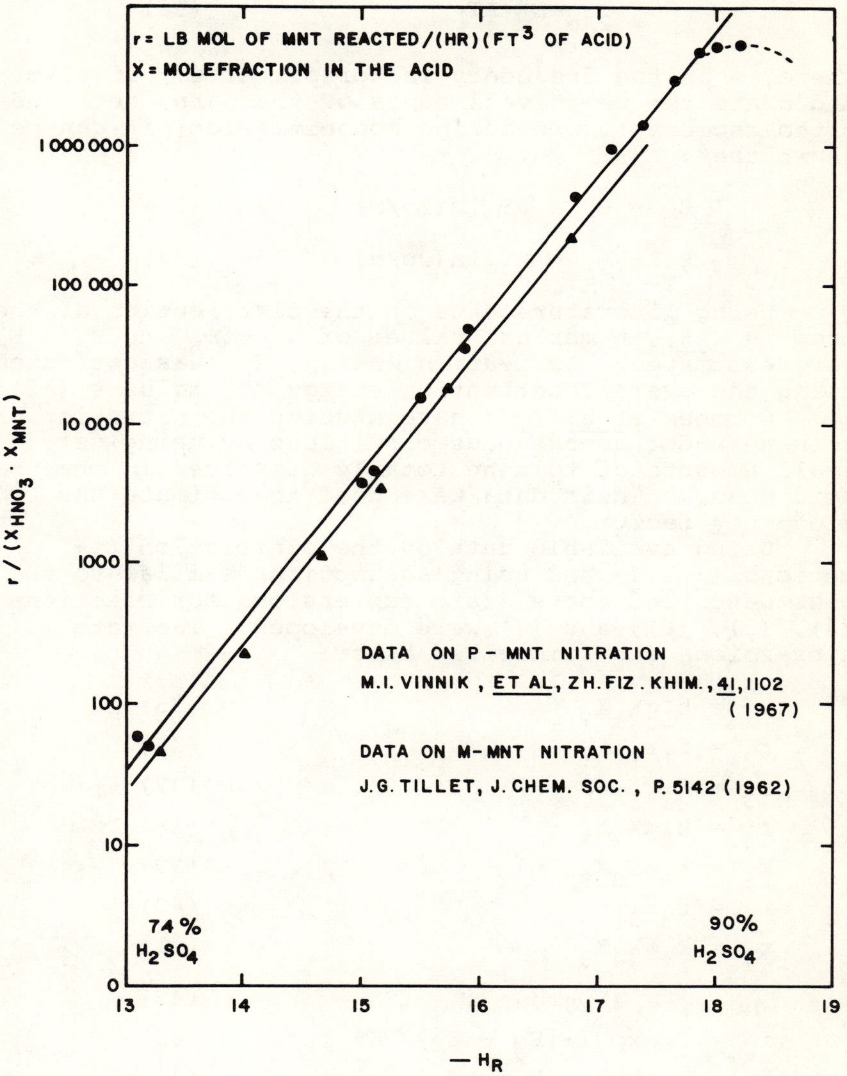

Figure 7. Dependence of $r/(X_{HNO_3} \cdot X_{ArH})$ on H_R. r = lb mol of MNT reacted/(HR) (ft³ of acid); X = mole fraction in the acid (3, 4).

$$k_o = 2Ae^{-E_o/RT} \qquad (30)$$
$$k_p = Ae^{-E_p/RT} \qquad (31)$$
$$k_m = 2Ae^{-E_m/RT} \qquad (32)$$

where, A is the frequency factor per site. If p, m. and o are the relative amounts of the para, meta and ortho isomers formed during mononitration, it can be shown that:

$$E_m - E_p = RT \ln(2p/m) \qquad (33)$$
$$E_o - E_p = RT \ln(2p/o) \qquad (34)$$

Using literature data on the distribution of the isomers (26), numerical values of $E_m - E_p$ and $E_o - E_p$ were estimated. Activation energy, E_p, was estimated using the overall activation energy for toluene (12).

Coombes et al (27) have studied the nitration of toluene under homogeneous conditions by using very small amounts of toluene totally dissolved in the acid phase. Their data were used to estimate the frequency factor.

Using available data on the dinitration reactions (3,5,4) and using assumptions similar to the ones described above, rate expressions for reactions (4), (5), (6), and (7) were developed. The rate expressions are summarized below:

$$r_1 = k_1 g X_1 X_9 \qquad (35)$$
$$r_2 = f_1 r_1 \qquad (36)$$
$$r_3 = f_2 r_1 \qquad (37)$$
$$r_4 = k_2 g X_2 X_9 \qquad (38)$$
$$k_5 = k_3 g X_3 X_9 \qquad (39)$$
$$r_6 = f_3 r_5 \qquad (40)$$
$$r_7 = k_4 g X_4 X_9 \qquad (41)$$

where, $g = 10^{-H_R}$ and (42)

$$f_1 = 2 \exp(-[E_o - E_p]/RT) \qquad (43)$$
$$f_2 = 2 \exp(-[E_m - E_p]/RT) \qquad (44)$$
$$f_3 = \exp(-[E_{26} - E_{24}]/RT) \qquad (45)$$

Solution of Equations

It is not possible to analytically solve the set of equations formulated in the above sections. However, it is possible to obtain a numerical solution corresponding to a specific set of reactor conditions. The model described here can be used to simulate any one stage of the nitration reactors.

Allied Chemical's process for DNT production consists of several interconnected nitrators and phase separators. The overall process simultion was accomplished using the so-called "building block" approach. This approach consists of developing subroutines for calculating the output of process vessels, given the operating conditions and the input streams, and then developing a main program which calls these subprograms in a certain sequence determined by the process topology (flow scheme) and converge on the recycle streams. After convergence, an energy balance is performed around each stage to permit quick evaluation of temperature-controlability.

Verification of the Model

A development project was undertaken to verify the model developed here and to study the mass transfer parameter k_oa as a function of phase ratio (organic/acid volumetric ratio) in the reactor, agitation, internal configuration of the reactor, etc. Results of laboratory runs could be explained by the model using k_oa as the only adjustable parameter. Table 1 shows an example of the excellent agreement between model predictions and laboratory data. In each run, at a different agitator speed, k_oa was determined by matching toluene content of the reactor effluent. The close agreement between the mononitrotoluene and dinitrotoluene isomer content of the actual reactor effluent with those predicted by the model verifies the accuracy and the adequacy of the model.

Admittedly, in the strictly theoretical sense, such a verification is a necessary but not sufficient requirement for the model to be the true model. It should be obvious to readers familiar with current research work on the nitration of aromatic compounds that the assumptions and mechanisms on which this model is based are under debate, albeit generally accepted. Exhaustive testing and verification of a model is usually not justifiable in a business

Table I

COMPARISON OF MODEL PREDICTIONS VS EXPERIMENTAL DATA FOR LAB CFSTR NITRATION

	Weight Percent in Reactor Effluent (Acid-Free Basis)					
	Agitation = 1500 RPM		Agitation = 1900 RPM		Agitation = 2300 RPM	
Compound	Measured	Calculated	Measured	Calculated	Measured	Calculated
Toluene	0.49	0.49	0.35	0.345	0.28	0.28
p-MNT	30.61	30.52	30.71	30.48	32.16	32.22
o-MNT	39.89	39.92	40.38	39.81	42.24	43.35
m-MNT	2.86	4.77	3.37	4.77	3.07	4.97
2,4-DNT	20.11	18.21	19.35	18.44	17.04	24.36
2,6-DNT	5.59	5.46	5.42	5.53	4.86	4.35
Other DNT	0.45	0.62	0.42	0.63	0.36	0.47

environment where the model is used primarily to provide direction in optimizing a multivariable process.

Remarks

The simulation program has been extensively used for process optimization studies as it permits accurate prediction of isomer distribution and heat release. It offers theoretical explanations for isomer control practices arrived at through several years of plant operating experience. The model was used in designing laboratory experiments to study mass transfer under various process conditions and reactor configuration. Since mass transfer and chemical kinetics are simultaneously important in this process, a model is necessary to "filter out" the kinetics effects for mass transfer correlations. The results of our laboratory studies will be presented in future papers.

NOMENCLATURE

a = activity or interfacial area as explained in text
A = frequency factor
c = concentration
e = extent of reaction
E = Activation energy
G = Organic molar flowrate
H_R = Acidity function, defined by eqn (10)
J_i = Interphase mass transfer rate for the i-th species
k = chemical velocity or transport parameter
K = Equilibrium constant
P_i = rate of production of i-th species/unit volume of acid phase
Q = molar flowrate for inorganic species
r_i = rate of i-th reaction, lb mol/hr ft
R = gas law constant
T = absolute temperature
V_A = volume of acid phase
W = weight percent sulfuric on nitric free basis
X = molefraction in the acid phase
y = molefraction in the organic phase (acid free)

GREEK LETTERS

γ = activity coefficient

SUBSCRIPT FOR SPECIES

1	=	toluene	2	=	p-Nitrotoluene
3	=	o-Nitrotoluene	4	=	m-Nitrotoluene
5	=	2,4 - Dinitrotoluene	6	=	2,6 - Dinitro-toluene
7	=	Ortho Dinitrotoluenes	8	=	water
9	=	nitric acid	10	=	sulfuric acid

Literature Cited

1. Norman, R.O.C., Taylor, R., "Electrophilic Substitution in Benzenoid Compounds", Elsevier Publishing Company, Amsterdam, 1965.

2. De La Mare, P.B.D., Ridd, J.H., "Aromatic Substitution, Nitration and Halogenation", Academic Press, New York, 1959.

3. Vinrik, M.I., Grabovskaya, Zh.E., Arzamaskova, L.N., Zh. Fiz. Khim., 1967, 41, 1102.

4. Tillet, J.G., J. Chem. Soc., 1962, p 5142.

5. Kobe, K.A., Fortman, J.T., Ind. Eng. Chem., 1961, 53, 269.

6. Hanson, C., Marsland, J.G., Wilson, G., Chem. Ind., 1966, p 675.

7. McKinley, C., White, R.R., Trans. AIChE, 1944, 40, 143.

8. Barduhn, A.J., Kobe, K.A., Ind. Eng. Chem., 1956, 48, 1305.

9. Miller, R.C., Noyce, D.S., Vermeulen, T., Ind. Eng. Chem., 1964, 56, 43.

10. Albright, L.F., Ind. Eng. Chem., 1965, 57, 53.

11. Giles, J., Hanson, C., Ismail, H.A.M., ACS Symposium Series, 1976, 22, 190.

12. Cox, P.R., Strachan, A.N., Chem. Eng. Sci., 1972, 27, 457.

13. Cox, P.R., Strachan, A.N., Chem. Eng. J., 1972, 4, 253.

14. Aris, R., "Introduction to the Analysis of Chemical Reactors", Prentice-Hall, 1965.

15. Ismail, H.A.M., Ph.D. Thesis, University of Bradford, U.K., 1973.

16. Redlich, O., Kister, A.T., Ind. Eng. Chem., 1948, 40, 345.

17. Ellis, S.R.M., J. Appl. Chem., 1957, 7, 152.

18. Renon, H., Prausnitz, J.M., AIChE J, 1968, 14, 135.

19. Cerfontain, H., Telder, A., Recueil, 1965, 84, 545.

20. Grabovskaya, Z.E., Vinnik, M.I., Zh. Fiz. Khim., 1966, 40, 1221.

21. Bennett, G.M., Brand, J.S.D., James, D.M., Saunders, T.G., Williams, D., J. Chem. Soc., 1947, p 474.

22. Deno, N.C., Jaruzelski, J.J., Schriesheim, A., J. Am. Chem. Soc., 1955, 77, 3044.

23. Gold, V., Hawes, B.W.V., J. Chem. Soc., 1951, p 2102.

24. Hammett, L.P., "Physical Organic Chemistry", McGraw Hill, 1940.

25. Arnett, E.M., Bushick, R.D., J. Am. Chem. Soc., 1964, 86, 1564.

26. Olah, G.A., Kuhn, S.J., Flood, S.H., Evans, J.C., J. Am. Chem. Soc., 1962, 84, 3687.

27. Coombes, R.G., Moodie, R.B., Schofield, K., J. Chem. Soc. (B), 1968 p 800.

RECEIVED November 5, 1979.

24

Modeling, Parameter Identification, and Adaptive Control of Anticoagulant Drug Therapy

THOMAS J. McAVOY

Department of Chemical Engineering, Goessman Laboratory, University of Massachusetts, Amherst, MA 01003

The scope of the medical problems which require the use of anticoagulant drugs cannot be overstated. Arterial thrombosis is a major contributor to the number one (acute myocardial infarction), number three (stroke) and number four (renal) causes of death in the United States (1). Venous thromboembolism is the most common non-surgical cause of death in patients hospitalized for major orthopedic procedures, the most frequent non-obstetrical cause of postpartum death, and a major cause of death in patients with chronic cardiac and pulmonary disease (1). Venous thrombosis is estimated to lead to the hospitalization of approximately 300,000 patients annually in the United States, of which more than 50,000 die (1). The average hospital stay for the 250,000 survivors is a week to 10 days and the cost of this hospitalization has been estimated as $750 million (1). If a one day reduction in hospital stay could be achieved through optimum dosing of anticoagulants, then the potential savings per year can be estimated as $50 million. It has further been estimated (2) that through improved administration of anticoagulants 8,000 to 12,000 lives could be saved each year in the United States alone.

The purpose of this paper is to illustrate how mathematical modeling and control theory can be applied to the problem of optimizing the administration of the anticoagulant drug heparin. The approach presented has the potential of solving some of the economic and life saving problems discussed above.

Mechanism of Clot Formation and Drug Action. Blood clotting is an extremely complicated process which involves an enzyme cascade. To date 13 enzymes, called clotting factors, have been identified and these are normally present in blood in their inactive state. Through an injury of some type one of the early enzymes is activated. This enzyme in turn catalyses the activation of the next enzyme in the cascade. Throughout the process a large amplification occurs. One molecule of a clotting factor is capable of activating many molecules of the next enzyme in the sequence and so forth. The final step in blood clotting involves

0-8412-0549-3/80/47-124-425$05.00/0

the polymerization of fibrinogen to fibrin. This reaction is catalysed by the clotting factor thrombin. Obviously the ability of blood to clot is extremely important when one has a wound or cut. However when clots form inside a vein or an artery a serious problem exists. Anticoagulant drugs are given to prevent this latter situation.

Recent experimental evidence (3,4) has demonstrated that heparin's anticoagulant effect results from its ability to catalyse the deactivation of several of the clotting factors. Of particular importance and the primary reason for heparin's anticoagulant effect is the drug's ability to deactivate thrombin by catalysing the reaction between thrombin and its inhibitor antithrombin III (AT-III). Because it acts directly to deactivate clotting factors, heparin has a very fast anticoagulant effect.

Laboratory Measurement and Goals of Anticoagulant Therapy. For in-hospital patients receiving heparin, a blood sample is drawn, at least daily. The red cells are centrifuged from the sample leaving only plasma. An agent that activates the clotting cascade is added to a fixed amount of plasma. The time required for a clot to form, called the clotting time, is then measured. In addition to the patient's plasma, normal plasma is also tested. The goal of anticoagulant therapy is to have the patient's clotting time be 2 to 2-1/2 times that for the normal plasma. For heparin the most common clotting test is called the partial thromboplastin time, PTT.

Modeling and Optimal Control of Heparin

Modeling Heparin Itself. For heparin it is reasonable to use the CSTR model shown in Figure 1. Drug can be inputted to the reactor either by a bolus injection (impulse forcing) or by continuous intravenous infusion. Within the reactor drug is continuously removed from the blood. However, the mechanism of this removal is at present unknown. As discussed by McAvoy (5) there are some interesting dynamic properties associated with the response of heparin itself. Figure 2 shows a plot of experimental data on heparin together with the response of one model that has been proposed for the drug (5). To generate the experimental data, three different bolus injections of heparin were given to healthy volunteers. Blood samples were then drawn periodically and heparin concentrations measured. When the logarithm of heparin concentration is plotted versus time, an apparently linear response is obtained. However, the slope of the response changes with each bolus. This change in slope indicates that an unusual nonlinear phenomena takes place when humans eliminate heparin. Stated mathematically, the initial condition inside the CSTR somehow is stored and it affects the entire transient response of the drug.

The model response plotted in Figure 2 is based on the following metabolite-inhibition mechanism (6)

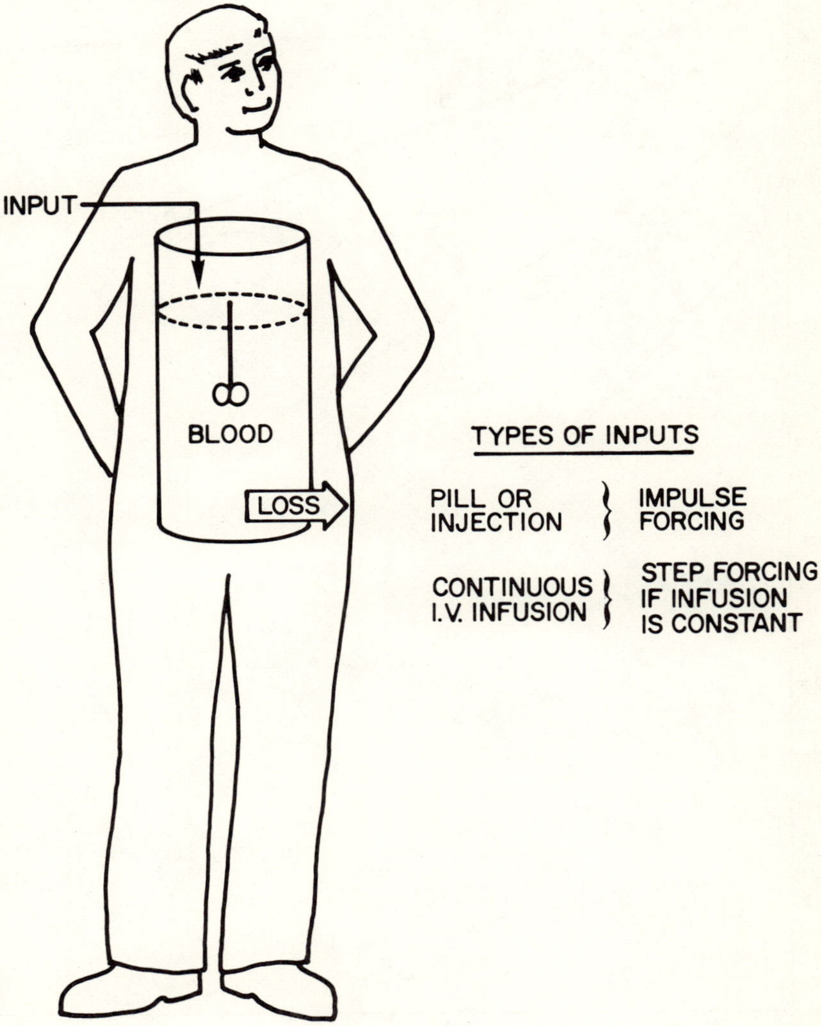

Figure 1. CSTR model for heparin

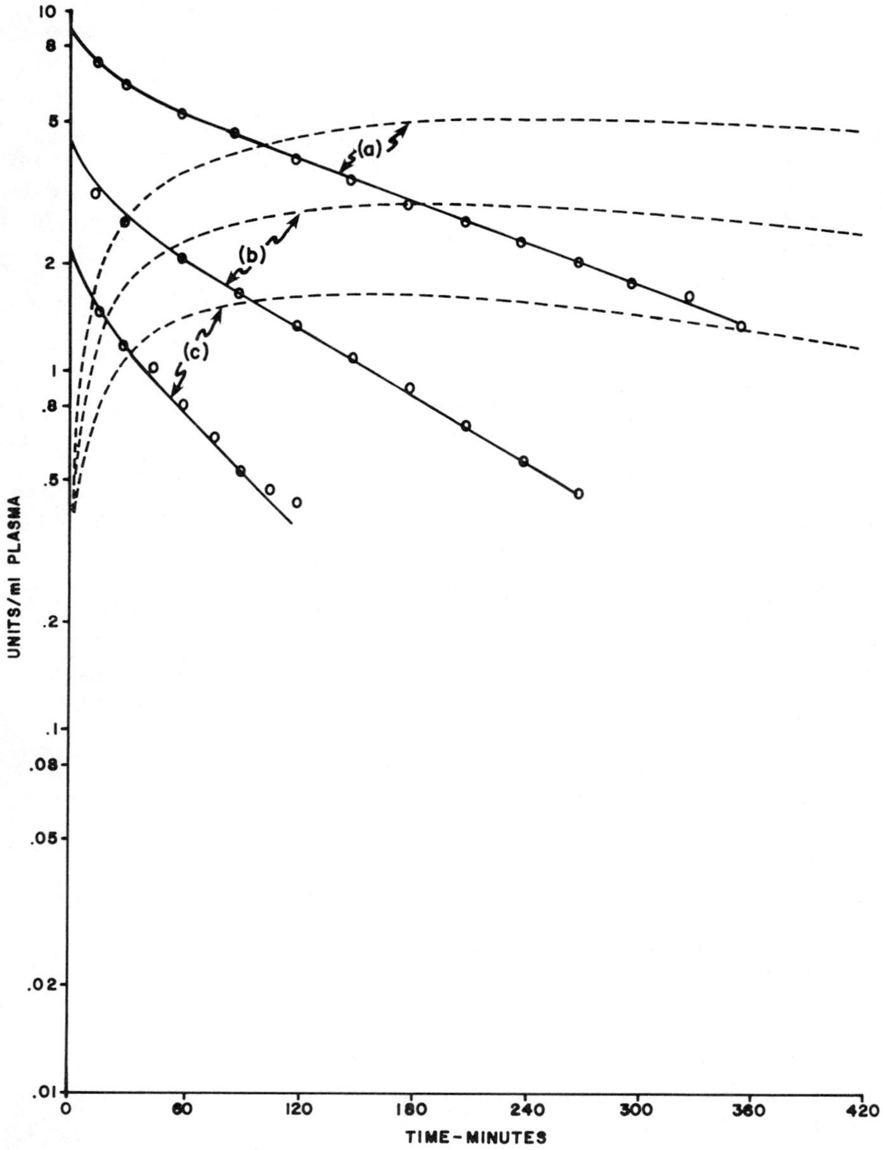

Figure 2. Heparin data and metabolite-inhibition model for three separate injections of drug. (——) heparin concentration; (○) experimental data; (– – –) metabolite concentration. (a) 400 μ/kg bolus; (b) 200 μ/kg bolus; (c) 100 μ/kg bolus (5).

$$H + E \rightleftarrows HE \rightarrow M_1 + E \qquad (1)$$

$$M_1 + E \rightleftarrows M_1E \qquad (2)$$

$$M_1 \rightarrow M_2 \qquad (3)$$

Heparin, H, is assumed to be metabolized by enzyme E to product M_1 which competes with the drug for binding sites on E. In addition, M_1 is biotransformed to M_2 which is non-competitive. If the mechanism given by equations 1-3 is assumed and balances on H and M_1 are made for the reactor shown in Figure 1, the following model for heparin results (6)

$$\frac{dH}{dt} = I - V_m H/(H + K_m(1 + M_1/K_P)) \qquad (4)$$

$$\frac{dM_1}{dt} = V_m H/(H + K_m(1 + M_1/K_P)) - k_{ep}M_1 \qquad (5)$$

where I is the rate of drug infusion divided by the reactor volume. Equations 4 and 5 contain four constants, V_m, K_m, K_P and k_{ep}. These constants were fitted to the experimental data shown in Figure 2 by numerically integrating the differential equations and then using a least square simplex approach on the results (7). The values obtained are: $V_m = .938$, $K_m = 15.7$, $K_P = .385$, and $k_{ep} = .00187$. The responses predicted by equations 4 and 5 are also shown in Figure 2. Note that I = 0 for these responses. As can be seen, equations 4 and 5 do an excellent job of fitting the experimental data. The M_1 response is given as the dashed line in Figure 2. As can be seen, M_1 builds up very quickly and achieves an almost constant value for most of the H transient. In fact it is M_1 which "remembers" the initial condition in the reactor. In effect M_1 blocks an approximately constant number of binding sites on E for each response and this produces the apparent first order response. Since the number of blocked sites increases with the initial condition for H, the slope of each of the H responses is different. In addition to the metabolite-inhibition model, McAvoy also presents an alternative model for heparin involving two nonlinear differential equations and based on a phagocytosis elimination mechanism (5).

Modeling the Clotting Time. Since the main objective of heparin therapy is to control a patient's clotting time (normally the PTT), it is necessary to have a model for the clotting time. Both the nature of the clotting cascade and published experimental results (8) indicate that clotting time is not a simple function of heparin concentration alone. Other variables which affect the clotting time include the concentration and degree of activation of the various clotting factors and the concentration of other proteins in the blood, such as antithrombin III. There is a degree of uncertainty at present as to what models will ultimately prove useful in describing heparin's effect on the clotting time because the necessary experimental data have not yet been gathered. The author is involved in a research program to generate these data. To illustrate the type of modeling approach that will be studied as well as the control problem which results, one specific example involving antithrombin III will be discussed in detail. This example is typical of the modeling which will be carried out.

Rosenberg's data (4) indicate that heparin binds to AT-III, while those of Marciniak (9) show that such binding does not take place. Both authors, however, agree that heparin's main route of action is via the enhancement of the inhibitory reaction which takes place between AT-III and thrombin. Marciniak and Gockerman (10) have also recently published data on AT-III for 24 patients receiving heparin infusions. These authors found that heparin therapy produced an exponential like transient decay in AT-III levels. After 3 days of heparin therapy AT-III levels approached a constant value which was 30% lower than normal. After cessation of heparin, AT-III levels returned to normal in about 3 days.

Since AT-III is so important to heparin's action, it is proposed to model its dynamic behavior, and to attempt to incorporate its concentration into an equation for clotting time. Thus, one equation which is proposed for describing clotting time is

$$E = g\ (H,\ AT\text{-}III) \qquad (6)$$

where E, the relative extension of the clotting time, is defined as

$$E = (PTT/PTT_n - 1) \qquad (7)$$

and PTT is the clotting time and PTT_n is the normal clotting time. The reason for considering an algebraic relationship between E, AT-III, and H is that the results of Rosenberg (4) indicate that heparin almost instantaneously catalyses the deactivation of thrombin by AT-III. Free thrombin levels in turn affect E. Thus, changes in both heparin level and AT-III level should have a very rapid effect on E.

In order to make equation 6 more specific it will be assumed that heparin binds to antithrombin III

$$H + AT\text{-}III \rightleftarrows (H\cdot AT\text{-}III) \qquad (8)$$

Further, it will be assumed that E is proportional to the concentration of the heparin antithrombin III cofactor

$$E = m\ (H\cdot AT\text{-}III) \qquad (9)$$

If thermodynamic equilibrium is assumed for equation 8, equation 9 can be written as

$$E = m K_{eq} (H)(AT\text{-}III) \quad (10)$$

where K_{eq} is the equilibrium constant for equation 8. To complete the model it is necessary to have an equation for antithrombin III. One model which Marciniak and Gockerman's data (10) suggest will be appropriate is

$$\frac{d}{dt}(AT\text{-}III) = - k_A (AT\text{-}III - AT\text{-}III_o) - k_B H \quad (11)$$

where k_A and k_B are constants and $AT\text{-}III_o$ is the normal blood concentration of antithrombin III. At steady state with no heparin present equation 11 forces AT-III to return to its steady state value. From Marciniak and Gockerman's data a value of $AT\text{-}III_o$ can be gotten as 36 mg/dℓ. To achieve a 3 day transient response for AT-III a reasonable value of k_A to use is .001 min. Lastly, if a steady state heparin concentration of .5 units/ml plasma produces the 30% reduction in AT-III levels then k_B can be calculated as .0216 ((mg/dℓ)/(units/ml min)) by setting d(AT-III)/dt = 0.

Optimal Control of Heparin Therapy. One complete model for a patient's response to heparin is given by equations 4, 5, 10 and 11. All of the parameters in the model have been specified except $m K_{eq}$ in equation 10. A reasonable value for $m K_{eq}$ can be gotten by assuming that E is 1.0 when H is .5 and AT-III is 70% of $AT\text{-}III_o$. This assumption gives $m K_{eq}$ as .0794. Using this model and modern control theory it is possible to formulate an optimum dosing policy. Prior to heparin being administered the initial conditions in the CSTR are that AT-III equals $AT\text{-}III_o$ and both H and M_1 are 0. One clinically reasonable performance index is to choose the heparin infusion, I, to minimize the time that it takes to get from this initial state to an anticoagulated state where the clotting time is twice normal (E = 1). Mathematically one should minimize the performance index

$$P.I. = \int_0^{\theta_f} dt \quad (12)$$

subject to

$$E(\theta_f) = 1 = .0794 (H)(AT\text{-}III) \quad (13)$$

In addition, I would be constrained to be less than some maximum clinically acceptable value I*

$$I \leq I^* \quad (14)$$

A reasonable value for I* is .968 units/hr ml plasma. This value results from a maximum infusion rate of 3000 units/hr into a typical plasma volume of 3100 ml. The problem posed involves time optimal control to a region. The approach which can be used to solve this problem has been discussed by Denn (11). By appending 3 adjoint equations to the state equations, one can start at the final time, θ_f, and numerically integrate backwards in time to solve for the optimal policy. To carry out the integration it is necessary to estimate $H(\theta_f)$ and $M_1(\theta_f)$ and iterate on these values until the initial conditions on M_1 and AT-III are satisfied. The time optimal policy calls for an initial injection of 1420 units of heparin with no infusion (I = 0) for 5.25 minutes. Once a patient is brought to an anticoagulated state it is necessary to keep him there and still not violate equation 14. McAvoy (5) has presented an infusion policy which instantaneously brings a patient to an anticoagulated state but which requires subsequent infusion rates about 1.5 times greater than $I^* = .968$. By solving the state equations 4, 5, and 11 subject to equations 13 and 14 it is possible to calculate the infusion policy necessary to keep a patient anticoagulated after the initial time optimal period. Indeed it is this latter infusion period where E = 1 and equation 14 holds which determines the 5.25 minute duration of the time optimal period. As I^* becomes larger the duration of the time optimal period becomes shorter and it eventually becomes 0 (solution given by McAvoy (5)). At 5.25 minutes I jumps to I^* and then begins to taper in an exponential-like manner as shown in Figure 3. As can be seen, I tapers over a period of 500 minutes and eventually reaches a steady state value of .257 units/hr ml plasma. Also shown in Figure 3 is the response of E to the optimal infusion policy.

It is interesting to compare the policy shown in Figure 3 to current clinical practice since reasonable parameter values were used in the heparin model. Very often it is found that a patient's need for heparin is greater during the early stages of therapy than later on. In the medical literature this has been referred to as initial resistance to heparin therapy. Generally heparin infusion rates are lowered over a period of several days. The infusion policy shown in Figure 3 does exhibit a tapering transient but the time involved, 500 minutes, is shorter than that observed clinically. This discrepancy may indicate the necessity to refine the model possibly by having the heparin response tied to the slower response of antithrombin III.

Clinical Use of the Model

In presenting the pharmacokinetic model for heparin some parameters were fitted to data on healthy volunteers while other parameters were arbitrarily chosen. To improve heparin therapy the pharmacokinetic modeling approach discussed here must be used in a prospective, predictive manner. By taking additional blood samples early on in heparin therapy data for each patient can be generated. One would measure such variables as heparin level, AT-III level, PTT, etc. Then, by using the model presented here or a more accurate model, parameters can be fitted (7) and the

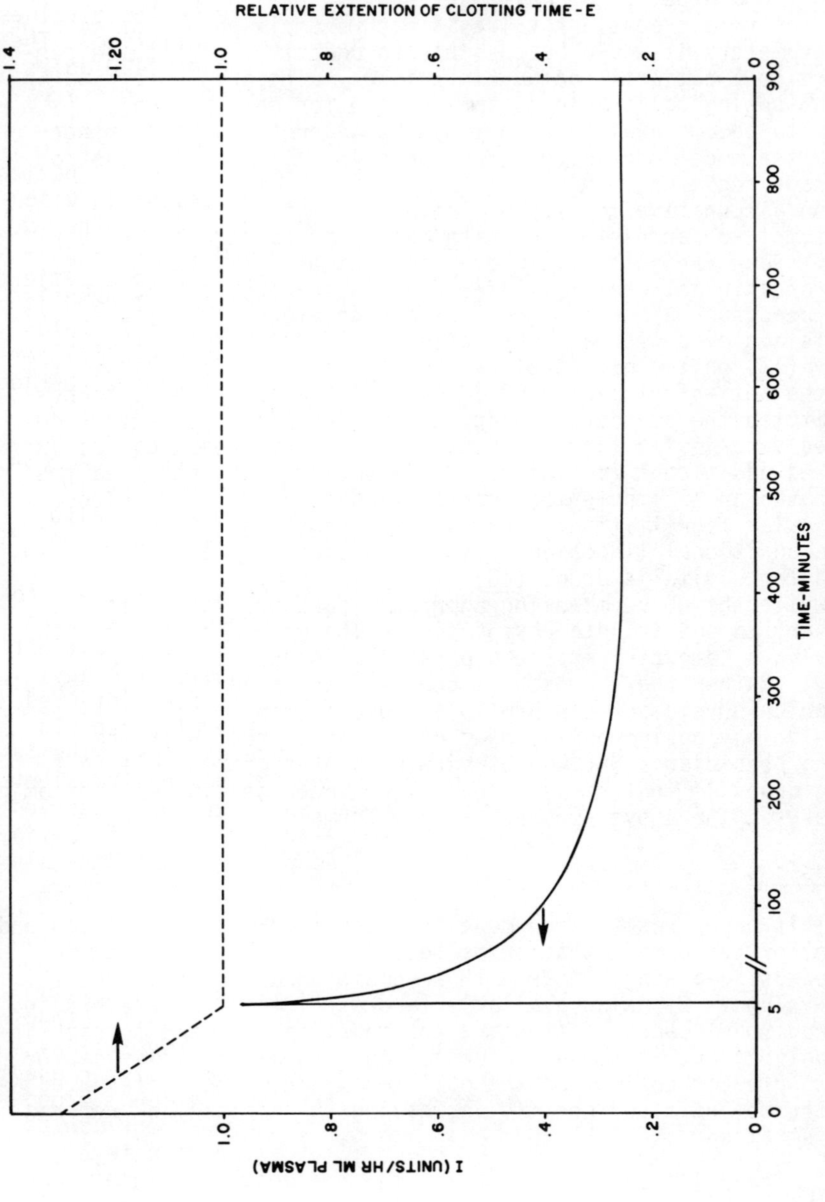

Figure 3. Optimum dosing policy and clotting time response for heparin model. (——) optimum dosing policy; (– – –) relative extension of clotting time

model tailor-made to each individual patient. The present model contains 8 parameters. For prospective clinical use it would be impractical to try to adaptively fit this many parameters. No doubt it would be necessary to set many of the parameters in the model to average constant values. With a small number of remaining parameters it would be possible to adaptively fit these paramaters to a patient's data. In this way a heparin model and an optimum dosing policy can be tailor-made for each patient. As more data become available they can be incorporated to further refine the model and in this fashion allow for adaptive control of dosage regimens.

An alternative to simply fitting parameters in the model is to attempt to correlate them with easily measured clinical variables. Such variables include: age, height, sex, hematocrit, serum creatinine, etc. Jelliffe and his co-workers have successfully used such a correlation approach to improve the therapeutic results achieved with a number of drugs. For example, they reported (12) on two groups of patients treated at Carmel Hospital with the anti-arrythmic drug lidocaine. A group of 68 patients constituted the standard therapy group. Of these patients 8 developed ventricular fibrillation, 5 had a toxic reaction and 33 required additional lidocaine. Of 71 patients who received the drug based on a computerized correlation approach, 2 developed ventricular fibrillation, 1 had a toxic reaction and only 2 required additional lidocaine. Similar success has also been reported for digitalis drugs (13).

While the above modeling approach looks promising, it is necessary to put it into perspective. The approach is not aimed at having a computer replace a physician in deciding dosing policy. Rather the approach is proposed as an additional, useful tool which physicians can use to improve therapy. The final decision on any dosing policy must of necessity rest with the attending physician. One way of viewing a physician's role in giving drugs is that he is a control engineer dealing with dynamic processes. The above approach simply formalizes this concept.

Conclusions

This paper has treated modeling, parameter identification and control of the administration of heparin. A model has been developed using a single CSTR with a volume equal to a patient's blood volume. By using available laboratory data, parameters in the drug model can be fitted and the model tailor-made to each individual patient. Once an accurate model is available it can be used to predict optimum dosing policies. The methods presented have the potential of greatly increasing the efficacy of heparin administration.

Abstract

A pharmacokinetic model which describes a patient's response to the widely used anticoagulant drug, heparin, is discussed. The model contains several unknown parameters. A patient's own past data can be used to identify these parameters and thus tailor-make a model for each individual under treatment. Once such a model is available it can be used to calculate optimum dosing policies to achieve a specified therapeutic goal.

Literature Cited

1. Wessler, S. "Current Dilemmas in the Clinical Use of Heparin," Fed. Proc., 1977, 36, 66-69.
2. Wessler, S. "A Heparin Symposium," Current Ther. Res., 1975, 18, 3-5.
3. Rosenberg, R. "Actions and Interactions of Antithrombin and Heparin," N. Engl. J. Med., 1975, 292, 146-151.
4. Rosenberg, R. "Chemistry of the Hemostatic Mechanism and Its Relationship to the Action of Heparin," Fed. Proc., 1977, 36, 10-19.
5. McAvoy, T. J. "Pharmacokinetic Modeling of Heparin and Its Clinical Implications," J. Pharm. Biopharm., 1979, 7, 331-354.
6. Perrier, D.; Ashley, J.; Levy, G. "Effect of Production Inhibition on Kinetics of Drug Elimination," J. Pharm. Biopharm., 1973, 1, 231-242
7. D'Argenio, D. Z.; Schumitzky, A. "A Program for Simulation and Parameter Estimation in Pharmacokinetic Systems," Comput. Programs in Biomed., 1979, 9, 115-134.
8. Hirsh, J.; VanAken, W. G.; Gallus, S. G.; Dollery, C. T.; Cade, J. F.; Yung, W. L. "Heparin Kinetics in Venous Thrombosis and Pulmonary Embolism," Circulation, 1976, 53, 691-695.
9. Marciniak, E. "Binding of Heparin in Vitro and In Vivo to Plasma Proteins," J. Lab. Clin. Med., 1974, 84, 344-356.
10. Marciniak, E.; Gockerman, J. "Heparin-Induced Decrease in Circulating Antithrombin III," Lancet, 1977, II, 581-584.
11. Denn, M. "Optimization By Variational Methods"; McGraw-Hill: New York, 1969, pp. 189-191.
12. Jelliffe, R.; Rodman, J.; Kolb, E. "Clinical Studies with Computer-Assisted Lidocaine Infusion Regimens," Circulation, 1976, 54, II-211.
13. Jelliffe, R.; Buell, J.; Kabala, R. "Reduction of Digitalis Toxicity by Computer-Assisted Glycoside Dosage Regimens," Ann. Int. Med., 1972, 77, 891-906.

RECEIVED November 5, 1979.

25

Linking Process Simulators to a Refinery Linear Programming Model

ANDRE W. POLLACK and W. DONALD LIEDER[1]

Suntech, Inc., P.O. Box 1135, Marcus Hook, PA 19061

The use of linear programming to optimize the flow of process streams through a petroleum refinery began in the mid-1950's (Symonds, 1955: Manne, 1956). Now, almost twenty-five years later, it is safe to say that one half of U.S. refining capacity is represented by linear programming or LP models which are routinely optimized to schedule operations, evaluate feedstocks, and study new process configurations.

The database for these refinery LP models is a mixture of economic and technical inputs. Economic inputs include the availability and price of refinery raw materials, the variable cost of operating the individual process units, and the demand and price for refinery products. Technical inputs include refinery product specifications as well as the operating constraints, usage of equipment and utilities, product yields, and product properties for each process unit.

The economic inputs are of vital concern to a petroleum refining company and various groups within the company are charged with monitoring and forecasting this information. For this reason, economic inputs are probably the easiest values to maintain and update in the refinery LP database. And for the same reason, specifications on refinery products are also easy to maintain and update.

Plant operating constraints are readily obtained from design data and operating history. There is usually little need to update these constraints unless the plant is debottlenecked or we want to study some new process configuration.

Unfortunately, the remaining technical inputs which characterize plant performance are extremely difficult to maintain and update. For whether we measure the usage of equipment and utilities, the product yields, and the product properties directly from a plant survey or whether we compute these inputs using a process simulator fitted to the plant, one fact is uncomfortably clear. The values are good only for the feed and operating con-

[1] Current Address: Sun Petroleum Products Company, Toledo, OH.

0-8412-0549-3/80/47-124-437$05.00/0

ditions at the time of the survey or specified in the computer run. That is why the refinery LP database often contains inputs for more than one plant operation and the LP model "blends" these operations to approximate plant performance on a new feed or at a new operating level.

To facilitate the maintenance and updating of plant performance inputs, we have developed and implemented an LP preprocessor. This preprocessor automatically generates and stores in the LP database the usage of equipment and utilities, the product yields, and the product properties for six process units at Sun Petroleum Products Company's Toledo Refinery. Linked to the preprocessor are three already existing process simulators: a fluid catalytic cracker or FCC simulator, a hydrocracker simulator, and a catalytic reformer simulator.

In this paper, we describe the preprocessor, the process simulators, and how we linked them together. We also discuss some LP results made possible through the use of the preprocessor.

The Preprocessor

The operation of the preprocessor is shown schematically in Figure 1. There are five basic steps:

- Select crude assay data
- Build and report input tables
- Generate crude data
- Generate process data
- Build and access LP data tables

In the first step, the preprocessor accesses the disc file which contains all of Sun Petroleum Products Company's crude assays. The preprocessor extracts assay data for those crudes which the user has identified by card input as part of the base crude mix or to be made available to the LP model as an incremental refinery feed. The user can identify up to ten crudes any five of which can be designated as incremental.

In the second step, the preprocessor builds and prints out tables showing the extracted crude assay data. The preprocessor also prints out in tabular form all information which the user has specified by card input for the following process units:

- Crude distillation unit
- Propane deasphalter
- FCC
- Gas oil hydrocracker
- Motor reformer
- BTX reformer

Base operating conditions and unit parameters for these six units are stored in the preprocessor. The user, through card input, may alter any of the base values to define a new base operation or to add one or more alternate operations.

In the third step, the preprocessor generates the yields and properties of the products obtained by crude distillation and

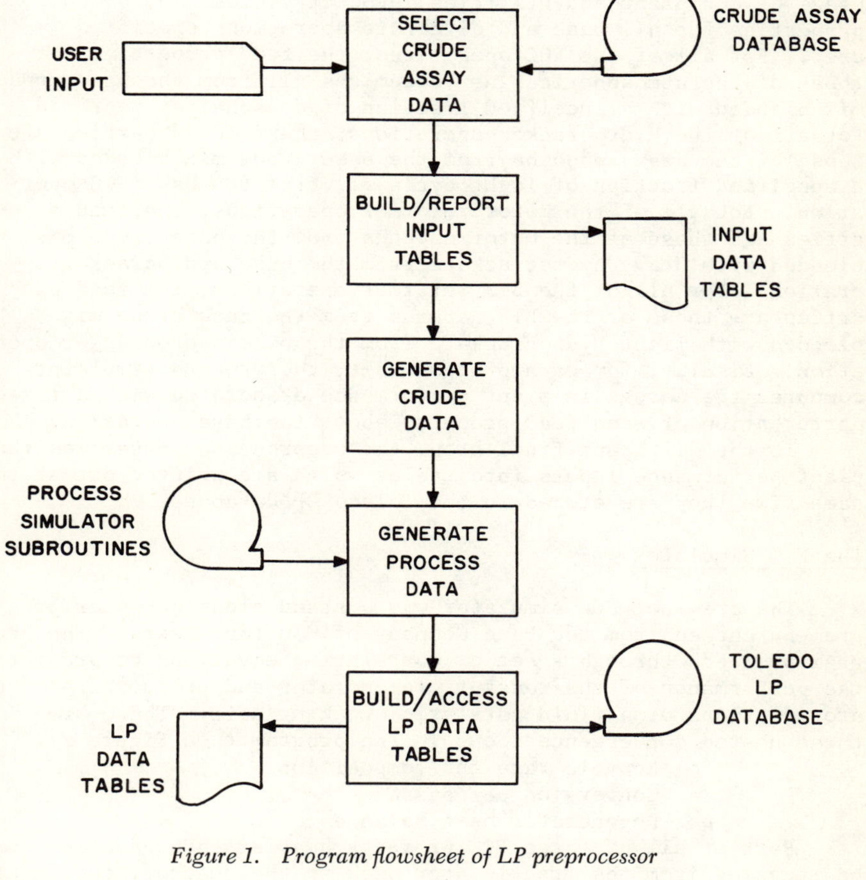

Figure 1. Program flowsheet of LP preprocessor

propane deasphalting of the base crude mix and of each incremental refinery feed. The preprocessor uses only simple correlations and interpolation procedures to transform the laboratory crude assay data into results applicable to the distillation and deasphalting operations in the refinery.

In the fourth step, the preprocessor generates plant performance data for the FCC, gas oil hydrocracker, motor reformer and BTX reformer. For each of these process units, the preprocessor calls the appropriate process simulator which computes the usage of equipment and utilities, product yields, and product properties for all base and alternate operations specified by the user. For all of the FCC operations, the feed properties are those of the atmospheric plus vacuum gas oil from the base crude mix blended with a specified fraction of deasphalter overhead. For all of the hydrocracker operations, the feed properties are those of the heavy naphtha from the base crude mix blended with a specified fraction of light cycle oil from the base FCC operation. For all of the motor reformer operations, the feed properties are those of the motor naphtha from the base crude mix blended with heavy hydrocrackate from the base hydrocracker operation. For all of the BTX reformer operations, the feed properties are those of the BTX naphtha from the base crude mix blended with light hydrocrackate from the base hydrocracker operation. Finally, for each process unit, the process simulator computes the change in plant performance associated with a fixed perturbation of each feed property about the base operation.

In the fifth and final step, the preprocessor assembles the plant performance inputs into tables which are printed out at the same time they are stored in the Toledo LP database.

The FCC Simulator

The original FCC simulator was a stand-alone computer program purchased from the Pace Company of Houston, Texas. The program proceeds through a set of correlating equations to predict the performance of the reactor, regenerator and product fractionator sections of a fluid catalytic cracking unit. There are three nested convergence loops in the program (see Figure 2):

- o Recycle rate and composition
- o Conversion per pass
- o Regenerator heat balance

Part or all of three FCC products heavier than gasoline can be recycled from the fractionator back to the reactor: the light cycle oil which at Toledo is a prime hydrocracker feed, the heavy cycle oil which is usually recycled, and the bottoms or slurry oil, part of which must be recycled to return entrained catalyst to the unit. The program offers six recycle options. For any option, the user specifies five out of the following eight recycle variables:

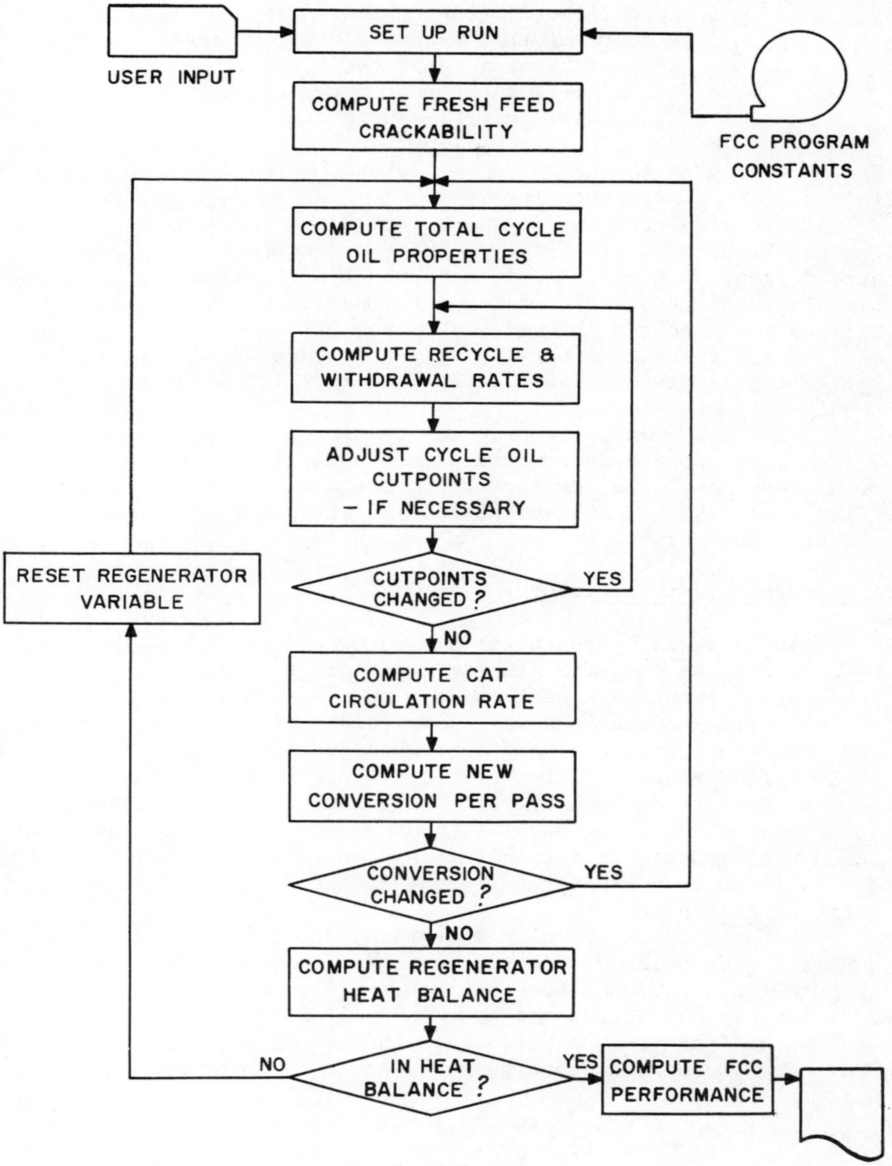

Figure 2. Program of FCC simulator

- Withdrawal rate of light cycle oil
- Recycle rate of light cycle oil
- Initial boiling point of heavy cycle oil
- Withdrawal rate of heavy cycle oil
- Recycle rate of heavy cycle oil
- Final boiling point of heavy cycle oil
- Withdrawal rate of slurry oil
- Recycle rate of slurry oil

The remaining three variables are determined by trial-and-error in the inner convergence loop.

Conversion per pass is the principal correlating variable in the equations predicting recycle rate and composition. In the second loop, this variable is resolved by computing the catalyst circulation rate which simultaneously satisfies the reactor heat balance and supports the conversion per pass.

In the third and outer loop, the program adjusts a user-selected regenerator variable to satisfy the heat balance in the catalyst regenerator.

The FCC simulator program was converted to subroutine form a few years ago and incorporated into a nonlinear programming model representing a complex of process units in the Toledo refinery. It is this subroutine version which has been linked with the LP preprocessor.

The Hydrocracker Simulator

The original hydrocracker simulator was developed in-house as a stand-alone computer program. The program is based on a fundamental reaction kinetic model and predicts the performance of two multi-bed reactors with inter-bed quench zones, high and low pressure separators, and a product fractionator. Figure 3 is a simplified flowsheet of the Toledo hydrocracker. The hydrocracker feed is defined by twenty-six components which can undergo hydrocracking, ring opening, hydrodealkylation, hydrogenation and denitrogenation reactions. Reaction rate expressions reflect the dual function nature of the catalyst and the inhibiting effects of adsorption.

The program numerically integrates the differential component and heat balances for the combined feed and recycle gas through the individual beds of both reactors accounting for the addition of cold quench gas between reactor beds and the recycling of fractionator bottoms to the second reactor inlet. There are two nested convergence loops in the program (see Figure 4):

- Recycle rate and composition
- Conversion per pass

In the inner loop, the rate and composition of fractionator bottoms recycle are determined by successive substitution sometimes accelerated by a secant method. The rate and composition of recycle plus makeup gas routed to the reactor inlets and

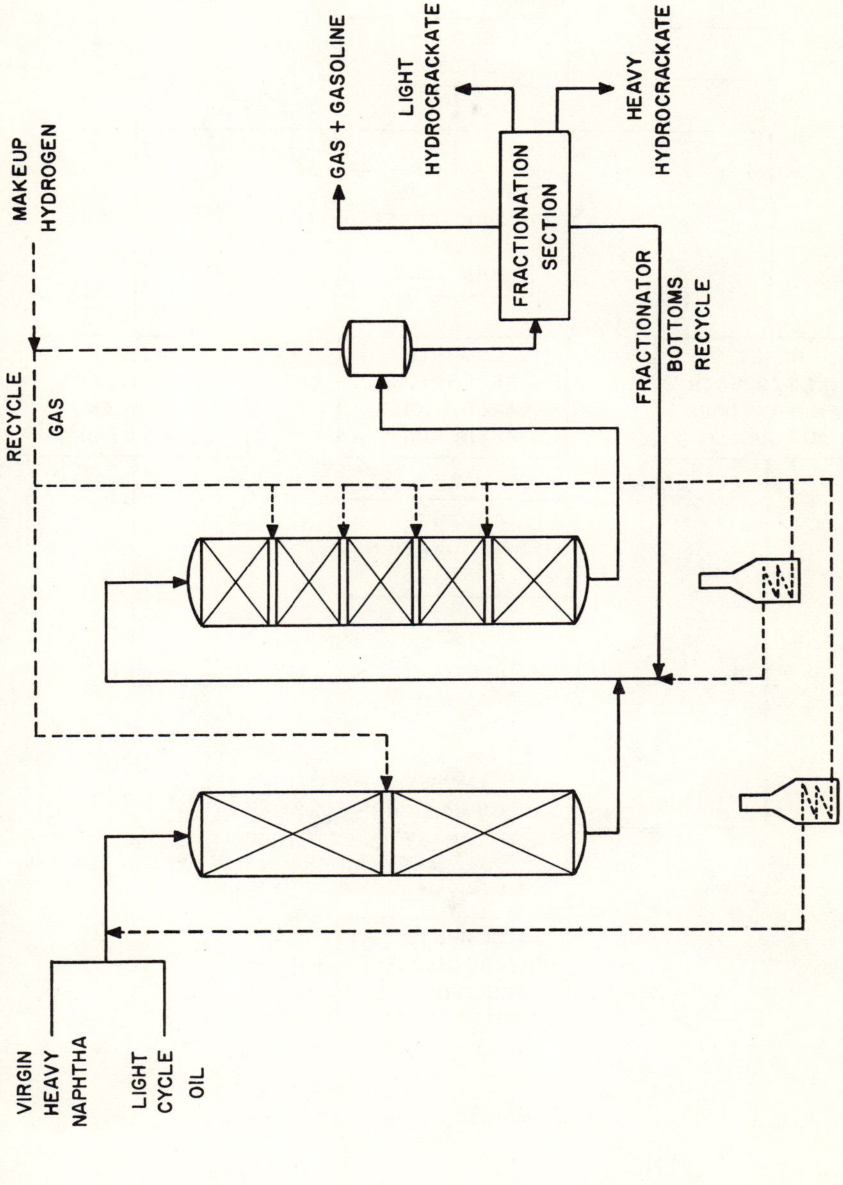

Figure 3. Process flowsheet of Toledo hydrocracker

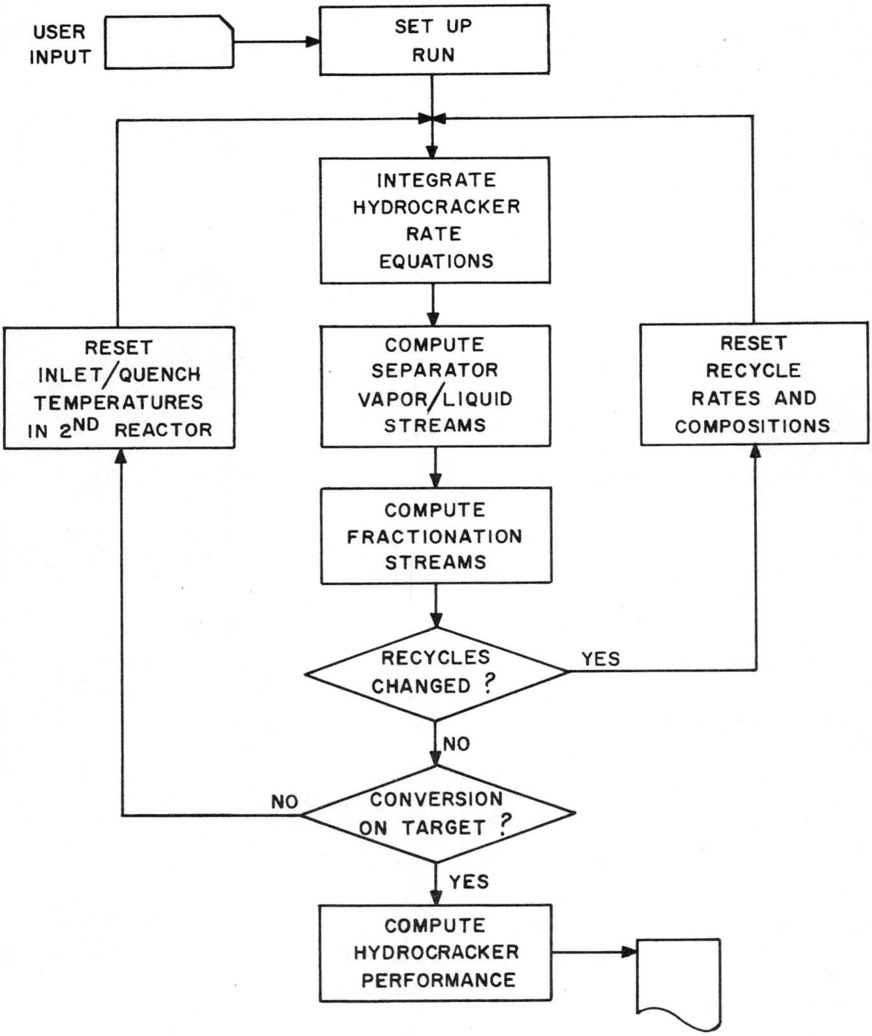

Figure 4. Program flowsheet of hydrocracker simulator

quench zones are also determined in the same inner loop using successive substitution.

In the outer loop, bed inlet temperatures in the second reactor are all adjusted by equal increments as the program converges on a target conversion per pass.

The hydrocracker simulator was also converted to subroutine form for inclusion in the nonlinear programming model of the Toledo process complex. The subroutine was considerably simplified, however, to save computer time and memory. The major differences are: (1) the fractionation section is represented by correlations instead of by a multi-stage separation model, (2) high pressure flash calculations use fixed equilibrium K-values instead of re-evaluating them as a function of composition, and (3) the beds in each reactor are treated as one isothermal bed, eliminating the need for heat balance equations.

The Reformer Simulator

The original reformer simulator was a stand-alone computer program purchased from the Pace Company of Houston, Texas. The program is based on a reaction kinetic model and predicts the performance of up to five fixed bed reactors with interheaters and a high pressure flash separator. In its present version, the stand-alone program handles a feed defined by thirty-nine components undergoing dehydrogenation, dehydrocyclization, hydrocracking, hydrodealkylation, and isomerization reactions. Reaction rate expressions reflect the dual function nature of the catalyst, but adsorption effects are neglected. The program numerically integrates the differential component and heat balances for the combined feed and recycle gas through the reactors, performs the flash calculations in the high pressure separator, and computes the properties of the C_5+ portion of the reformer product.

There are two nested loops in the program (see Figure 5):

- o Composition of the recycle gas
- o Octane number of the C_5+ reformate

In the inner loop, the composition of the hydrogen recycle gas is determined by successive substitution. If a target reformate octane is specified, an outer loop adjusts the inlet temperatures to all the reactors by equal increments until the target is reached.

The reformer simulator was converted to subroutine form for inclusion in nonlinear programming models of two refinery complexes. To save computer time and memory, the subroutine uses a linearized version of the original kinetic model, with 28 components and 33 reactions. Instead of numerical integration, the linearized model is solved analytically at constant temperature, pressure, and total mols using special subroutines to find the eigenvalues and eigenvectors of the reaction rate constant matrix.

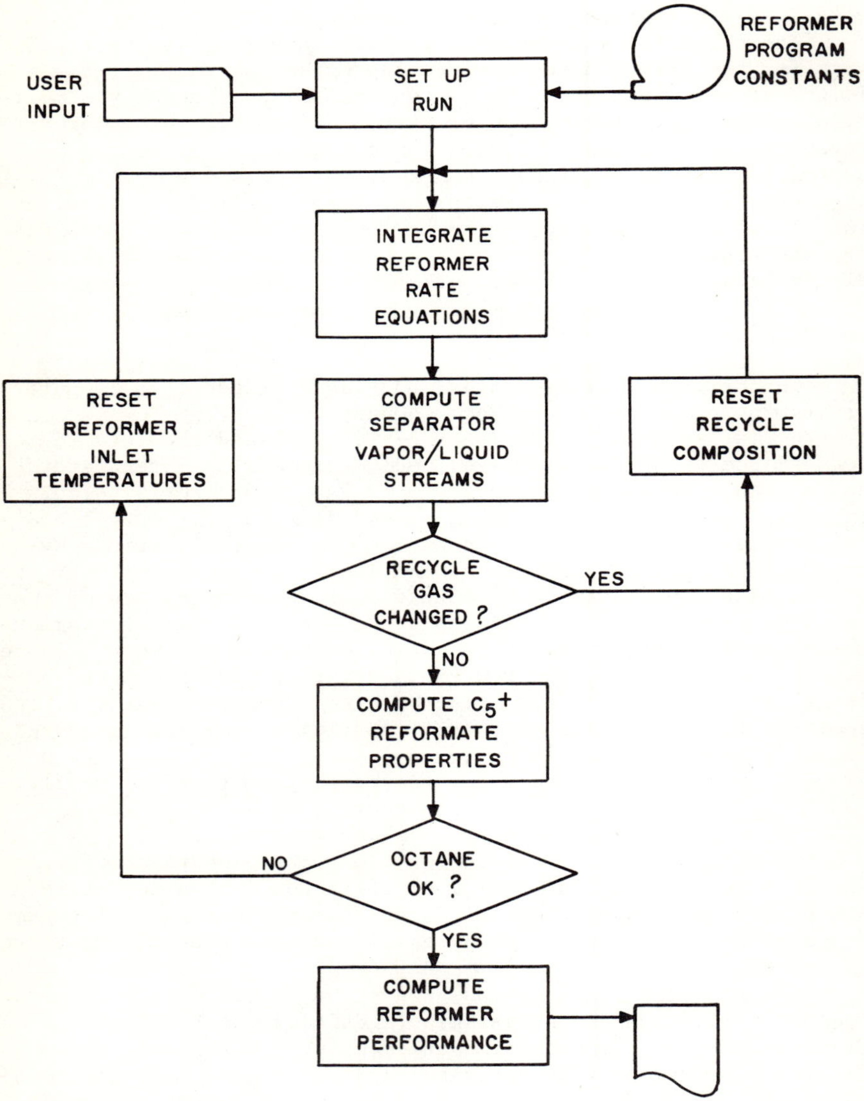

Figure 5. Program flowsheet of reformer simulator

Linking The FCC Simulator To The Preprocessor

The FCC simulator requires a description of the feed in terms of measured properties such as gravity, distillation, and Ramsbottom carbon, as well as properties derived from physical measurements such as molecular weight and percent carbon in aromatic and cycloparaffin rings. All of the necessary properties are reported in the crude assay for two standard gas oil cuts. The preprocessor creates the properties of the base FCC feed by (1) adjusting the standard cuts of each crude in the base mix according to the cutpoints specified for the atmospheric and vacuum gas oils from the crude distillation unit and (2) computing the blended properties of the base FCC feed composed of atmospheric and vacuum gas oils from each crude according to its proportion in the base mix along with a specified fraction of deasphalter overhead. Adjustment of the properties of the standard crude assay cuts is done by linear interpolation using the midpoint between initial and final boiling points as the independent variable. Properties are blended according to volume so that measurements expressed in weight percent are multiplied by specific gravity to put them on a weight per unit volume basis. Instead of blending molecular weight, the preprocessor blends specific gravity divided by molecular weight.

The FCC simulator also requires a description of the unit and its operation in terms of equipment dimensions and constraints, operating variables and unit parameters. Standard or default values for all of these are stored in the preprocessor database. The user, however, can change the values of any of the following in specifying the base and alternate FCC operations:

- o Seven operating variables including reactor bed temperature and fresh feed rate
- o Twenty-seven unit parameters including reactor efficiency and gasoline selectivity
- o Steam and operating cost factors
- o Maximum regenerator bed temperature

When the FCC simulator was linked to the preprocessor and tested on alternate operations and feed perturbations, it was discovered that the simulator would sometimes fail to converge in either the recycle loop or the conversion per pass loop. In case this happens, the user can change the initial estimates of the final boiling point of the heavy cycle oil and the conversion per pass for the base or any alternate FCC operation. To assist the user in selecting new estimates, the preprocessor prints out the following whenever the FCC simulator fails to converge:

- o Identity of operation or feed property change when failure occurred
- o Values of all operating and feed variables at the time of failure, including the last-computed heavy cycle oil endpoint and conversion per pass

To avoid convergence problems during feed perturbation runs of the FCC simulator, we have made adjustments to the size of each feed property change and reset all initial estimates before each feed perturbation to the values computed by the simulator for the base FCC operation.

Linking The Hydrocracker Simulator To The Preprocessor

The feed to the hydrocracker consists of the heavy naphtha cut from the base crude mix and light cycle oil from the base FCC operation. The hydrocracker simulator requires a description of each of these feeds in terms of the hydrocarbon components shown in Table I. Since these components are not directly measured in the crude assay nor are they predicted by the FCC simulator, special techniques were developed to estimate them from available data.

For the heavy naphtha cut from the base crude mix, the following technique is used by the preprocessor:

1. From each crude assay, the density, molecular weight, and average aromatic and cycloparaffin rings per molecule are obtained for both the saturate and aromatic fraction from each of the two gas oil cuts distilled in the laboratory.
2. These data are linearly extrapolated to the heavy naphtha cut specified for the crude distillation unit. Again, midpoint between initial and final boiling points is the independent variable.
3. Component mol fractions in each heavy naphtha cut are then determined by solution of a set of equations. The equations include aromatic and cycloparaffin ring balances as well as empirical equations relating ring distribution to ring average. Coefficients for the empirical equations were derived from mass spectrometer analytical data on a number of virgin hydrocracker feeds.
4. Blended component fractions are computed for the heavy naphtha cut from the base mix.

For the light cycle oil from the base FCC operation, a correlation predicts the component mol fractions based on various properties computed by the FCC simulator. These properties include density, molecular weight, and percent carbon as aromatic ring and cycloparaffin ring. The correlation is based on mass spectrometer data on FCC cycle oils.

The hydrocracker simulator also requires a description of the unit and its operation in terms of equipment dimensions and constraints, operating variables, and unit parameters. Standard or default values are stored in the preprocessor. The user can change the values of any of the following in specifying the base

TABLE I

HYDROCARBON COMPONENTS RECOGNIZED BY HYDROCRACKER SIMULATOR

Hydrogen
Methane
Ethane
Propane
i/n Butanes
C_5/C_6 Paraffins
Benzene
Methylcyclopentane
C_7 - C_{12} Paraffins
Single Ring Cycloparaffins
Single Ring Aromatics
$C_{13}+$ Paraffins
Double Ring Cycloparaffins
Double Ring Aromatic Cycloparaffins
Double Ring Aromatics
Multi Ring Cycloparaffins
Multi Ring Aromatic Cycloparaffins
Multi Ring Aromatics

and alternate operations:

- o Eight operating variables including conversion per pass and feed rate
- o Five unit parameters including catalyst metal and acid activities in both reactors
- o Fuel, steam and operating cost factors

No convergence problems were found in linking the hydrocracker simulator to the preprocessor and running on a wide range of alternate operations and feed property changes.

Linking The Reformer Simulator To The Preprocessor

The reformer simulator also requires a description of the feed in terms of hydrocarbon components. These are shown in Table II. Fortunately, most of these components are measured in the crude assay and are predicted by the hydrocracker simulator.

For the motor and BTX naphthas from the base crude mix, the preprocessor (1) determines the amount of each component present in the crude, (2) computes the fraction distilled overhead for each component at the naphtha cutpoints specified for the crude distillation unit, (3) constructs for each naphtha the composition of the material remaining between the initial and final boiling points, and (4) computes the blended compositions of the motor and BTX naphthas from the base crude mix.

For the heavy and light hydrocrackates, the hydrocracker simulator includes a product fractionation subroutine which distributes components between adjacent fractionator cuts using a Fenske-type formulation.

Standard or default values for the motor and BTX reforming operations are stored in the preprocessor. The user can change the values of any of the following in specifying the base or any alternate operation:

- o Six operating variables for the motor reformer, nine for the BTX reformer, including target reformate octane and feed rate
- o Three unit parameters including metal and acid activities of the catalyst in the idealized single reactor
- o Fuel, steam, operating cost, and capacity factors

To speed the operation of the preprocessor, the convergence loops in the linked reformer simulator have been modified. In the inner loop, after a certain number of iterations, the mole fractions of C_4 and heavier in the recycle gas are held constant. It was found that slight variations in these components (which are assumed to react in the reforming kinetic model) slowed down the rate of convergence without materially improving the accuracy of the results.

TABLE II

HYDROCARBON COMPONENTS RECOGNIZED BY REFORMER SIMULATOR

Hydrogen	Toluene
Methane	C_8 Paraffins
Ethane	C_8 Cycloparaffins
Propane	C_8 Aromatics
i-Butane	C_9 Paraffins
n-Butane	C_9 Cycloparaffins
Pentanes + Cyclopentane	C_9 Aromatics
Hexanes	C_{10} Paraffins
Cyclohexane	C_{10} Cycloparaffins
Methylcyclopentane	C_{10} Aromatics
Benzene	C_{11}+ Paraffins
Heptanes	C_{11}+ Cycloparaffins
Methylcyclohexane	C_{11}+ Aromatics
C_7 Cyclopentanes	

For the outer loop, the preprocessor adjusts isothermal temperature by secant method to satisfy the target octane.

Structure Of The Preprocessor-Generated LP Model

The preprocessor generates the following activities in the refinery LP matrix:

1. Activities in the crude distillation unit submatrix represent the base crude mix operation and up to five incremental crude operations. To preserve their properties for downstream processing or product blending, the distillation cuts from each crude unit activity enter separate and distinct stream balance rows.
2. Activities in the propane deasphalting (PDA) unit submatrix represent the operations on vacuum tower bottoms from the base crude mix and from the incremental crudes. The deasphalter overhead streams from all activities enter one stream balance row with common properties. Each deasphalter bottoms enter a separate row for No. 6 fuel oil blending.
3. Activities in the FCC, hydrocracker, motor and BTX reformer submatrices represent (1) feed streams from each of the crude operations and from other process units, (2) material transfers between process units simulating a change in crude distillation cutpoint, (3) base and alternate operations, and (4) changes in each feed property.

For these activities, the preprocessor computes coefficients in the following refinery LP rows:

- Fuel and steam balance rows
- Operating cost row
- Equipment usage rows limiting equipment capacity
- Stream balance rows
- Property blending rows to meet product specifications

In addition, for each of the major downstream processing units (FCC, hydrocracker, motor and BTX reformers), the preprocessor generates a set of feed property balance rows. It is through these rows and the corresponding feed property change activities that the unit performance can respond to changes in crude mix and distillation cutpoint as well as changes in properties or proportion of feed from other process units.

To see how this is accomplished, let us examine the general structure of a downstream processing unit submatrix (Figure 6).

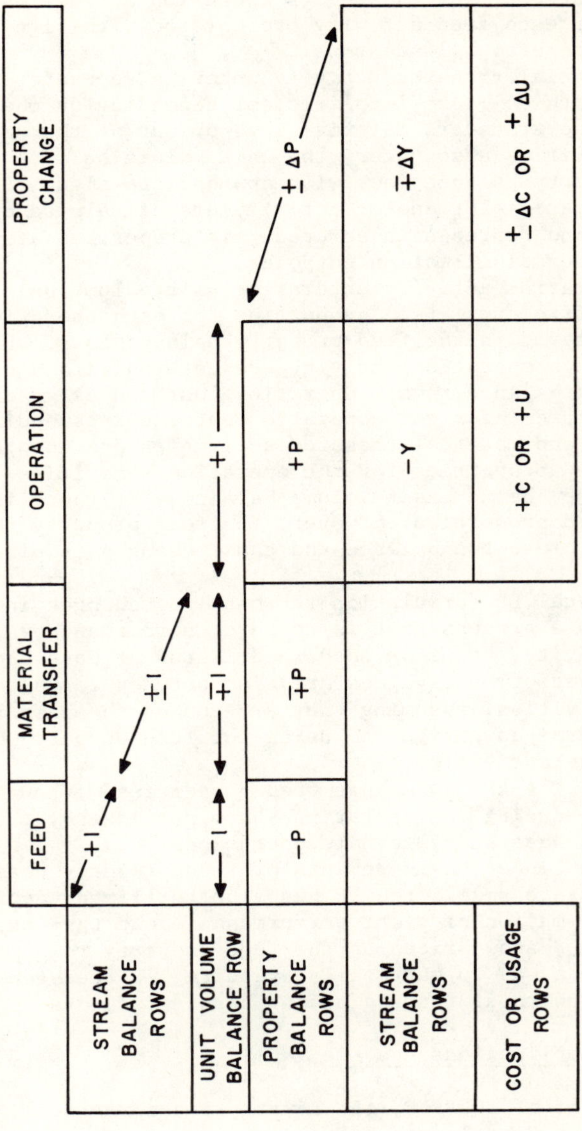

Figure 6. LP submatrix for downstream processing unit

Each feed activity enters a separate stream balance row with a +1 coefficient signifying that it is a destination for that particular stream produced elsewhere in the refinery model. It also enters the unit volume balance row with a -1 coefficient as a source of material for the unit's operation activities. The properties for each feed activity are entered with sign reversed in the feed property balance rows.

Each material transfer activity enters a separate stream balance row with a +1 or -1 coefficient depending on whether the cutpoint change transfers material into or out of the unit. Each of these activities also enters the unit volume balance row and the feed property balance rows with appropriate sign. Feed properties for a material transfer activity are calculated by the preprocessor and represent base crude mix properties at the corresponding crude distillation cutpoint.

Each operation activity enters the unit volume balance row with a +1 coefficient. Feed properties for each operation activity are entered in the feed property balance rows. Product yields from each operation activity are entered with a negative sign in stream balance rows. Operating cost and usage of fuel, steam and equipment for each operation enter corresponding rows.

The feed and material transfer activities provide the material and feed properties for the operation activities. There must, of course, be an exact volume balance. If there is an imbalance in feed properties, however, the feed property change activities must close the balance and cause changes in plant performance.

A more usual LP formulation for handling changes in feed composition to a process unit is to enter performance values for each feed activity, assuming each feed is run at base operating conditions. Performance values are also entered for alternate operation activities, assuming they are run on the base feed. There are several advantages to using the feed property balance rows and change activities:

- The LP may change feed properties without reducing the operating space provided by the base and alternate operations.
- Since the increments of feed property change are small, the LP has a better linear approximation of plant performance about the base feed.
- Shadow prices on the feed property rows provide a way to marginally evaluate new feedstocks without running the process simulators.

Preprocessor Applications And Use

Development of TOLPREP, the LP preprocessor system, was sponsored by Sun Petroleum's Toledo refinery as an improved method to perform a variety of economic studies. Since downstream refinery unit performance is affected by variations in crude type, the preprocessor system is chiefly viewed as a method of incorporating these variations into economic evaluations. Obviously, the most applicable tasks will involve crude oil allocations, evaluation and other problems involving optimum raw material slates at the refinery.

Specific crude oil evaluations have been successfully performed by the use of the system. Other applications are being developed and will be tried shortly. Although we have just started to use the system, the following points have been established:

1. Reliability and usability of the system are acceptable, although some improvements are suggested between the preprocessor and LP steps.
2. Evaluations of one specific case indicated an optimum rate for a crude as an incremental refinery charge. This rate was a unique development of the preprocessor system and accompanying LP run and would not have been properly ascertained by a conventional LP approach. Preliminary evaluations indicate that the peaking of refinery profit as incremental crude rate increases is realistic.
3. Use of the system for future raw material decisions will improve refinery profitability and/or indicate specific expansion/revamp areas to be evaluated. In general terms, the results are acceptable and believable and additional applications of the system will develop as it is put into extensive use at Toledo.

An Example

Initially, uses of the TOLPREP system have centered on crude oil evaluation. Here is a specific example of such an evaluation.

A preprocessor case was developed with capability of running both Toledo's base crude mix and the incremental under consideration. This base prepared by the preprocessor was then run as an LP case.

Incremental crude was initially precluded from the refinery charge in the first LP case and a reasonable LP representation of the Toledo refinery was generated. In this case, there was a dual price associated with the precluded incremental crude indicating that it had a significant value as a refinery charge, for the first increment of the crude.

Alternative cases were developed by using a parametric LP approach to replace part of the base crude mix with incremental crude. This approach allowed the evaluation of the incremental crude, as a refinery charge, at a series of increasing rates to the refinery and gave some surprising results. The most important result was that optimum profitability of the crude occurred at less than the maximum possible rate for that crude.

Refinery profit as measured by the LP objective function increased as the incremental crude was charged to the refinery. At a point equal to about 50% of the available incremental crude, however, the profit function began to decline. The profit at maximum incremental crude rate actually ended up at a lower value than the initial LP case value using this parametric approach.

The reasons for the rise and fall in profit were complex, but involved factors which could only have been realistically assessed by an LP model whose downstream processing yields responded to the following changes:

1. Changes in the virgin/catalytic charge ratio to the hydrocracker.
2. Changes in the crude tower cut temperatures, between reforming, hydrocracking, and FCC feeds.
3. Changes in the ratio of crude-derived to hydrocracker-derived charge to both catalytic reformers.

Properly representing the yield effects of these changes without the preprocessor-generated LP model would have been such an enormous task in terms of manpower and computer time that, frankly, it would not have been attempted.

Further Plans

The Toledo refinery plans to use the preprocessor system in a similar manner on other studies. However, the system will be implemented using an evolutionary approach. An immediate use of the preprocessor for all studies and work in this area would not be realistic.

Routine crude evaluations are presently accomplished at Toledo by a method which does not accurately consider the rate of the incremental crude. Problems can develop with the capacity and operating flexibility of various downstream units when the incremental crude rate reaches significant levels. We plan to use the preprocessor to expose and avoid such problems.

Expansion of this preprocessor-based evaluation procedure to include investigation of several crudes by the parametric method is the next appropriate application and is underway now.

Literature Cited

Manne, Alan S., "Scheduling of Petroleum Refinery Operations," Harvard Economic Studies, Vol. 48, Harvard University Press, Cambridge Mass. 1956.

Symonds, Gifford, H., "Linear Programming: The Solution Of Refinery Problems," Esso Standard Oil Company: New York, 1955

RECEIVED November 5, 1979.

26

Optimal Design of Batch Ultrafiltration-Diafiltration Process

ELMER H. HSU, STUART BACHER, and CARLOS B. ROSAS
Merck, Sharp, & Dohme Research Laboratories, Rahway, NJ 07065

Ultrafiltration, which uses selective membranes to separate materials on the basis of different molecular sizes, has become a valuable separation tool for a wide variety of industrial processes, particularly in the separation of dispersed colloids or suspended solids. In many cases where a high degree of separation is desired, a batch ultrafiltration process is used because it is the most economical in terms of membrane area.

A comprehensive mathematical analysis of batch ultrafiltration coupled with diafiltration is presented. The time cycle of the ultrafiltration-diafiltration has been correlated with the volume initially charged, percent of solute recovered, membrane area and flux. The optimum diafiltration volumes which result in the minimum cycle time or the minimum membrane area were solved for in terms of the operating conditions.

For a product recovery of 96 percent, optimum solutions were obtained and are presented graphically via design charts. The design charts plot the optimum diafiltration volume and total time cycle as a function of other operating conditions, i.e., initial volume, recovery, membrane area and flux. For a recovery other than 96 percent, the optimum solution can be obtained using the equations developed in this paper in a similar manner.

Introduction

Numerous studies relating to the application of ultrafiltration have been presented in the literature. For example, protein ultrafiltration has been studied by Kozinski (1972). Separations of complex aqueous suspensions and organic solutions have been reported by Bhattacharyya (1974, 1975). Industrial applications have been reviewed by Klinkowski (1978). Theoretical aspects of ultrafiltration have been discussed by Michaels (1968), Porter (1972), Shen (1977) and others.

Often where a high degree of separation is desired, a

0-8412-0549-3/80/47-124-457$05.00/0

batch ultrafiltration process is preferred because it is the most economical in terms of membrane area. However, due to the decrease in membrane flux as the solids concentration increases, the batch separation is normally carried out in two stages. First the dilute batch is concentrated to a specific point. Then, water is added continuously while the filtration continues at nearly constant flux. This latter filtration stage, when water is added to maintain a constant flux, is referred to as diafiltration. Proper choice of the diafiltration starting time can minimize the required membrane area, which is often the major part of the capital cost in an ultrafiltration process. Due to the complication of diafiltration, determination of the optimum ultrafiltration cycle normally requires time consuming experimental work or tedious calculations.

In this paper, complete mathematical formulations for correlating the time cycles with other operating conditions are presented. The optimum diafiltration cycle (in terms of volume fraction), and the total cycle time are solved as functions of membrane area, flux, initial volume and recovery. Convenient charts, which can be used as a guide in designing or modifying an ultrafiltration process, are provided.

Mathematical Formulation

The schematic of a typical batch ultrafiltration process used for separating suspended solids is shown in Figure 1. The operating tank is charged initially with a fixed volume of slurry. Then, the slurry is circulated continuously through the membrane at a high flow rate. A high degree of turbulence is maintained so that the concentration polarization film thickness on the membrane surface is minimized and the highest possible flux is attained, (Klinkowski, 1978). As the permeate is continuously removed, the slurry volume in the operating tank decreases. Thus, the solid concentration increases, and the flux drops accordingly. To avoid having a flux too low to be practical at very low operating volume, diafiltration is adopted toward the latter stage of the filtration. During the diafiltration phase, the rate of water added is kept equal to the flux so that the solids concentration and, thus, flux can be maintained nearly constant.

A material balance on the solution in the operating tank during diafiltration gives the following equation

$$-V_D dC = J_D A C dt \qquad (1)$$

Where V_D = volume during the diafiltration stage
C = Solute concentration
J_D = flux during diafiltration
A = membrane area
t = time

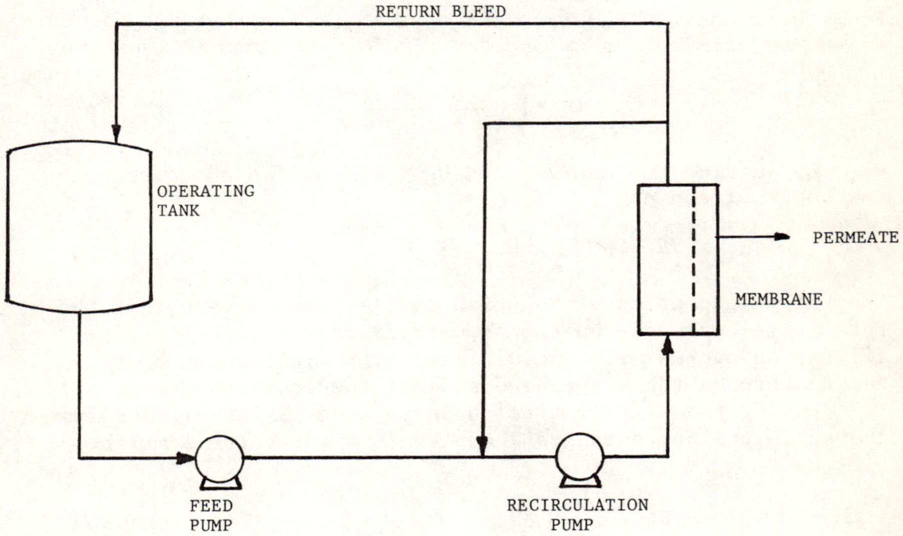

Figure 1. Flow diagram of batch ultrafiltration process

Integration of Eq. (1) from the beginning of the diafiltration phase, Tu, to the end, T, gives

$$C_f = C_o \text{ EXP } \left[\frac{A\ J_D\ (Tu - T)}{V_D}\right] \tag{2}$$

Where C_o = initial solute concentration
C_f = final solute concentration

The fraction recovery of the solute, R, is defined as

$$R = 1.0 - \frac{C_f V_D}{CoVo} \tag{3}$$

Where Vo = initial volume
From Equations (2) and (3), the total time cycle, T, can be solved by eliminating C_f and Co.

$$T = Tu + \frac{V_D}{AJ_D} \ln \left[\frac{V_D}{(1 - R)\ Vo}\right] \tag{4}$$

By letting K = AJ_D/Vo, and U_D = V_D/Vo, the above equation can be rewritten as

$$T = Tu + \frac{U_D}{K} \ln \left[\frac{U_D}{(1 - R)}\right] \tag{5}$$

From inspection of Equation (5), it can be seen that the total time cycle is the sum of ultrafiltration and the diafiltration cycles with the diafiltration cycle given by the second term on the right-hand side of the equation.

During the ultrafiltration phase, the differential volume change in the operating tank can be related to the membrane flux by

$$- dV = J\ A\ dt \tag{6}$$

The membrane flux, J, is in general a logarithmic function of the suspended solids concentration in the slurry (Michaels, 1968)

$$J = m \ln \frac{S}{So} + b \tag{7}$$

The slope, m, and the intercept, b, are constant for a given ultrafiltration process. Since the product of solids concentration and total volume in a given batch is always a constant, the flux can also be expressed as

$$J = m \ln \frac{Vo}{V} + b \tag{8}$$

Substituting the above relationship for J in Equation (6), one obtains

$$-\frac{dv}{dt} = m\ A\ \ln\frac{Vo}{V} + bA \tag{9}$$

There are four parameters (m, b, A, and Vo) in the above equation. Letting U = V/Vo, P = mA/Vo, and Q = bA/Vo, Equation (9) is reduced to a two-parameter equation

$$-\frac{dU}{dt} = P\ \ln\left(\frac{1}{U}\right) + Q \tag{10}$$

By substituting $X = P\ \ln\left(\frac{1}{U}\right) + Q$ and $\frac{dx}{dt} = -P\frac{1}{U}\frac{dU}{dt}$, Equation (10) can be transformed to

$$e^{(-Q/P)}\ P\ dt = \frac{1}{X}\ e^{(-X/P)}\ dX \tag{11}$$

Integration between the beginning of the operation (t = 0; V = Vo) to the end of the ultrafiltration phase (t = Tu; V = V_D) yields the expression for the ultrafiltration time cycle.

$$Tu = \frac{1}{P}\ e^{(Q/P)}\left[\left[\ln\left[P\ \ln(Vo/V_D) + Q\right] - \frac{P\ \ln(Vo/V_D) + Q}{P} + \frac{\left[P\ \ln(Vo/V_D) + Q\right]^2}{2\cdot 2!\ \cdot\ P^2} - \frac{\left[P\ \ln(Vo/V_D) + Q\right]^3}{3\cdot 3!\cdot\ P^3} + \ldots\right] - \left[\ln Q - \frac{Q}{P} + \frac{Q^2}{2\cdot 2!\ \cdot\ P^2} - \frac{Q^3}{3\cdot 3!\ \cdot\ P^3} + \ldots\right]\right] \tag{12}$$

Since the initial volume, recovery, membrane area and flux parameters (m, b) are all constant and known or specified, the total batch cycle time is a function of the relative diafiltration volume (Vo/Vd) only. The optimum can be determined by differentiating Equation (5) with respect to U_D and setting the result to zero, i.e.

$$\frac{\partial T}{\partial U_D} = \frac{\partial Tu}{\partial U_D} + \frac{\partial T_D}{\partial U_D} = 0 \tag{13}$$

OR $$D_1 + D_2 = 0 \tag{14}$$

Where

$$D_1 = \frac{1}{F}\ e^{(Q/P)} - \left[\frac{P}{KU_D} + \frac{1}{U_D} - \frac{2K_D(P/U_D)}{2\cdot 2!\ \cdot\ P^2} + \frac{3K_D^{\,2}(P/U_D)}{3\cdot 3!\ \cdot\ P^3} - \ldots\right] \tag{15}$$

$$D_2 = \frac{1}{K}\ \ln\left[(1-R)/U_D\right] - \frac{P}{K^2}\ \ln\left[(1-R)/U_D\right] + \frac{1}{K} \tag{16}$$

$$K = P\ \ln\frac{1}{U_D} + Q \tag{17}$$

Equation (14) is an implicit algebraic equation of the optimum relative diafiltration volume, U_D. It can be solved numerically by any one of a number of methods, e. g. Newton, Raphson Technique, (Lapidus, 1962). Once the value of U_D is determined, the optimum time cycles of the ultrafiltration and diafiltration stages, Tu and T_D, can be calculated readily from Eqs. (12) and (5).

Use of Design Charts

It has been shown that the relative diafiltration volume and the total time cycle can be solved for in terms of three parameters, P, Q, and R (Eqs. (14) and (5)). Thus, at a given recovery, R, the values of U_D and T can be obtained for various values of P and Q. Considering R as a parameter, plots of U_D and T as functions of P and Q can be made. The curves for batch ultrafiltration within the following operating ranges (P = - 0.02 to - 0.38; Q = 0.1 to 0.60; R = 0.96) are presented in this paper (Figures 2 and 3).

The use of the charts is straight forward. For each recovery, there are two corresponding charts. One determines the optimum time cycle and the other determines the optimum diafiltration volume. For the case where the initial volume, membrane flux, desired recovery, and the time cycle are specified or known, the required membrane area can be determined from the corresponding Time Cycle Chart. The procedure is to first calculate P and Q based on an assumed area. Then, the time cycle is found from the chart. Finally, the area is adjusted until the time cycle read from the chart matches the specified time cycle. Once the area is determined, the optimum relative diafiltration volume can be found from the corresponding relative diafiltration Volume Chart.

For the case when the membrane area and recovery are known, the optimum cycle time and the relative diafiltration volume, for any amount of initial charge with any flux rate can be read directly from the chart. Thus, the charts not only assist the design of new processes, they also provide a quick quide to the existing plant in adjusting the proportion of the diafiltration cycle when operating conditions are changed or the membrane flux is altered due to long-term fouling or deterioration. They eliminate the need for tedious calculation and minimize the amount of experimental work required to provide an ultrafiltration process design.

Examples

Use of the design charts is illustrated by the following two examples:

Case 1. Determination of Membrane Area. Assume a batch,

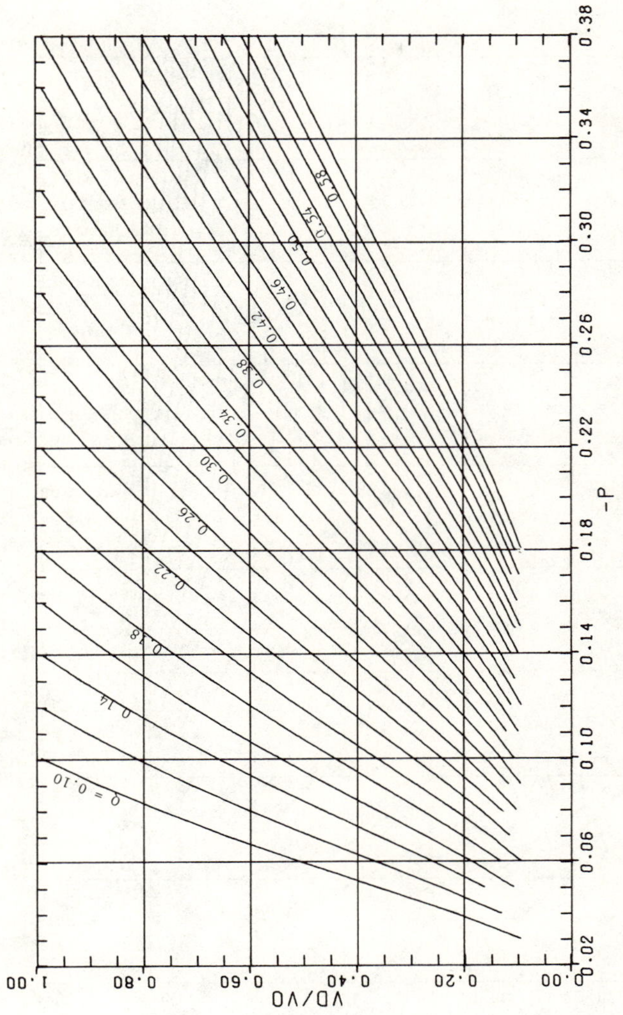

Figure 2. Volume chart (R = 0.96)

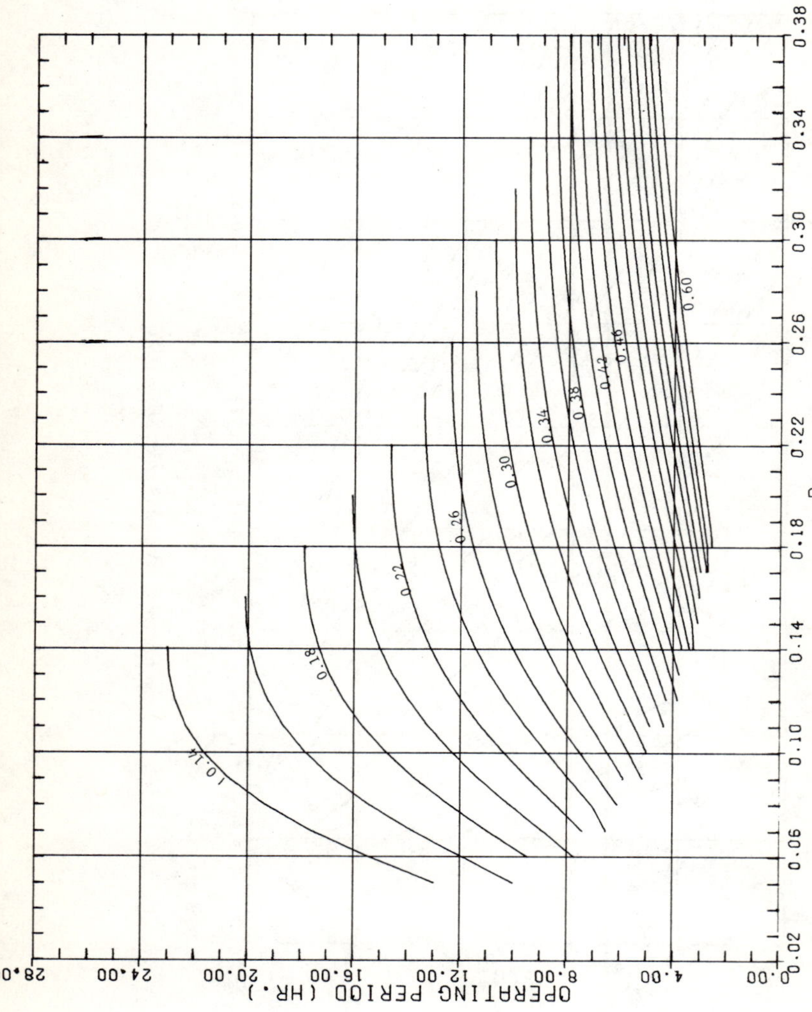

Figure 3. Time chart (R = 0.96)

11,500 gallons in volume, to be processed by the ultrafiltration-diafiltration technique. If 96% is to be recovered in 8 hours, and the flux curve follows $J = -0.32 \ln (Vo/V) + 0.88$ gal/hr - ft^2, then, the membrane area required can be determined as follows:

1) An arbitrary area is picked, $A = 2300$ ft^2.

2) P and Q are calculated

$$P = mA/Vo = -0.064$$

$$Q = bA/Vo = 0.176$$

3) T is obtained using Figure 3 , T = 11 hours

4) Since T is longer than 8 hours, the area required must be more than 2300 ft^2 (Pick A = 4600 ft^2).

5) P and Q are recalculated.

$$P = -0.128$$

$$Q = 0.352$$

6) T is obtained using Figure 3 T = 5.8 hours

7) Now, T is shorter than 8 hours, A is decreased. (Pick A = 3000).

8) Again, P and Q are calculated.

$$P = -0.0835$$

$$Q = 0.23$$

9) T is obtained using Figure 3 T = 8 hours

Therefore, the required membrane area is 3000 ft^2. Now, with P = -0.0835 and Q = 0.23, using the volume chart, (Figure 2), the optimum diafiltration volume is found to be 17%.

Case II: Determination of Optimum Cycle Time & Diafiltra-Volume. If the same process equipment is used to process the same amount of broth (Vo = 11,500 gal; A = 3,000 ft^2), but the flux has dropped 20% (as a result of fouling), then for the same recovery (96%), the total processing time and the optimum diafiltration volume can be found directly from Figures 3 and 2.

Given: $R = 0.96$
$Vo = 11{,}500$
$m = -0.32 \times 0.8 = -0.256$
$b = 0.88 \times 0.8 = 0.704$

Calculated: $P = mA/Vo = -0.0668$
$Q = bA/Vo = 0.184$

Found from Fig. 3: $T = 10.3$ hours
from Fig. 2: $V_D/Vo = 0.18$

Conclusions

The optimum time cycle and the relative diafiltration volume in the ultrafiltration-diafiltration process can be expressed as a function of three variables, P, Q, and R. P and Q are simple functions of the initial volume, membrane area, and flux ($P = mA/Vo$, $Q = bA/Vo$), and R is the solute recovery. From these, the time cycle and relative diafiltration volume (V_D/Vo) can be solved at various values of m, b, Vo, A, and R (m and b are respectively the slope and intercept of the flux, $J = m \ln Vo/V + b$). At a fixed recovery, the optimum time cycle and the relative diafiltration volume become functions of only two variables P and Q. Thus, the optimum operating condition can be simply plotted as function of P and Q. These plots, providing convenient and sufficient information, can be used as a guide in the design and operation of the ultrafiltration process.

The design charts and the examples provided in this paper illustrate the simple procedure of solving a common ultrafiltration problem. In general, when P, Q and R fall beyond the covered ranges, additional charts can be readily prepared by solving the implicit equations presented in this paper.

Nomenclature

A = Membrane area
b = Intercept of flux curve
c = Solute concentration
Cf = Final solute concentration
Co = Initial solute concentration
D_1 = Derivitive of ultrafiltration time cycle with respect to U_D
D_2 = Derivitive of diafiltration time cycle with respect to U_D
J = Flux
J_D = Flux during diafiltration phase
K = J_DA/Vo
m = Slope of flux curve
P = mA/Vo
Q = bA/Vo
R = Fraction recovery of solute

S	=	Suspended solid concentration
So	=	Initial suspended solid concentration
t	=	Time
T	=	Total time cycle
T_D	=	Time cycle of diafiltration phase
Tu	=	Time cycle of ultrafiltration phase
U	=	Volume fraction remained (V/Vo)
U_D	=	Volume fraction remained in diafiltration phase (V_D/Vo)
V	=	Solid free liquid volume in operating tank
V_D	=	Operating volume in diafiltration phase
Vo	=	Initial liquid volume
X	=	JA/Vo

Literature Cited

Bhattacharyya, Dibakar, et. al., AIChE J., 20, 1206 (1974)
Bhattacharyya, Dibakar, et. al., AIChE J., 21, 1057 (1975)
Klinkowski, P.R., Chemical Engineering, May 8, 165 (1978)
Kozinski, A.A., Lightfoot, E.N., AIChE J., 18, 1030 (1972)
Lapidus, L., "Digital Computation for Chemical Engineers", Chapter 6, McGraw-Hill, New York, 1962.
Michaels, A.S., Chem. Eng. Progr. 64 (12), 31 (1968)
Porter, M.C., Ind. Eng. Chem., Prod. Res. Dev. 11, 234 (1972)
Shen, Joseph J.S., Probstein, Ronald F., Ind. Eng. Chem., Fundam. 16, 459 (1977)

RECEIVED November 5, 1979.

Yield Optimization in a Tube-Wall Reactor

DANIEL TA-JEN HUANG and ARVIN VARMA

Department of Chemical Engineering, University of Notre Dame, Notre Dame, IN 46556

A tube-wall reactor, in which the catalyst is coated on the tube wall, is conceptually ideally suited for highly exothermic and equilibrium-limited reactions because the heat generated at the wall can be rapidly taken away by the coolant. Previous work (1) has numerically demonstrated that for highly exothermic selectivity reactions, the optimized tube-wall reactor is superior from both steady state production and dynamic points of view to the fixed-bed reactor. Also, the tube-wall reactor is being advanced as a possible reactor for carrying out methanation in coal gasification plants (2). From a reaction engineering point of view, it therefore seems appropriate to analyze the reactor for the analytically resolvable case of complex first-order isothermal reactions.

In the case of methanation, the tube-wall reactor (3,4) has the advantage of being able to accommodate feeds containing as much as 25% CO without recycle and yet operate under nearly isothermal conditions which greatly eases reactor control. Mathematically speaking, an isothermal model is one which uses only the material balance equations and drops the energy balance equation. As shown by Smith and Carberry (5), using the tube-wall reactor for the highly exothermic oxidation of naphthalene on V_2O_5 catalyst, dropping the energy balance equation still provides a rather accurate picture of the reactor conversion and yield behavior. Also Senkan et al. (6) and Schehl et al. (3) have shown that for methanation, the material balance equation can be solved independently of the energy balance equation in diffusion-limited cases, because the effects of temperature variation on gas properties essentially cancel each other. It is therefore justified to consider the isothermal model for the purpose of yield optimization.

It is our purpose to optimize the performance of a tube-wall reactor with the reaction rate constants as parameters, and also with respect to some other key design parameters.

Tube-Wall Reactor Model

The following assumptions have been made:

0-8412-0549-3/80/47-124-469$05.00/0

(1) Steady state operation is achieved.

(2) To enhance mass transfer to and from the wall surface, the tube is packed with inert pellets. Thus, plug-flow in the axial direction is assumed and only radial dispersion is considered, which are valid not only for highly turbulent flow but even for relatively low Reynolds number flow compared to the empty tube.

(3) The catalyst activity is assumed to be constant. Because of excellent temperature control achievable in tube-wall reactors, thermal sintering of the catalyst is less likely than in a fixed-bed reactor. Some methods of avoiding catalyst deactivation or reducing the deactivation rate for methanation are described by Mills and Steffgen (7).

(4) The physical properties of the gas and the mass transfer coefficient are assumed to be independent of position in the reactor.

(5) Since the catalyst film coated is usually very thin, the catalyst may be considered to be only an active superficial surface and thus the gas/catalyst interface is assumed to coincide with the wall.

The following first-order consecutive-parallel reaction scheme is used, which is in accord with the oxidation of naphthalene (1):

$$A \xrightarrow{k_1} B \xrightarrow{k_2} C \quad ; \quad A \xrightarrow{k_3} D$$

If C and D are the same, it is a triangular reaction, e.g., in the silver-catalyzed partial oxidation of ethylene. By setting k_3 equal to zero, we get a consecutive reaction:

$$A \xrightarrow{k_1} B \xrightarrow{k_2} C$$

And by setting k_2 equal to zero instead, we get a parallel reaction:

$$A \xrightarrow{k_1} B \quad ; \quad A \xrightarrow{k_3} D$$

In all cases, B is the desired product. In the following derivations, only the consecutive-parallel reaction scheme is considered, since the other two are just special cases of this more general scheme. With the list of symbols given at the end,

At the Catalyst Surface

$$C_{wA} = \frac{C_A}{1 + (k_1+k_3)/k_{gw}} \tag{1}$$

$$C_{wB} = \frac{C_B + (k_1/k_{gw})C_{wA}}{1 + k_2/k_{gw}} \tag{2}$$

Material Balance

$$v \frac{\partial C_i}{\partial z^*} = D_{eri} \left(\frac{\partial^2 C_i}{\partial r^{*2}} + \frac{1}{r^*} \frac{\partial^2 C_i}{\partial r^*} \right) \quad ; \quad i = A, B \tag{3}$$

$$\text{at} \quad z^* = 0, \quad C_i = C_{i0} \tag{4a}$$

$$\text{at} \qquad r^* = 0, \qquad \frac{\partial C_i}{\partial r^*} = 0 \tag{4b}$$

at the gas/catalyst interface, i.e., at $r^* = R$,

$$- D_{eri} \frac{\partial C_i}{\partial r^*} = k_{gw}(C_i - C_{wi}) \tag{4c}$$

With the dimensionless variables defined in the list of symbols at the end, and substituting the expressions (1) and (2) for C_{wA} and C_{wB} into the boundary condition (4c), we obtain the following dimensionless equations:

For species A

$$\frac{\partial u_A}{\partial z} = \frac{1}{Pe_A} \left(\frac{\partial^2 u_A}{\partial r^2} + \frac{1}{r} \frac{\partial u_A}{\partial r} \right) \tag{5}$$

$$\text{at} \qquad z = 0, \qquad u_A = 1 \tag{5a}$$

$$\text{at} \qquad r = 0, \qquad \frac{\partial u_A}{\partial r} = 0 \tag{5b}$$

at $r = a$ (gas/catalyst interface),

$$\frac{\partial u_A}{\partial r} + K u_A = 0 \tag{5c}$$

where

$$K = \frac{k_{gw} d_p}{D_{erA}} \left[1 - \frac{1}{1 + (k_1 + k_3)/k_{gw}} \right] \tag{6}$$

For species B

$$\frac{\partial u_B}{\partial z} = \frac{1}{Pe_B} \left(\frac{\partial^2 u_B}{\partial r^2} + \frac{1}{r} \frac{\partial u_B}{\partial r} \right) \tag{7}$$

$$\text{at} \qquad z = 0, \qquad u_B = u_{BO} \tag{7a}$$

$$\text{at} \qquad r = 0, \qquad \frac{\partial u_B}{\partial r} = 0 \tag{7b}$$

$$\text{at} \qquad r = a, \qquad \frac{\partial u_B}{\partial r} + K_1 u_B = K_2 u_A \tag{7c}$$

where

$$K_1 = \frac{k_{gw} d_p}{D_{erB}} \left(1 - \frac{1}{1 + k_2/k_{gw}} \right) \tag{8}$$

$$K_2 = \frac{k_1 d_p}{D_{erB}(1 + k_2/k_{gw})[1 + (k_1 + k_3)/k_{gw}]} \tag{9}$$

The correlation for the pressure drop through the packed bed is the modified Ergun equation by Tallmadge, as described by Smith and Carberry (1):

$$\Delta P = \frac{\rho v_s^2 (1-\varepsilon) b M^*}{\varepsilon^3} \left[\frac{150(1-\varepsilon)M^*}{Re_s} + 4.2 \left(\frac{1-\varepsilon}{Re_s} \right)^{1/6} \right] \quad \text{for } 10^{-1} < Re_s < 10^5 \tag{10}$$

where

$$M^* = 1 + \frac{1}{3(1-\varepsilon)a} \tag{10a}$$

The mass transfer coefficient at the wall, k_{gw}, is calculated by the Yagi and Wakao (8) correlation:

$$k_{gw} = \frac{0.2 v_s}{Sc^{2/3} Re_s^{1/5}} \quad \text{for } Re_s > 40 \tag{11}$$

$$k_{gw} = \frac{0.6 v_s}{Sc^{2/3} Re_s^{1/2}} \quad \text{for } Re_s < 40 \tag{12}$$

Care should be taken here because plug-flow has been assumed in the model which may not hold for small Reynolds numbers, or when $R/d_p \lesssim 4$.

Analytic Solutions

Reactant (A) Conversion

The method of Finite Fourier Transform (9) is applied to solve Eq. (5) with boundary conditions (5a), (5b), and (5c). The resulting dimensionless concentration profile of reactant A is:

$$u_A(z,r) = \sum_{n=1}^{\infty} \frac{2KJ_o(\sqrt{\lambda_n}\, r)}{a(K^2+\lambda_n) J_o(\sqrt{\lambda_n}\, a)} \exp\left(- \frac{\lambda_n z}{Pe_A} \right) \tag{13}$$

where λ_n is the n-th eigenvalue of the problem

$$\frac{d}{dr}\left(r \frac{d\phi}{dr}\right) = - \lambda r \phi \tag{14}$$

$$\text{at} \quad r = 0, \quad \frac{d\phi}{dr} = 0, \quad \text{or} \quad \phi \text{ finite} \tag{14a}$$

$$\text{at} \quad r = a, \quad \frac{d\phi}{dr} + K\phi = 0 \tag{14b}$$

and satisfies the transcendental equation

$$K J_0 (\sqrt{\lambda_n}\, a) = \sqrt{\lambda_n}\, J_1 (\sqrt{\lambda_n}\, a) \qquad n = 1, 2, \ldots\ldots \tag{15}$$

where $J_k(x)$ is the Bessel function of the first kind of order k. The radially averaged conversion at the reactor exit, z = b, is

$$X(b) = 1 - \frac{4K^2}{a^2} \sum_{n=1}^{\infty} \frac{1}{(K^2+\lambda_n)\lambda_n} \exp\left(-\frac{\lambda_n b}{Pe_A}\right) \tag{16}$$

Product (B) Yield

Substituting Eq. (13) into the boundary condition (7c), we get:

$$\text{at} \quad r = a, \quad \frac{\partial u_B}{\partial r} + K_1 u_B = K_2 \sum_{n=1}^{\infty} \frac{2K}{a(K^2+\lambda_n)} \exp\left(-\frac{\lambda_n z}{Pe_A}\right) \tag{17}$$

The method of Finite Fourier Transform with step and impulse responses (9) is used to solve Eq. (7) with boundary conditions (7a), (7b), and (17). The resulting dimensionless concentration profile of the desired product B is:

$$u_B(z,r) = \frac{4KK_1K_2}{a^2 Pe_B}\left(u_{BO} - \frac{1}{K_1}\right) \sum_{n=1}^{\infty}\sum_{m=1}^{\infty} \frac{\mu_m}{(K_1^2+\mu_m)(K^2+\lambda_n)\left(\frac{\mu_m}{Pe_B} - \frac{\lambda_n}{Pe_A}\right)}$$

$$\times \frac{J_o(\sqrt{\mu_m}\, r)}{J_o(\sqrt{\mu_m}\, a)} \left[\exp\left(-\frac{\mu_m z}{Pe_B}\right) - \exp\left(-\frac{\lambda_n z}{Pe_A}\right)\right] \tag{18}$$

where μ_m satisfies

$$K_1 J_o(\sqrt{\mu_m}\, a) = \sqrt{\mu_m}\, J_1(\sqrt{\mu_m}\, a) \qquad m = 1, 2, \ldots. \tag{19}$$

The radially averaged yield at the reactor exit is

$$Y(b) = \frac{8KK_1^2K_2}{a^3 Pe_B}\left(u_{BO} - \frac{1}{K_1}\right) \sum_{n=1}^{\infty}\sum_{m=1}^{\infty} \frac{\exp\left(-\frac{\mu_m b}{Pe_B}\right) - \exp\left(-\frac{\lambda_n b}{Pe_A}\right)}{(K_1^2+\mu_m)(K^2+\lambda_n)\left(\frac{\mu_m}{Pe_B} - \frac{\lambda_n}{Pe_A}\right)} \tag{20}$$

Numerical Results and Discussion

A numerical example was used to examine the desired product yield versus reactant conversion behavior in a tube-wall reactor described above. Values assumed for the physical and chemical properties and the operating conditions are:

$Pe_A = 10$	$Der_i = d_p v/Pe_i$; $i = A,B$
$Pe_B = 10$	$D = 0.2\ cm^2\ sec^{-1}$
$\mu = 2.48 \times 10^{-4}\ g\ cm^{-1}\ sec^{-1}$	$\varepsilon = 0.5$
$\rho = 9.22 \times 10^{-4}\ g\ cm^{-3}$	$k_1 = 1\ cm\ sec^{-1}$
$d_p = 0.635$ cm (0.25 in)	$u_{BO} = 0$

Where the values for μ and ρ are taken as those of air at 200°C and 1 atm. u_{BO} is assigned a value of zero which means that no product B exists in the feed stream. The plug-flow velocity, v, is determined by the choice of the Reynolds number. The values of k_2 and k_3 depend on the chosen ratios of these rate constants with respect to k_1.

In the calculations, the following controlling policy was used for evaluating the infinite series in the expressions of X(b) and Y(b): For the series $\sum_{n=1}^{\infty} C_n$, the computation was continued till

$$\left| \frac{C_{N+1}}{\sum_{n=1}^{N} C_n} \right| < \text{tolerance (e.g., } 10^{-8}\text{)}.$$

It was found that the convergence is usually very rapid. The yield Y(b) versus conversion X(b) functionality is thus determined and is described below for the three different cases.

The Consecutive-Parallel Reaction:

$$A \xrightarrow{k_1} B \xrightarrow{k_2} C; \; A \xrightarrow{k_3} D$$

The parameters of the base case, which are fixed for every figure except the specific parameter varied, are assigned the following values:

$$\frac{k_2}{k_1} = 1 \;, \quad \frac{k_3}{k_1} = 1 \;, \quad a \equiv \frac{R}{d_p} = 10, \quad Re_s = 100$$

The following observations can be made from the calculations:

1. From Figures 1a and 2a, as k_3/k_1 decreases Y_{max} increases at the expense of smaller conversion, and also at the expense of a longer reactor and higher pressure drop. From Eq. (10), pressure drop is a linear function of the reactor length b, and a non-linear function of Re_s. Thus if Re_s is fixed, ΔP_m is a linear function of b_m, and longer reactors mean higher pressure drop, where ΔP_m denotes ΔP at b_m.

2. From Figures 1b and 2b, reducing k_2/k_1 will increase Y_{max} largely with an increase of X_m until a point is reached, from which further reducing k_2/k_1 does not increase Y_{max} much, but makes X_m so close to 1 that the required reactor length b_m (and hence ΔP_m) are very large.

3. From Figures 1c and 2c, we note that as R/d_p decreases, Y_{max} increases slightly with very small increase of X_m but at the benefit of largely reducing b_m. Results of the type shown in Figure 2c are valuable in making practical judgements concerning reactors, for they show the interplay between reactor diameter, length, pressure drop and achievable product yield.

4. From Figures 1d and 2d, as we increase Re_s, Y_{max} increases slightly and rapidly reaches a point, from which further increase of Re_s will only add to the reactor length needed and largely increase ΔP_m. This is because as Re_s increases the gas residence time decreases relatively, and thus a longer reactor is required to achieve equivalent conversion.

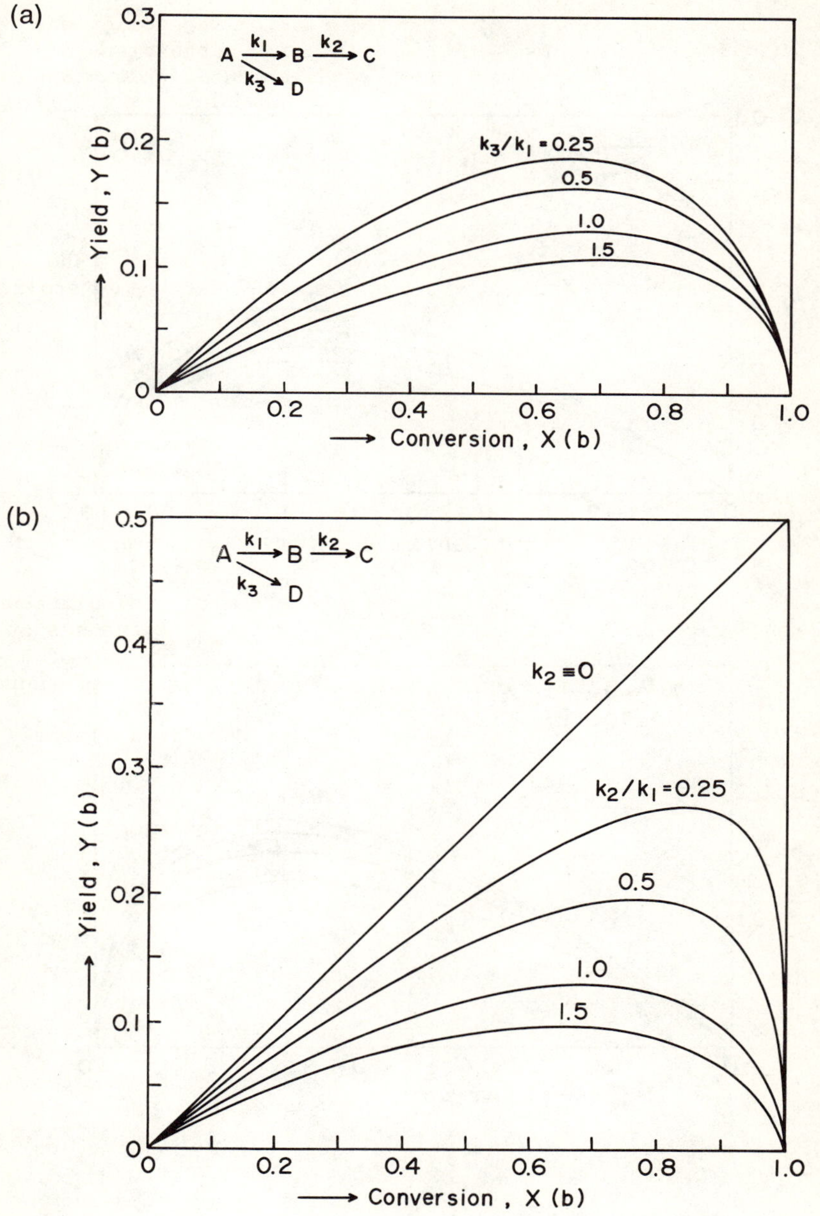

Figure 1. Product (B) yield versus reactant (A) conversion, varying (a) k_3/k_1, (b) k_2/k_1, (c) R/d_p, and (d) Re_s

(c)

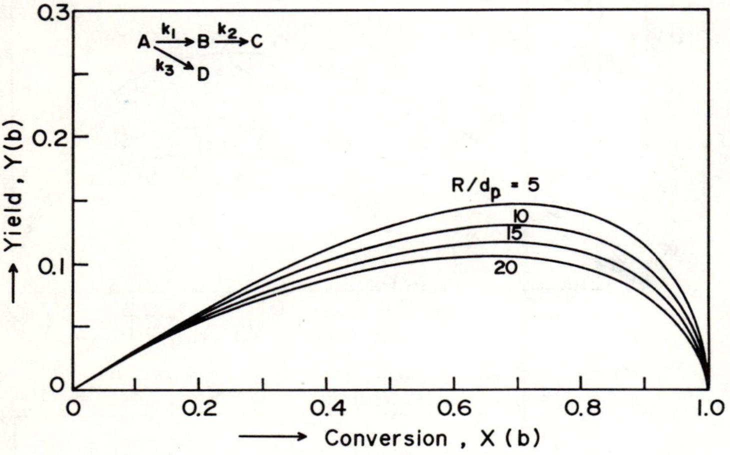
0.3
0.2
0.1
0
A
k1
B
k2
C
k3
D
Yield, Y(b)
R/dp = 5
10
15
20
0
0.2
0.4
0.6
0.8
1.0
Conversion, X (b)

(d)

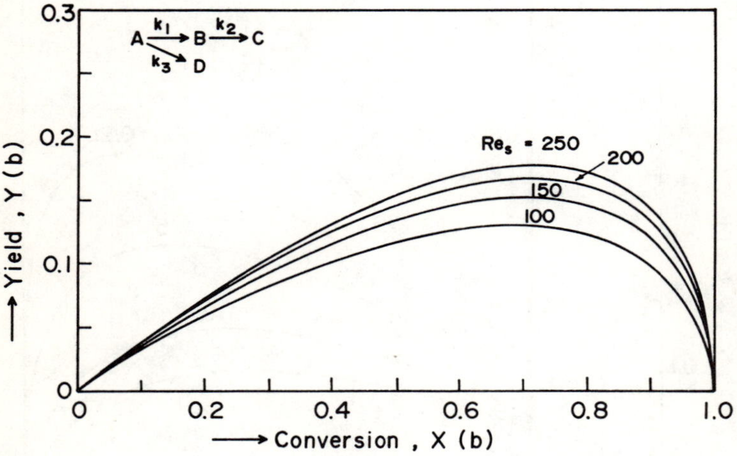
0.3
0.2
0.1
0
A
k1
B
k2
C
k3
D
Yield, Y (b)
Res = 250
200
150
100
0
0.2
0.4
0.6
0.8
1.0
Conversion, X (b)

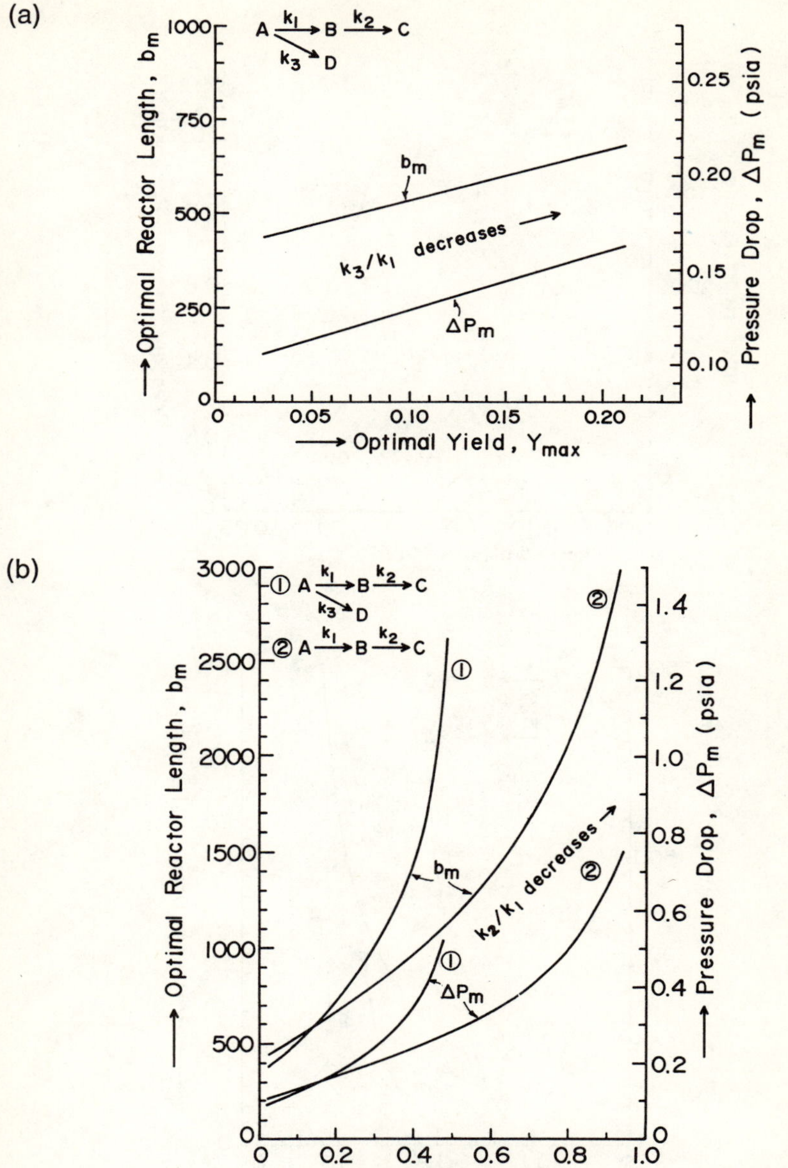

Figure 2. Optimal reactor length and corresponding pressure drop versus optimal yield, (a) k_3/k_1, (b) k_2/k_1, (c) R/d_p, and (d) Re_s as implicit parameter

(c)

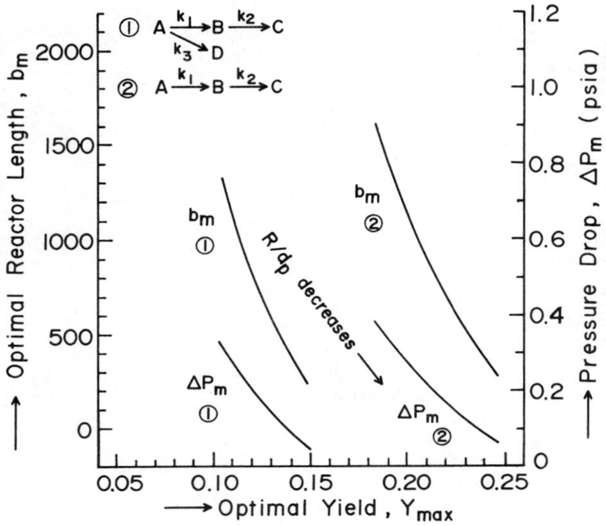
① A —k1→ B —k2→ C
k3 → D
② A —k1→ B —k2→ C
2000
1500
1000
500
0
→ Optimal Reactor Length , bm
1.2
1.0
0.8
0.6
0.4
0.2
0
→ Pressure Drop , ΔPm (psia)
bm ①
bm ②
R/dp decreases
ΔPm ①
ΔPm ②
0.05
0.10
0.15
0.20
0.25
→ Optimal Yield , Ymax

(d)

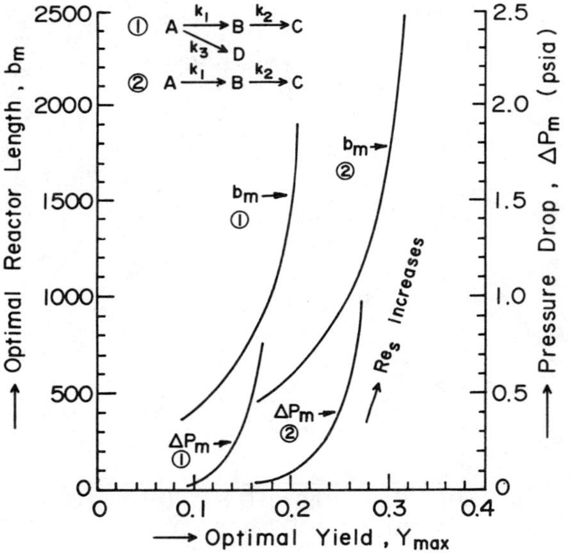
① A —k1→ B —k2→ C
k3 → D
② A —k1→ B —k2→ C
2500
2000
1500
1000
500
0
→ Optimal Reactor Length , bm
2.5
2.0
1.5
1.0
0.5
0
→ Pressure Drop , ΔPm (psia)
bm ①
bm ②
ΔPm ①
ΔPm ②
Res Increases
0
0.1
0.2
0.3
0.4
→ Optimal Yield , Ymax

It is thus clear that larger maximum desired product yield, Y_{max}, may be obtained by reducing k_3/k_1, k_2/k_1, or R/d_p, and increasing Re_s. But it is better to keep Re_s within a reasonable range from the point of reactor length and pressure drop. It should be noted that although higher Re_s will provide better cooling effect, it is at the expense of a longer reactor. Also note that plug-flow has been assumed, and in a packed reactor turbulent flow occurs for Re_s greater than 100, and for the transition zone, $10 < Re_s < 100$.

It is thus found that the most benefit of a higher Y_{max} is obtained by reducing k_2/k_1, then by reducing R/d_p and k_3/k_1, and relatively little from increasing Re_s. Since k_2/k_1 and k_3/k_1 are usually functions of temperature, the selection of an optimal operating temperature is very important for yield optimization.

The Consecutive Reaction: $A \xrightarrow{k_1} B \xrightarrow{k_2} C$.

This is a special case of the consecutive-parallel reaction, and its behavior is similar to it -- as shown in Figures 2b,c,d. However, the values of R/d_p and Re_s have little effect on X_m in this case.

The Parallel Reaction: $A \xrightarrow{k_1} B$; $A \xrightarrow{k_3} D$

For this case, since the desired product B does not react further, its yield versus conversion depends only on the ratio k_3/k_1:

$$\frac{Y(b)}{X(b)} = \frac{k_1}{k_3} \tag{21}$$

We therefore have to solve only for X(b), which is exactly the same as Eq. (16). We see that yield increases as k_3/k_1 decreases, and have found that the values of R/d_p and Re_s have no effect on the yield versus conversion functionality. It is apparent that no local maximum occurs on the Y(b) versus X(b) plot.

Concluding Remarks

Analytic solutions for reactant (A) conversion and desired product (B) yield for the first-order reaction schemes:

$A \rightarrow B$, $A \rightarrow C$	Parallel
$A \rightarrow B \rightarrow C$	Consecutive
$A \rightarrow B \rightarrow C$, $A \rightarrow D$	Consecutive-parallel

occurring isothermally in a packed tube-wall reactor are reported. These solutions take the form of rapidly convergent infinite series of Bessel functions. The yield versus conversion functionality is thus deduced, from which the reactant conversion (and hence the reactor length) leading to an optimum yield of B is identified. It is found that selection of the operating temperature is very important for the purpose of yield optimization. The design of the optimal ratio of the reactor diameter (2R) to the inert particle diameter is also important for both the consecutive-parallel and consecutive reactions. This ratio can be varied by changing either particle or reactor diameter. Obviously, larger packing particles will lead to lower pressure drop but poorer effective radial diffusivity.

List of Symbols

a	R/d_p
b	L/d_p
C	concentration
d_p	pellet diameter
D_{er}	effective radial mass diffusivity
D	molecular diffusivity
k_i	specific rate constant
k_{gw}	gas-wall mass transfer coefficient
L	reactor length
Pe	Peclet number, $d_p v/D_{er}$
r^*	radial coordinate
r	r^*/d_p
R	inside tube radius
Re_s	Reynolds number, $d_p v_s \rho/\mu$, at reactor inlet
Sc	Schmidt number, $\mu/\rho D$
v	interstitial gas velocity
v_s	superficial velocity, εv
u	C/C_{A0}
X	$\int_0^a [1-u_A(z,r)]rdr/\int_0^a rdr$
Y	$\int_0^a u_B(z,r)rdr/\int_0^a rdr$
Y_{max}	maximum Y at reactor exit
z^*	axial coordinate
z	z^*/d_p

Greek Symbols

ε	void fraction
μ	viscosity
ρ	gas density

Subscripts

0	inlet of reactor, z = 0
A	reactant A
B	desired product B
m	corresponding to Y_{max}
w	wall

Acknowledgment

The Union Oil Fellowship in Reaction Engineering and a Reilly tuition scholarship for D. T.-J. Huang are gratefully acknowledged.

Literature Cited

1. Smith, T. G.; Carberry, J. J. Chem. Eng. Sci., 1975, 30, 221.
2. Haynes, W. P.; Schehl, R. R.; Weber, J. K.; Forney, A. J. Ind. Eng. Chem. Proc. Des. Dev., 1977, 16, 113.
3. Schehl, R. R.; Weber, J. K.; Kuchta, M. J.; Haynes, W. P. Ind. Eng. Chem. Proc. Des. Dev., 1977, 16, 227.
4. Pennline, W. H.; Schehl, R. R.; Haynes, W. P. Ind. Eng. Chem. Proc. Des. Dev., 1979, 18, 156.
5. Smith, T. G.; Carberry, J. J. Chem. Eng. Sci., 1976, 31, 1071.
6. Senkan, S. M.; Evans, L. B.; Howard, J. B. Ind. Eng. Chem. Proc. Des. Dev., 1976, 15, 184.
7. Mills, G. A.; Steffgen, F. W. Catalysis Reviews, 1973, 8, 159.
8. Yagi, S.; Wakao, N. AIChE J., 1959, 5, 79.
9. Amundson, N. R. Ind. Eng. Chem., 1956, 48, 26.

RECEIVED November 5, 1979.

Modeling, Simulation and Control of an Extractive Distillation Column

E. D. GILLES, B. RETZBACH, and F. SILBERBERGER

Universität Stuttgart, Institut für Systemdynamik und Regelungstechnik, Stuttgart, West Germany

The large scale distillation column shown in Figure 1 is part of an extractive distillation plant employed in the waste water refinement process at HOECHST AG, Frankfurt/Main.

The binary azeotrope isopropanol/water F_A is separated by using the extractant glycol F_E. Under normal operating conditions, the distillate D has to be of pure isopropanol, whereas the bottom product A must not contain any of it. This column was first investigated by Kunstmann (1), who carried out extensive experiments on the plant. In these experiments a high temperature gradient, a so-called "temperature front" was observed within a small area of the column. It could be shown that a strong dependency exists between the product compositions and the locus of the temperature front. Therefore, three temperatures were measured at the steady-state locus of the front, averaged and fed to a PID-controller which effected the heating steam flow rate u.

Because of its high sensitivity to feed disturbances, this control system was not reliable under all operating conditions and the demand for a better configuration arose. The development of this new control concept is treated in the following.

Mathematical Model for Simulation

Digital simulation based on an appropriate mathematical model is the best tool in order to understand the dynamic behaviour of the column. We used a set of equations which models every stage as an heterogeneous two-phase system consisting of the two totally mixed homogeneous phases vapour and liquid (Figure 2).

Assuming constant molal holdup n', the balance equations of mass and enthalpy h on the k-th stage read

$$0 = L_{k-1} - L_k - \sum_{i=1}^{I} j_{ik} + F_k \tag{1}$$

$$n'\frac{dx_{ik}}{dt} = L_{k-1}x_{ik-1} - L_k x_{ik} - j_{ik} + F_k x_{Fik} \; ; \quad i=1,\ldots,I-1 \tag{2}$$

0-8412-0549-3/80/47-124-481$05.00/0

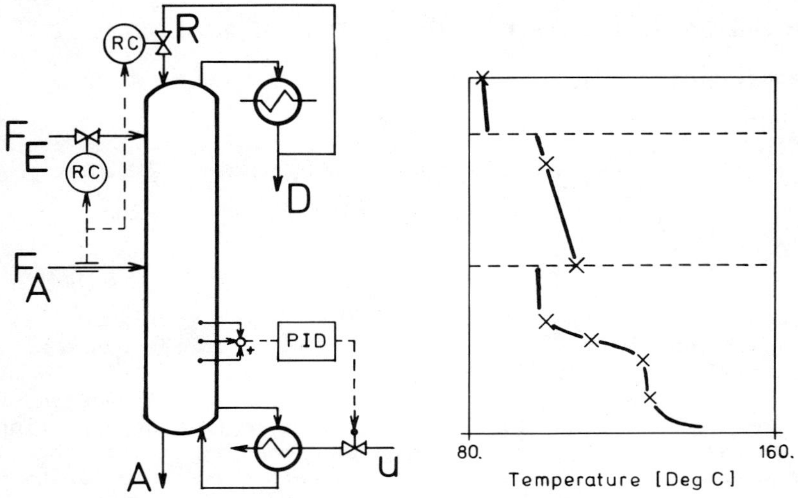

Figure 1. Extractive distillation column and its temperature profile

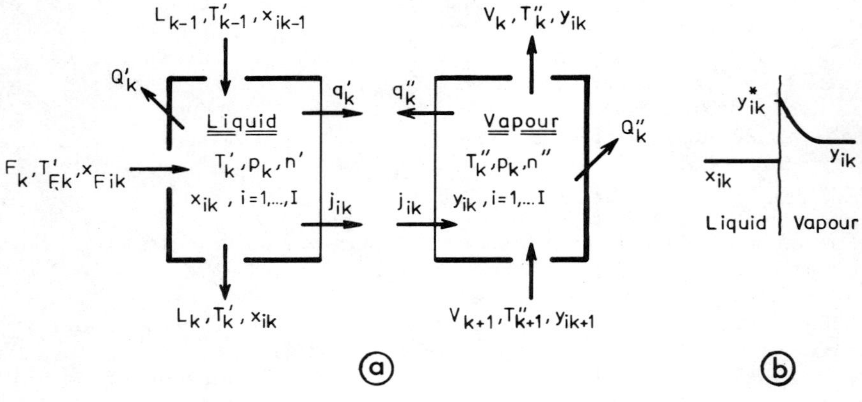

Figure 2. Heterogeneous two-phase system (a) and assumed mole fractions near phase boundary (b)

$$n'\frac{dh'_k}{dt} = L_{k-1}h'_{k-1} - L_k h'_k - \sum_{i=1}^{I} j_{ik}\overline{h}'_{ik} + F_k h'_{Fk} - q'_k - Q'_k \quad , \tag{3}$$

where the j_{ik}, $i=1,\ldots,I$ denote the interphasial mass transfer rates and q'_k, Q'_k stand for the heat transfer rates.

The molal holdup of the vapour phase is very small, $n'' \ll n'$, therefore

$$0 = V_{k+1} - V_k + \sum_{i=1}^{I} j_{ik} \tag{4}$$

$$0 = V_{k+1}y_{ik+1} - V_k y_{ik} + j_{ik} \qquad i=1,\ldots,I-1 \tag{5}$$

$$0 = V_{k+1}h''_{k+1} - V_k h''_k + \sum_{i=1}^{I} j_{ik}\overline{h}''_{ik} - q''_k - Q''_k \quad . \tag{6}$$

The pressure drop Δp_k of a stage is mainly effected by the vapour flow V_k. Taking only this flow effect into consideration, the pressure of the k-th stage is approximately

$$p_k = p_o + \xi \sum_{m=1}^{k} V_m^2 \quad , \tag{7}$$

where p_o denotes the top pressure.

Because our main interest is concerned with the mass exchange between vapour and liquid, further simplifications are introduced in the energy balance equations (3) and (6). The molal enthalpy dh'_k is expressed by dT'_k, dp_k and dx_{ik}. Now using the mass balance equations and neglecting the heats of mixing and the effect of pressure on the molal enthalpy yields

$$n'c'_{pk}\frac{dT'_k}{dt} = L_{k-1}c'_{pk-1}(T'_{k-1}-T'_k) + F_k c'_{pFk}(T'_{Fk}-T'_k) - q'_k - Q'_k \tag{8}$$

$$0 = -V_{k+1}c''_{pk+1}(T''_k - T''_{k+1}) - q''_k - Q''_k \quad . \tag{9}$$

Energy- and mass transfer between the phases are connected by the energy balance equation of the phase boundary

$$q''_k + q'_k = \sum_{i=1}^{I} (\overline{h}''_{ik} - \overline{h}'_{ik}) j_{ik} = \sum_{i=1}^{I} r_{ik} j_{ik} \quad . \tag{10}$$

Using the concept of irreversible thermodynamics, the mass transfer rates j_{ik} are formulated as linear functions of temperature gradient and gradients of chemical potentials. These quantities are considered driving forces. A simplified expression is obtained if the heat exchange is assumed to be perfect and if all mass transfer resistances may be concentrated on the vapour phase (Figure 2). With these assumptions it is $T''_k = T'_k = T_k$ and the stage temperature may be computed from the liquid composition by the well known boiling point procedure using WILSON-equation to determine liquid phase activity. This procedure also yields the vapour equilibrium composition y^*_{ik}, i=1,...,I. Now the change of energy stored in the holdup is a function of composition changes. This amount is neglected because it is small compared with those impressed from the surroundings or interchanged between the phases. With these simplifications using eqs.(8),(9) the left hand side of eq.(10) $\Delta E_k = q'_k + q''_k$ can be obtained as

$$\Delta E_k = L_{k-1}c'_{pk-1}\Delta T_{k-1} - V_{k+1}c''_{pk+1}\Delta T_k + F_k c'_{pFk}\Delta T_{Fk} - Q_k \quad . \qquad (11)$$

Here the ΔTs denote the temperature differences $\Delta T_j = T_j - T_{j+1}$. The loss of heat to the surroundings is summarized by $Q_k = Q'_k + Q''_k$. The mass transfer is caused by the concentration gradients $y^*_{ik} - y_{ik}$, i=1,...,I and the surplus of energy ΔE_k. Therefore, the mass transfer rates are written assuming a diagonal matrix of phenomenological coefficients α_i

$$j_{ik} = \alpha_i(y^*_{ik} - y_{ik} + \Delta F_{Ek}) \qquad i=1,\dots,I \quad . \qquad (12)$$

where ΔF_{Ek} represents the influence of energy and is computed from eqs.(10), (11).

Under normal operating conditions and with good thermal isolation ΔE_k is very small except around a feed stage. This allows us to neglect this quantity and also ΔF_{Ek} on every stage except the feed stages. Hence everywhere else ΔE_k and ΔF_{Ek} may be taken as zero and we obtain

$$\sum_{i=1}^{I} r_i\alpha_i(y^*_{ik} - y_{ik}) = 0 \quad . \qquad (13)$$

Equation (13) gives a relationship between the phenomenological coefficients. Because of $\Sigma y = 1$ only one of them can be chosen independently, that means $\alpha_i = r_I\alpha_I/r_i$, i=1,...,I. Taking into account a non-vanishing ΔE_k, the mass transfer rate is

$$j_{ik} = \frac{r_I}{r_i}\,\alpha_I(y^*_{ik} - y_{ik}) + \frac{\Delta E_k}{I\cdot r_i} \quad ; \qquad i=1,\dots,I \quad . \qquad (14)$$

This equation is consistent with the intuitive idea, that a surplus of energy on a stage leads to an acceleration or to a damping of the mass transfer.

The dynamic behaviour of the reboiler is described by two lag elements (2). One of the time lags is caused by the heat transfer through the heat exchanger

$$\frac{dQ_K}{dt} = -\frac{1}{\tau_1}Q_K - \frac{c_K}{\tau_1}u \quad . \qquad (15)$$

The other lag element represents the energy balance equation of the reboiler section

$$\frac{dV_K}{dt} = -\frac{1}{\tau_2}V_K - \frac{1}{\tau_2 r_K}Q_K \quad . \qquad (16)$$

Furthermore, the vapour V_K leaving the reboiler is assumed to be of equilibrium composition $y_{iK} = y^*_{iK}$, $i=1,\ldots,I$.

In the condenser the vapour is condensed totally. Thus the mole fractions in the distillate D and in the reflux R are $x_{i0} = y_{i1}$, $i=1,\ldots,I$.

The developed mathematical model is based on certain serious simplifications. However, measured data were matched at a sufficient accuracy by adjusting the parameters holdup n', vapour flow resistance ξ and exchange rate coefficient α_I (3).

Simulation Results

Below the feed F_A, in the steady state temperature profile (Figure 3) two regions of saturation, separated by the above mentioned temperature front, can be distinguished. The temperature front is caused by the interphasial exchange of isopropanol and water.

If a step disturbance reduces the feed rate F_A, the region of exchange and the temperature front are forced to move upward the column (Figure 4). The state variables of all other regions are only slightly altered.

A stepwise decrease in the flow rate of the heating steam forces the front to move downward (Figure 5).

Simplified Dynamic Model

Modern control synthesis procedures are often based on the state variable representation of the plant equations. In applying them to high order systems, such as distillation columns, practical reasons of controller design require order reduction. That means one has to replace the large number of state variables obtained by physical laws, thru the introduction of a smaller set of suitably chosen state variables.

Looking at the simulation results makes an appropriate choice apparent. The essential dynamic quality of the extractive column is the movement of the region of the temperature front, generated by the moving region of high mass transfer isopropanol/water. If the front is fixed sufficiently apart from both bottom and top of the column, the product specifications are insured. Therefore, the locus of the front is a suitable state variable.

To get the dynamic model of the movement of the front, the column is separated into two parts by cutting off below the thirtieth stage (Figure 6). It is assumed that the total amount of isopropanol leaves the liquid phase within the region of the temperature front. Furthermore, the vapour flow rates are supposed to be the same along the column $V_k = V$, $k=1,\ldots,K$. It is to be emphasized that the mole fractions of isopropanol remain almost uneffected in the upper part. As a result the reflux R consists of pure isopropanol and hence the overall mass balance equations of the upper part are

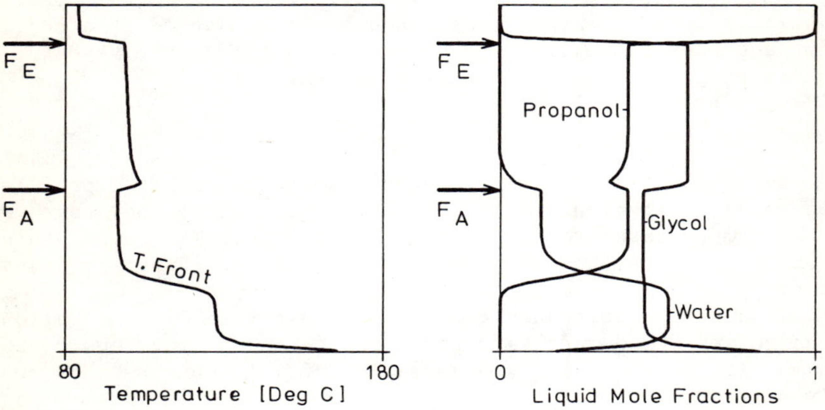

Figure 3. Steady-state temperature and mole fractions profiles

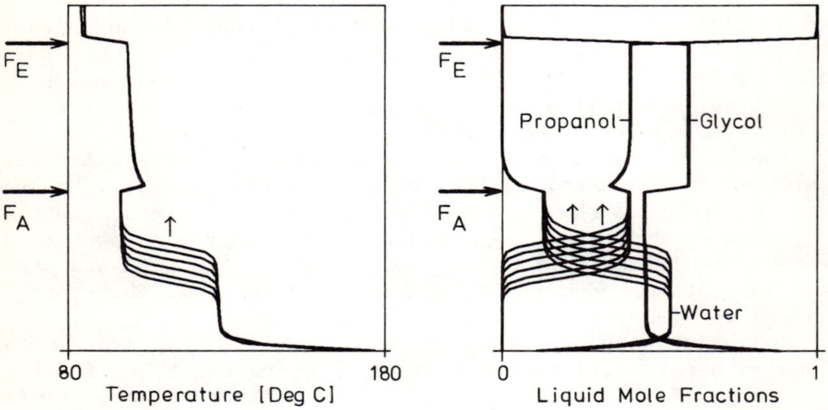

Figure 4. Dynamic behavior: stepwise decrease of feed rates

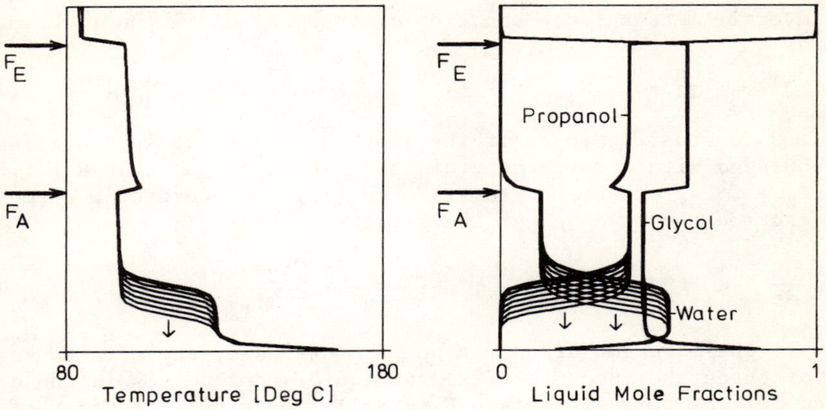

Figure 5. Dynamic behavior: stepwise decrease of heating steam flow rate

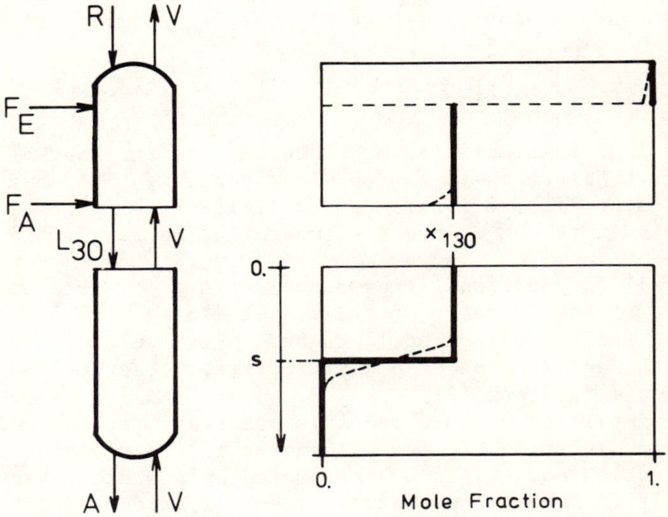

Figure 6. Column parts and approximated mole fraction profile of isopropanol

$$L_{30}x_{130} = F_A x_{FA_1} + R \quad (17)$$

$$L_{30} = F_A + F_E + R \ . \quad (18)$$

The overall mass balance equation of isopropanol in the lower part yields

$$n' x_{130} \frac{ds}{dt} = L_{30}x_{130} - V \ , \quad (19)$$

where s denotes the locus of the temperature front. As the column is equipped with a ratio control (see Figure 1), so that $R = f_1 F_A$ and $F_E = f_2 F_A$, we obtain from (17) to (19) an integrating element to describe the movement of the front

$$\frac{ds}{dt} = K_I [F_A (x_{FA} + f_1) - V] \ ; \quad K_I = \frac{1+f_1+f_2}{n'(x_{FA_1}+f_1)} \ . \quad (20)$$

If the movement of the front is measured by the average voltage of three thermocouples fixed at equidistant points in the desired area of this front (see Figure 1), the effective range of measurement is limited due to saturation. Within the limits, the averaged temperature changes nearly linearly with the deviation of the front. If the limits are exceeded, the output of the measuring device can be considered to be constant.

To complete the simplified dynamic model (Figure 7), we take into account the connection between vapour flow rate and heating steam flow rate as they are given by the lag elements (15) and (16).

New Control Configuration

By using the simplified model to analyse the control system according to Figure 1, it can be clearly shown, that as a consequence of stability requirements and limitations in the measurement, the gain factor of the PID controller is restricted to small values.

Installing additional thermocouples is a possible way to increase the effective range of the measurement and to improve the quality of the control. A still higher control efficiency is insured by the application of a modified state observer and of an optimal state feedback (2).

The simplified dynamic model is observable if the measured temperature deviation ΔT lies within the limits of saturation. Its state variables Δq, ΔV, Δs are estimated by a state observer using the simplified model. The observer feedback is turned off, if ΔT exceeds any of the limits so that the system becomes unobservable.

The correcting variable Δu is computed from the state variables using the optimal feedback law

$$\Delta u = -c_1 \Delta q - c_2 \Delta V - c_3 \Delta s - c_4 \int \Delta s dt \ .$$

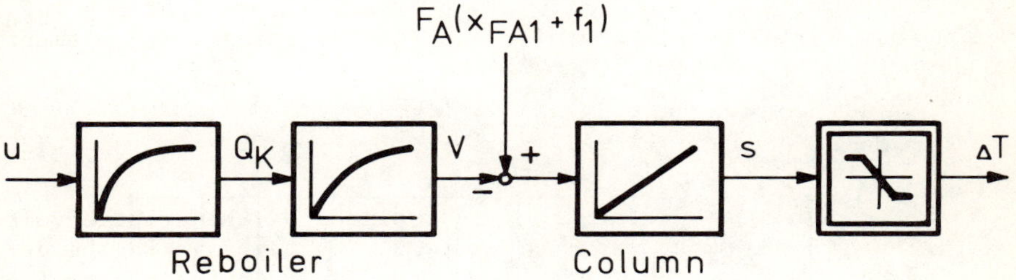

Figure 7. Block diagram of the simplified dynamic model

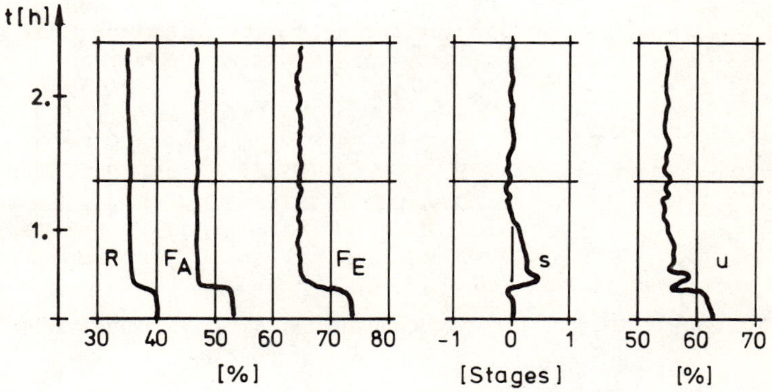

Figure 8. New control system: stepwise decrease of feed rates

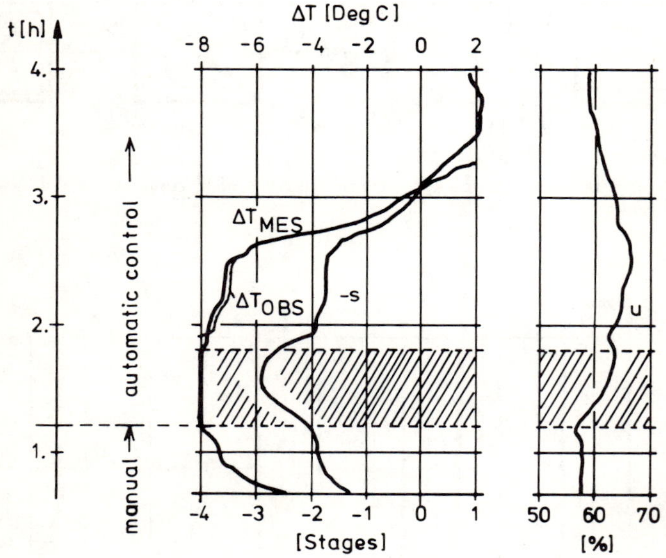

Figure 9. New control system: behavior in the unobservable region

The gain factors c_i, $i=1,\ldots,4$ are obtained by minimizing a time-invariant quadratic performance index by the solution of the steady state matrix RICCATI-equation.

Implementation and Field-Tests

An INTEL 8080 μ-processor was programmed and implemented in the large scale plant. Two key experiments with the new control concept shall be discussed here.

In Figure 4 no control was employed; a feed disturbance caused the front to move to the top of the column. Due to observer and optimal control the maximum of deviation was about half a plate (Figure 8) and the product specifications were within the prescribed limits.

In order to test the control system in an unobservable region of the plant (shaded in Figure 9), the heating steam flow rate was first controlled manually in order to force the front to move out of the observable region. When the measured temperature ΔT_M passed the lower limit of saturation, the observer feedback was turned off and the control loop was closed. The heating steam flow rate changed according to the simulated values of the state variables. This caused the front to return to the observable region, where the observer was again balanced by means of its feedback. Then the desired locus of the temperature front was attained.

The new control configuration has been in continuous operation for one year. It has proven itself to be highly reliable and efficient. The former difficulties have been surmounted.

Abstract

The digital simulation of an extractive distillation column was performed in order to understand the dynamic behaviour of the system. Based on this results a considerably simplified dynamic model of sufficient accuracy could be developed. This model was employed in the design of a state observer and of an optimal control. After implementation in the large scale plant this new control system has proved to be highly efficient and reliable.

Acknowledgement

The project was sponsored by the "Bundesministerium für Forschung und Technologie" as a part of the PDV (Data processing) Project within the 2nd and 3rd data processing programmes of the German Government (DV 5.505).

Literature Cited

(1) G.Kunstmann: Betriebsversuche zur Regelung einer Extraktiv-Destillation, Regelungstechnische Praxis und Prozess-Rechentechnik (1973) Heft 8, S.193-197.

(2) F.Silberberger: Modellbildung, Simulation und Regelung einer Extraktiv-Destillationskolonne. Dissertation, Universität Stuttgart, 1979.
(3) F.Silberberger: Dynamische Modelle und Simulation chemischer Prozesse, Teil II. Destillationskolonne. PDV-Ber.KFK-PDV 95, 1977.

RECEIVED November 5, 1979.

INDEX

INDEX

D

  mass and energy balances ... 155–156

E

Q

R

S

T

Packing List for Order #058-7896448-7039763

Shipping to

Rodolfo H Garanchon
9151 Paso Robles
N/A 9151 Paso Rob
Northridge, CA 91325
USA

Order Information

Payment Date: 06/29/2007
Buyer Name: Rodolfo H Garanchon
Shipping Speed: Standard

Item Details

Computer Applications to Chemical Engineering: Process Design and Simulation...
ASIN/ISBN: 0841205493
Quantity: 1
Listing ID: 0508F042244
Order-Item ID: 45136179156661
Condition: Used - Good
Comments: book inside clean, no marks, dustcover has marks and a few scratches

Seller Information

Shipped by: solo-saint-juste
Seller Email: elster@ohiou.edu

Thanks for buying on Amazon Marketplace. Please be sure to rate your experience with this seller by visiting www.amazon.com/feedback. View more information about this seller at www.amazon.com/seller/solo-saint-juste.

PROCESS SYSTEMS ENGINEERING

The "systems approach" is a fundamental, integral concept that is not explicitly addressed in the chemical engineering curriculum. The concept of analyzing a collection of components and processes as an overall system, rather than as individual components, is critical. Systems engineering equips the chemical engineering graduate to:

- create and understand mathematical descriptions of physical phenomena
- scale variables and perform order-of-magnitude analyses,
- structure and solve complex problems,
- manage large amounts of messy data, including missing data and information,
- resolve complex and sometimes contradictory issues of process design: sensitivity of solutions to assumptions, uncertainty in data, what-if questions, process optimization.

The systems component of the curriculum is the part that trains the students in the tools for synthesis, analysis, design, and manipulation of chemical processes, units and collections thereof. The systems education teaches the students how to convert scientific facts and principles of chemical systems into engineering decisions. The knowledge base of systems consists of methods for dynamic and steady-state simulation at multiple length and time scales, statistical analysis of data, sensitivity analysis, optimization, parameter estimation and system identification, design and analysis of feedback, methods for on-line monitoring and diagnosis, and methods for design of products and processes.

Systems Engineering tools will enable specific scientific knowledge to be harnessed for engineering purpose:

- **Overview.** Plant-wide and product viewpoints, exposure to multi-faceted, real-world problems, degrees-of-freedom analysis, computer programming concepts, and simple applications. The Chemical Process Industries: function and economic structure.

- **Engineering Systems.** Conservation laws for simple dynamic and steady-state systems, simple models for an experimental dynamic system (chemically reacting system), acquisition and analysis of noisy, complex, dynamic laboratory data, numerical simulation for simple models (single ODE), parameter estimation for simple models (one or two parameters estimated from one or two dynamic sensor measurements), equipment construction and sensor design.

- **Molecular Systems.** Random variables, probability and statistics, stochastic systems and molecular level reactions as systems, stochastic kinetic models and Monte Carlo models, simulation as an enabling technology, optimization principles for design, parameter estimation and decision making, general principles of experimental design for static and dynamic systems, use of models in predicting and understanding system behavior (synthesis).

- **Systems Integration and the Marketplace.** Multi-scale systems, separation and resolution of time and length scales, design and analysis of feedback control systems, frequency response and analysis of spectroscopic data, monitoring and fault detection, energy and mass integration, design for environment and process efficiency, network targeting concepts, process operations: planning, scheduling, and the supply chain, the design experience: economics and business skills, safety, marketing, environmental impact, life cycle analysis, ethics, intellectual property, globalization, social and national needs.